Exploring Autodesk Navisworks 2017

(4th Edition)

CADCIM Technologies

525 St. Andrews Drive
Schererville, IN 46375, USA
(www.cadcim.com)

Contributing Authors
Prof. Sham Tickoo

Purdue University Northwest
Department of Mechanical Engineering Technology
Hammond, Indiana, USA

Manash Chatterjee

BIM Implementation Specialist
CADCIM Technologies, USA

CADCIM Technologies

Exploring Autodesk Navisworks 2017, 4th Edition
Sham Tickoo

CADCIM Technologies
525 St Andrews Drive
Schererville, Indiana 46375, USA
www.cadcim.com

ISBN   978-1-942689-49-2

NOTICE TO THE READER

Publisher does not warrant or guarantee any of the products described in the text or perform any independent analysis in connection with any of the product information contained in the text. Publisher does not assume, and expressly disclaims, any obligation to obtain and include information other than that provided to it by the manufacturer.

The reader is expressly warned to consider and adopt all safety precautions that might be indicated by the activities herein and to avoid all potential hazards. By following the instructions contained herein, the reader willingly assumes all risks in connection with such instructions.

The publisher makes no representation or warranties of any kind, including but not limited to, the warranties of fitness for particular purpose or merchantability, nor are any such representations implied with respect to the material set forth herein, and the publisher takes no responsibility with respect to such material. The publisher shall not be liable for any special, consequential, or exemplary damages resulting, in whole or part, from the reader's use of, or reliance upon this material.

www.cadcim.com

Online Training Program Offered by CADCIM Technologies

CADCIM Technologies provides effective and affordable virtual online training on various software packages including Computer Aided Design, Manufacturing, and Engineering (CAD/CAM/CAE), computer programming languages, animation, architecture, and GIS. The training is delivered 'live' via Internet at any time, any place, and at any pace to individuals as well as the students of colleges, universities, and CAD/CAM/CAE training centers. The main features of this program are:

Training for Students and Companies in a Classroom Setting

Highly experienced instructors and qualified Engineers at CADCIM Technologies conduct the classes under the guidance of Prof. Sham Tickoo of Purdue University Northwest, USA. This team has authored several textbooks that are rated "one of the best" in their categories and are used in various colleges, universities, and training centers in North America, Europe, and in other parts of the world.

Training for Individuals

CADCIM Technologies with its cost effective and time saving initiative strives to deliver the training in the comfort of your home or work place, thereby relieving you from the hassles of traveling to training centers.

Training Offered on Software Packages

CADCIM Technologies provide basic and advanced training on the following software packages:

CAD/CAM/CAE*: CATIA, Pro/ENGINEER Wildfire, SolidWorks, Autodesk Inventor, Solid Edge, NX, AutoCAD, AutoCAD LT, Customizing AutoCAD, AutoCAD Electrical, EdgeCAM, and ANSYS*

Architecture and GIS*: Autodesk Revit (Architecture, Structure, MEP), AutoCAD Civil 3D, AutoCAD Map 3D, Autodesk Navisworks, Bentley STAAD. Pro, RISA 3D and Oracle Primavera P6*

Animation and Styling*: Autodesk 3ds Max, Autodesk Maya, Autodesk Alias, Pixologic ZBrush, and CINEMA 4D*

Computer Programming*: C++, VB.NET, Oracle, AJAX, and Java*

*For more information, please visit the following link: **http://www.cadcim.com***

Note

If you are a faculty member, you can register by clicking on the following link to access the teaching resources: ***http://www.cadcim.com/Registration.aspx***. The student resources are available at ***http://www.cadcim.com***. We also provide **Live Virtual Online Training** on various software packages. For more information, write us at *sales@cadcim.com*.

Table of Contents

Chapter 2: Exploring the Navigation Tools in Navisworks

Chapter 3: Selecting, Controlling, and Reviewing Objects

Chapter 4: Viewpoints, Sections, and Animations

Chapter 5: TimeLiner

Chapter 6: Working with Animator and Scripter

Chapter 7: Quantification

Chapter 8: Clash Detection

Chapter 9: Autodesk Rendering in Navisworks

Conversion Table

Conversion Table-Metric/English			
	Units	Multiply By (Factor)	To Obtain
Length	Inch	2.54	Centimeter
	Centimeter	0.393	Inch
	Feet	0.301	Meter
	Meter	3.281	Feet
	Kilometer	0.54	Nautical Mile
	Nautical Mile	1.852	Kilometer
	Feet	0.000304	Kilometer
Weight and Mass	Ounce	28.35	Gram
	Gram	0.0353	Ounce
	Pound	0.453	Kilogram
	Kilogram	2.205	Pounds
	Metric Ton	1.102	Ton
Liquid Measures	Fluid Ounce	0.0296	Liter
	Gallon	3.785	Liter
	Liter	0.264	Gallon
Thrust / Pressure	Pounds Force	4.448	Newton
	Newton	0.225	Pound
	Pound per square inch (psi)	6.895	Kilo Pascal
Temperature	Kelvin	1	Degree Celsius-273.15
	Degree Celsius	1.8	Degree Fahrenheit +32

Preface

Autodesk Navisworks 2017

Autodesk Navisworks 2017, developed by Autodesk Inc., is a powerful software used for reviewing integrated models and data. It has various tools that help in combining 3D models, navigating around them, reviewing them, creating 5D simulation, creating quantity takeoff, and rendering. This enables the users to harness the power of BIM with Autodesk Navisworks for their specific use. Autodesk Navisworks is an effective tool for architects, civil engineers, and construction professionals.

Autodesk Navisworks has interoperability with major design softwares such as Autodesk Revit Architecture / Structure / MEP, AutoCAD, AutoCAD Plant 3D, Autodesk Inventor, Microsoft Project, and Autodesk 3ds Max.

Exploring Autodesk Navisworks 2017 is a comprehensive textbook that has been written to cater to the needs of the students and professionals. The chapters in this textbook are structured in a pedagogical sequence, which makes the learning process very simple and effective for both the novice as well as the advanced users of Autodesk Navisworks. In this textbook, the author emphasizes on creating 4D simulation, performing clash detection, performing quantity takeoff, rendering, creating animation, and reviewing models through tutorials and exercises. In addition, the chapters have been punctuated with tips and notes, wherever necessary, to make the concepts clear, thereby enabling you to create your own innovative projects.

The highlight of this textbook is that each concept introduced in it is explained with the help of suitable examples to facilitate better understanding. The simple and lucid language used in this textbook makes it a ready reference for both the beginners and the intermediate users.

- **Tutorial Approach**
  The author has adopted the tutorial point-of-view and learn-by-doing approach throughout the textbook. This approach guides the users through various processes involved in creating and analyzing spatial data. At the end of each chapter, tutorials are provided to practice the concepts learned in the chapter.

- **Real-World Projects as Tutorials**
  The author has used about 17 real-world Navisworks projects as tutorials in this book. This will enable the readers to relate the tutorials to the real-world projects in the industry. In addition, there are about 9 exercises based on the real-world projects.

- **Tips and Notes**
  The additional information related to topics is provided to the users in the form of tips and notes.

- **Learning Objectives**
  The first page of every chapter summarizes the topics that are covered in that chapter.

- **Self-Evaluation Test, Review Questions, and Exercises**
  Every chapter ends with Self-Evaluation Test so that the users can assess their knowledge of the chapter. The answers to Self-Evaluation Test are given at the end of the chapter. Also, the Review Questions and Exercises are given at the end of each chapter and they can be used by instructors as test questions and exercises.

- **Heavily Illustrated Text**
  The text in this book is heavily illustrated with screen capture images.

Symbols Used in the Textbook

Note
The author has provided additional information related to various topics in the form of notes.

Tip
The author has provided a lot of information to the users about the topic being discussed in the form of tips.

Enhanced
This symbol indicates that the command or tool being discussed is enhanced in Navsworks 2017.

Unit System Followed in the Textbook
In this book, the Metric system has been used as the default unit system. In addition, the Imperial conversions of Metric units are also mentioned in parenthesis wherever required.

Formatting Conventions Used in the Textbook
Please refer to the following list for the formatting conventions used in this textbook.

- Names of tools, buttons, options, browser, palette, panels, and tabs are written in boldface.

  Example: The **Walk** tool, the **Enable Scripts** button, the **Project** panel, the **Home** tab, and so on.

- Names of dialog boxes, drop-downs, drop-down lists, list boxes, areas, edit boxes, check boxes, and radio buttons are written in boldface.

  Example: The **File Options** dialog box, the **Alignment** drop-down list, the **Position** edit box of the **Edit Viewpoint - Current View** dialog box, and so on.

- Values entered in edit boxes are written in boldface.

  Example: Enter **Buildings** in the **Name** edit box.

- Names of the files are italicized.

  Example: *c03_navisworks_2017_tut1*

Naming Conventions Used in the Textbook

Tool
If you click on an item in a panel of the ribbon and a command is invoked to create/edit an object or perform some action, then that item is termed as tool. For example: **Select** tool and **Zoom** tool.

Button
The item in a dialog box that has a 3d shape like a button is termed as Button. For example, **OK** button, **Cancel** button, **Apply** button, and so on.

Drop-down
A drop-down is the one in which a set of common tools are grouped together. You can identify a drop-down with a down arrow on it. These drop-downs are given a name based on the tools grouped in them. For example, **Select All** drop-down, **Windows** drop-down, and so on, refer to Figures 1 and 2.

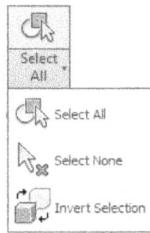

*Figure 1 The **Select All** drop-down*   *Figure 2 The **Windows** drop-down*

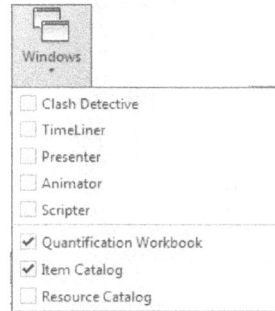

Drop-down List
A drop-down list is the one in which a set of options are grouped together. You can set the parameters using these options. You can identify a drop-down list with a down arrow on it. For example, **Model Takeoff** drop-down list, **Linear Units** drop-down list, and so on, refer to Figure 3.

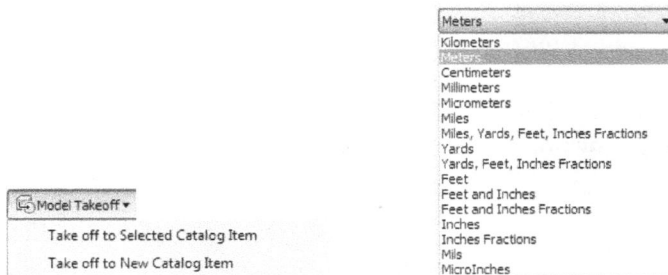

*Figure 3 The **Model Takeoff** and **Linear Units** drop-down lists*

Options

Options are the items that are available in shortcut menus, dialog boxes, drop-down lists, and so on. For example, choose the **Test Action** option from the shortcut menu displayed on right-clicking in the drawing area.

Free Companion Website

It has been our constant endeavor to provide you the best textbooks and services at affordable price. In this endeavor, we have come out with a Free Companion website that will facilitate the process of teaching and learning of Autodesk Navisworks 2017. If you purchase this textbook, you will get access to the files on the Companion website.

The resources available for the faculty and students in this website are as follows:

Faculty Resources

- **Technical Support**
  You can get online technical support by contacting *techsupport@cadcim.com*.

- **Instructor Guide**
  Solutions to all review questions and exercises in the textbook are provided in the instructor guide to help the faculty members test the skills of the students.

- **PowerPoint Presentations**
  The contents of the book are arranged in PowerPoint slides that can be used by the faculty for their lectures.

- **Part Files**
  The Navisworks files used in illustration, tutorials, and exercises are available for free download.

Student Resources

- **Technical Support**
  You can get online technical support by contacting *techsupport@cadcim.com*.

- **Part Files**
  The Navisworks files used in illustrations and tutorials are available for free download.

If you face any problem in accessing these files, please contact the publisher at *sales@cadcim.com* or the author at *stickoo@pnw.edu* or *tickoo525@gmail.com*.

Stay Connected

You can now stay connected with us through Facebook and Twitter to get the latest information about our textbooks, videos, and teaching/learning resources. To stay informed of such updates, follow us on Facebook (*www.facebook.com/cadcim*) and Twitter (*@cadcimtech*). You can also subscribe to our YouTube channel (*www.youtube.com/cadcimtech*) to get the information about our latest video tutorials.

Chapter 1

Introduction to Autodesk Navisworks 2017

Learning Objectives

After completing this chapter, you will be able to:
- *Understand the basic features of Autodesk Navisworks 2017*
- *Start Autodesk Navisworks 2017 software*
- *Use different components of the User Interface screen of Navisworks 2017*
- *Understand the Local and Global options in Navisworks*
- *Understand different file types in Navisworks*
- *Manage files in Navisworks*
- *Share data in Navisworks*
- *Work in different workspaces in Navisworks*
- *Access the Navisworks 2017 Help*

INTRODUCTION

In the construction industry, sometimes many teams working on different tasks are involved in a project. In such cases, the coordination between the team members and integration of the project data is important for seamless execution of the project. Autodesk Navisworks is designed to provide an integrated environment to coordinate, share, and manage project files between multiple teams. It helps them to view the whole project and make better design conclusions, accurate construction documents, and to anticipate performance of the design. It is a powerful software which takes construction and design to new levels. It holds a building information modelling project together. The use of Autodesk Navisworks 2017 provides a competitive advantage and a higher profitability to building industry professionals.

Autodesk Navisworks has three product verticals: Navisworks Freedom, Navisworks Simulate, and Navisworks Manage.

Navisworks Freedom is a free viewer provided by Autodesk for viewing NWD files. Navisworks Freedom comprises all the basic tools including the complete set of navigation tools. This software can be used to quickly open the NWD files, review objects, create animation, and for navigation purpose.

Navisworks Simulate includes the tools and options for reviewing, real-time navigation, simulation, and animation. This software can also be used to communicate the project information to others. Construction managers can use this software for simulating the construction schedule.

Navisworks Manage is a comprehensive software which includes all the advanced and basic tools for coordination, real-time navigation, simulation, animation, clash detection, and estimation required for the project. This software can be used by construction managers, BIM engineers, architects, and so on.

Navisworks allows you to open the Autodesk Revit files in their native format. It also supports other file formats from other software such as CATIA, SolidWorks, Autodesk ReCap, Autodesk Inventor, Faro 4.8, and so on.

In this book, you will learn about all the basic and advanced tools in Navisworks Manage.

BASIC FEATURES OF Navisworks Manage

Navisworks Manage is a complete project software which allows analysis, reviewing, coordination, and simulation. The basic features of Navisworks are discussed next.

Navigation and Reviewing Tools

Navigating and reviewing models in a scene are the essential features of Navisworks. In Navisworks, you can navigate within a model using tools such as **ViewCube**, **Pan**, **Zoom**, **Orbit**, **Look**, **Walk/Fly**, and so on. There are some additional features in Navisworks that will give you the real-time experiences while working on a project.

While navigating, you can review a model by using various selection methods, transformation tools, measure tools, and redline tools. You can also analyze the properties of an individual or a group of objects in a model. All these tools and methods will be discussed in detail in the later chapters.

TimeLiner

TimeLiner is used to understand the construction sequence of a model. It helps you to create 4D and 5D simulations and allows you to see the effect of a schedule on the project. Using the TimeLiner feature, you can compare between the planned dates (start and finish) and the actual dates (start and finish) of any activity in the project. This feature helps in avoiding the potential problems that can arise during construction before it starts. Thus, you can reduce the delay and the amount of rework in a construction project. You can import the schedules from other project management software such as Microsoft Project, Primavera, and Asta.

Quantification

Quantification is a process in which the quantity takeoff of model components or of the entire BIM model is performed. It helps you to perform automatic and manual takeoff for 2D and 3D models. You can also perform takeoff of various combined source files, which will include measuring length, width, height, and area. Using Autodesk BIM 360, the takeoff data can also be exported to excel and can be shared in cloud with other project members for reviewing.

Clash Detection

Clash Detection is used for detecting interferences between objects. Performing clash detection helps in identifying the errors and rectifying them. It thus reduces the risk of human errors and saves time and efforts. Using this feature, you can check the clashes and report them to the team members by exporting the clash reports.

Autodesk Rendering

Autodesk Rendering is used to create photorealistic images and visualizations by applying materials, real lighting effects, and background effects. Navisworks has several types of materials, effects, and rendering styles that you can use to render the project with photorealistic effect before delivering it to the clients.

Animator and Scripter

Animator is used for creating animation by capturing every movement. Scripter is used for adding interactivity to the animated objects in a model.

BENEFITS OF NAVISWORKS

There are various advantages of using Navisworks. Some of them are listed below:

1) Navisworks has an open format which supports a large variety of file formats.

2) Navisworks allows real-time navigation, clash detection, tolerance checking, design reviews, and project coordination with team members.

3) It improves productivity and quality by pre-visualizing the construction projects.

4) It allows rendering that helps in verifying the materials and textures that are appropriate for a proposed design.

5) It has the ability to import data from other applications to produce a single unified model.

STARTING Autodesk Navisworks 2017

You can start Navisworks by double-clicking on the Navisworks Manage 2017 icon on the desktop. Alternatively, choose the **Start > All Programs > Autodesk > Navisworks Manage 2017 > Manage 2017** from the taskbar (in Windows 7); the interface screen will be displayed, as shown in Figure 1-1.

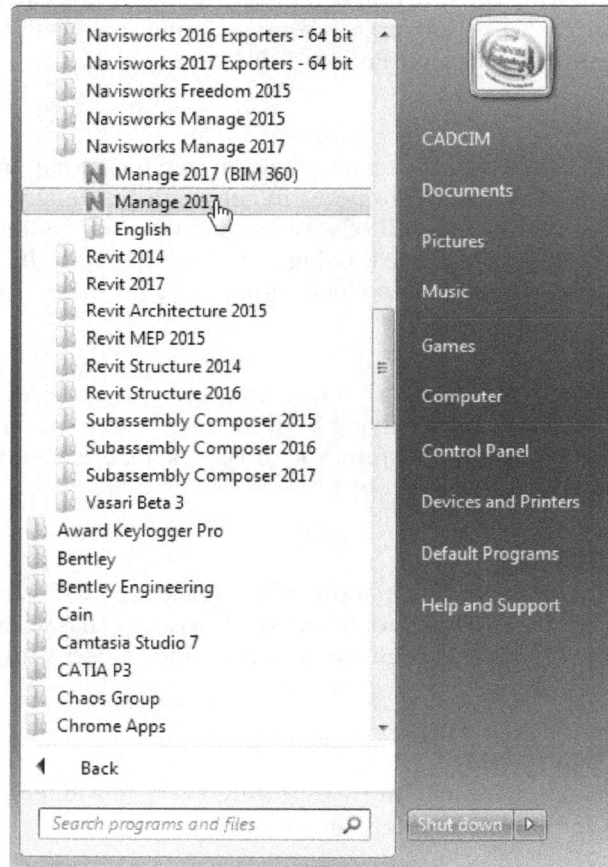

Figure 1-1 *Starting Autodesk Navisworks Manage from the taskbar*

USER INTERFACE

The user interface of Autodesk Navisworks comprises several elements such as the **Application Button**, **Ribbon**, **Quick Access Toolbar**, **Info Centre**, **Scene View**, **Navigation Bar**, **Dockable Windows**, and **Status Bar**, as shown in Figure 1-2. The interface elements are exclusively designed to provide an easy access to the tools and windows. These elements are discussed next.

Title Bar

The Title bar is placed at the top of the user interface. It displays the program's logo and name of the current project. It also contains the **Quick Access Toolbar** and **Info Centre** toolbar.

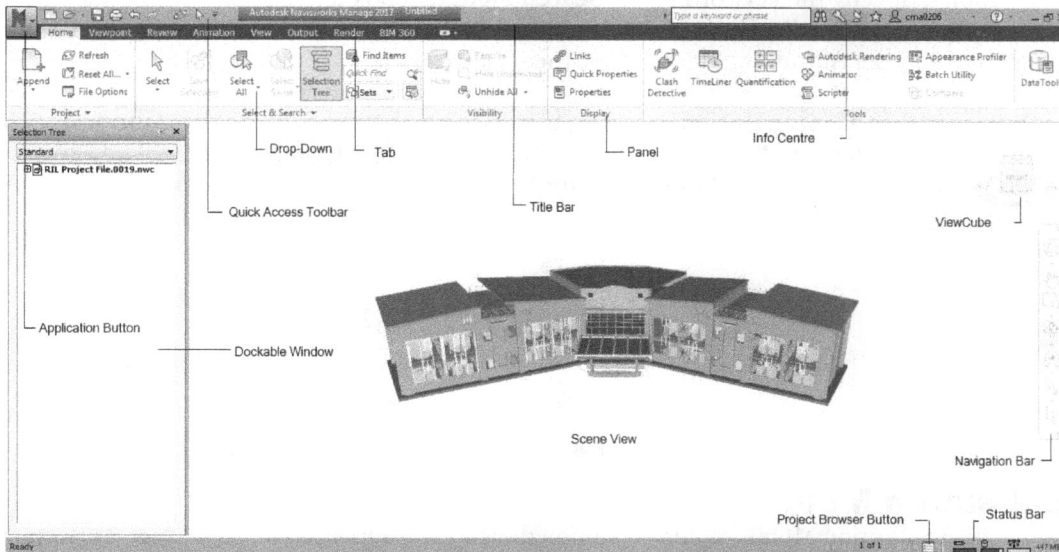

Figure 1-2 *The user interface of Autodesk Navisworks Manage 2017*

Quick Access Toolbar

The **Quick Access Toolbar** comprises several frequently used tools such as **New**, **Open**, **Save**, **Print**, **Undo**, **Redo**, **Refresh**, and **Select**. Figure 1-3 shows the **Quick Access Toolbar**. It is located at the top of the application window. You can add the required tools from the ribbon to the **Quick Access Toolbar**. You can also remove the unwanted tools. To add a tool from the Ribbon, choose the tab which contains that tool, and right-click on the tool; a shortcut menu will be displayed. Choose the **Add to Quick Access Toolbar** option from the shortcut menu; the selected tool will be added to the **Quick Access Toolbar**. To remove a tool from the **Quick Access Toolbar**, right-click on it; a shortcut menu will be displayed. Choose the **Remove from Quick Access Toolbar** option from the menu; the tool will be removed. You can place the toolbar below the Ribbon by choosing the **Show Below the Ribbon** option from the menu.

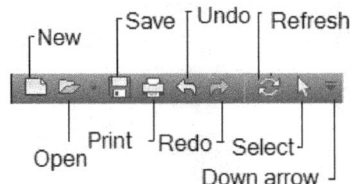

Figure 1-3 *The Quick Access Toolbar*

InfoCenter

The InfoCenter toolbar contains several options, which are used to access product related information. You can use the search option in the InfoCenter toolbar to find the information on various topics. Specify the keyword in the **Search Field** text box and choose the **Search** button; the search results will be displayed in the Autodesk Wikihelp website. The **Sign In** option is used for signing in to the Autodesk 360 to access the online services.

The **Subscription Center** option is used to subscribe the services of the Autodesk. The **Communication Center** option is used for product updates and announcements. The **Favorites** panel is used to access the saved topics. Figure 1-4 shows the tools in InfoCenter.

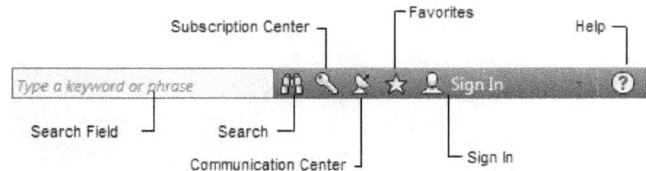

Figure 1-4   The InfoCenter

Application Menu

The Application Menu contains common tools such as **Open, Save, New, Save As, Export, Publish**. To access these tools, choose the **Application Button**; the Application Menu will be displayed, as shown in Figure 1-5.

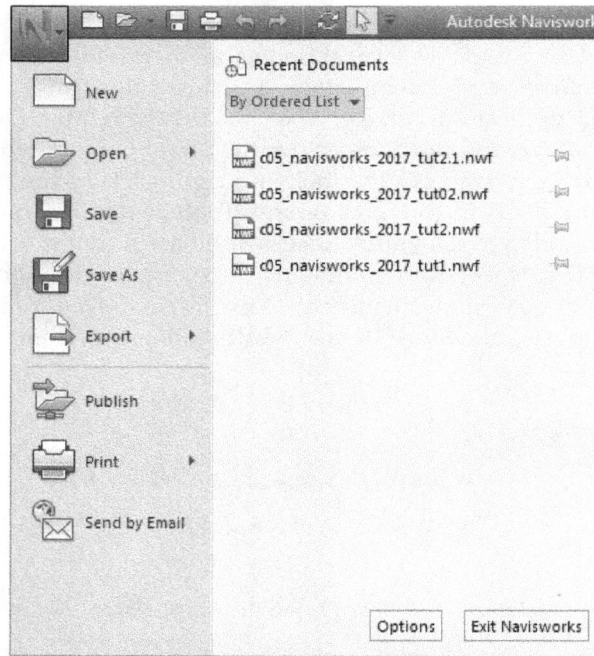

Figure 1-5   The Application Menu

In the Application Menu, you can view the recently used files. You can group and sort these files by selecting the options from the **Recent Documents** drop-down list, as shown in Figure 1-6. The default value for the maximum number of recently used files displayed in the Application Menu will be four. To change this number, go to **Options Editor > General > Environment** and then specify the desired value in the **Maximum Recently Used Files** edit box. The options in the Application Menu will be discussed later in this chapter.

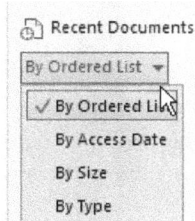

*Figure 1-6 The **Recent Documents** drop-down list*

Ribbon

The Ribbon is a palette in the interface that is displayed at the top of the screen, as shown in Figure 1-7. It comprises several tools and options necessary for creating a project. These tools and options are grouped together in a task-based series of panels within the tabs. These tabs and panels can be customized by the users. This can be done by moving the panels and changing the view of the Ribbon, as discussed next.

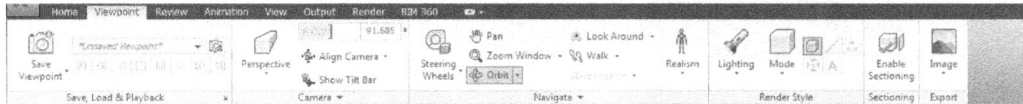

Figure 1-7 Partial view of the Ribbon

Moving the Panels

Navisworks allows the user to change the location of panels in the **Ribbon**. The panels can also be placed in the Scene View. To do so, click on the required panel in the **Ribbon**, press and hold the left mouse button and then drag the panel to some desired place on the screen.

Changing the View of the Ribbon

You can control the space taken by the Ribbon in the interface by changing its view. To do so, click on the second arrow on the right of the **BIM 360** tab; a flyout will be displayed, as shown in Figure 1-8. Select one of the four minimizing Ribbon options.

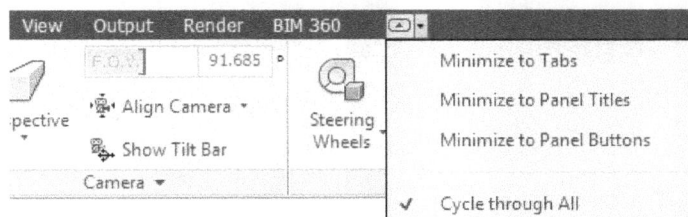

Figure 1-8 Various options in the flyout for changing the views of the Ribbon

To display only the tab titles, choose the **Minimize to Tabs** option from the menu. To display only tab and panel titles, choose the **Minimize to Panel Titles** option from the menu. To display only tab titles and panel buttons, choose the **Minimize to Panel Buttons** option from the menu. To cycle through all four Ribbon states, choose the **Cycle through All** option from the menu.

Scene View

This is the area where you can actually view and work with 3D models. By default, there is only one Scene View but Navisworks allows you to split the Scene View into multiple viewports by using the options available in the **Scene View** panel, as shown in Figure 1-9. Multiple views will help you to compare the views while rendering and creating animations. To split the Scene View horizontally, choose the **Split Horizontal** tool from the **View > Scene View > Split View** drop-down. Similarly, to split the scene vertically, choose the **Split Vertical** tool from the **View > Scene View > Split View** drop-down. These added Scene Views can be moved and docked. To do so, choose the **Show Title Bars** button from the **Scene View** panel in the **View** tab; all the added Scene Views will now have title bars, and they can be moved. You can display the current Scene View in the full screen mode. To do so, choose the **Full Screen** tool from the **Scene View** panel in the **View** tab; the current view will be displayed in full screen. You can use the **Background** tool from the **Scene View** panel to change the background settings such as color. The **Window Size** tool is used to adjust the size of the window for the desired scene view.

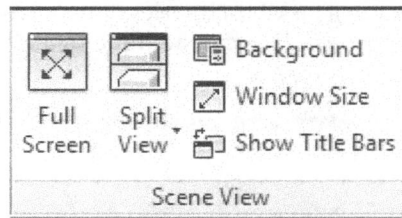

Figure 1-9   The Scene View panel

Navigation Bar

The **Navigation Bar** provides access to the tools used for navigation and orientation in a model. It contains **Steering Wheels**, **Pan**, **Zoom**, **Orbit**, **Look**, **Walk**, and **Fly** tools, as shown in Figure 1-10. You can customize the **Navigation Bar** and can also change its docking position in the Scene View, which will be discussed in the next chapter.

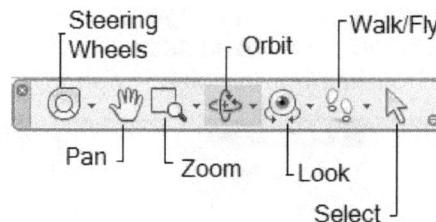

Figure 1-10   The Navigation Bar

ViewCube

The ViewCube is a navigation control which is used to set the orientation of a model. It is always displayed at the top right corner in the Scene View. It comprises a cube and a compass ring at the base, as shown in Figure 1-11.

Dockable Windows

Navisworks allows you to access most of its features from the dockable windows. These dockable windows can be moved, docked, and resized in the Scene View. To display these windows,

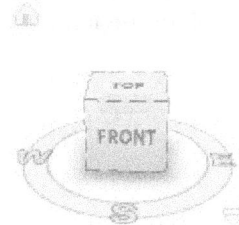

Figure 1-11 The ViewCube

click on the **Windows** drop-down in the **Workspace** panel of the **View** tab; various options will be displayed, as shown in Figure 1-12. Now, select the check box next to the required window in the list. For example, to display the **Animator** window, select the **Animator** check box; the **Animator** window will be displayed.

These windows can be customized. You can move, group, ungroup, and auto hide the windows. To move a window, click and drag the title bar of docking window and place it at the desired place. To group dockable windows, click and drag the title bar of the window to be added to the other window or group. To ungroup the window, click on the tab of the window you want to remove. Next, click and drag the window out of the group and place it at the desired place.

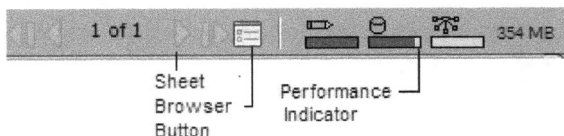

Status Bar

The Status Bar is located at the bottom of the interface screen, as shown in Figure 1-13. The Status Bar comprises the **Sheet Browser** button to navigate between models in the multi-sheet file. It also displays the progress of the project and the amount of memory being used by Navisworks.

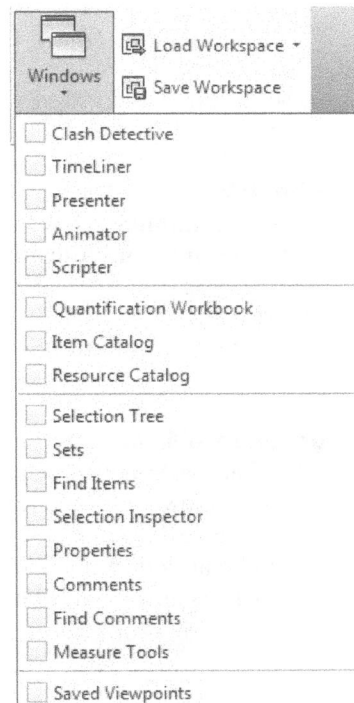

Figure 1-13 The Status Bar

*Figure 1-12 Partial view of the **Windows** drop-down*

Sheet Browser

The **Sheet Browser** is a dockable window. It contains all 2D sheets and 3D models currently opened in the file. To display this window, choose the **Sheet Browser** button in the Status Bar; the **Sheet Browser** window will be displayed, as shown in Figure 1-14. This window is divided into three areas. The top area displays the name of the current file. The middle area dislplays the list of all the loaded sheets. The bottom area displays the property of the selected sheet.

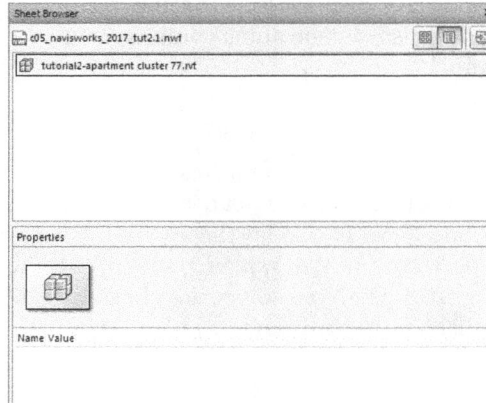

*Figure 1-14 The **Sheet Browser** window*

There are three buttons located at the top right corner in the window. These buttons are discussed next.

Thumbnail View

The **Thumbnail View** button is used to display the sheets and models as thumbnail images in the middle area of the window.

List View

The **List View** button is used to display the sheets and models in a list.

Import Sheets & Models

The **Import Sheets & Models** button is used to import sheets and models from external files. When you choose this button, the **Insert From File** dialog box will be displayed. Browse to the folder containing the required file and choose the **Open** button; the sheet/ model will be added in the list. Figure 1-15 shows the **Sheet Browser** window containing a list of 2D sheets and 3D models.

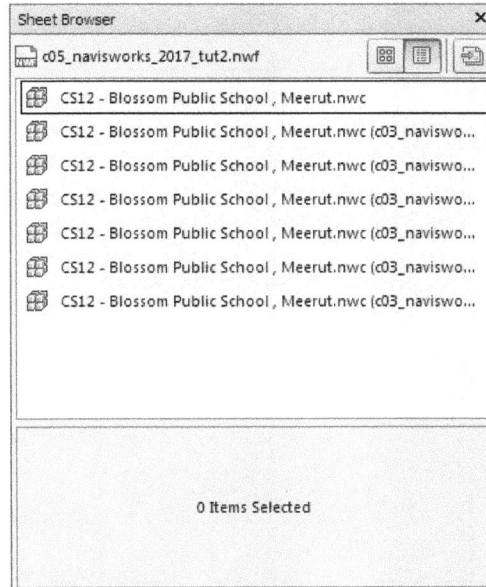

*Figure 1-15 The **Sheet Browser** window containing a list of added sheets and model*

After adding sheets or models, you will notice that not all the sheets and models are prepared to be used in Navisworks. These sheets/models are represented by the symbol. To prepare the sheets/models, you need to click on the prepare icon located next to the sheets/models, refer to Figure 1-15.

Now, to open the required sheet/model in the Scene View, double-click on it in the **Sheet Browser** window; the selected file will be loaded in the Scene View. You can also use the multi-navigation buttons available in the Status Bar.

There are some options that are used for managing the sheets/models added in the **Sheet Browser** window. To access these options, right-click on the required sheets/models in the **Sheet Browser** window; a shortcut menu will be displayed, as shown in Figure 1-16. The options in the menu are discussed next.

Open
The **Open** option is used to open the selected sheet/model in the Scene View.

Delete
The **Delete** option is used to delete the selected sheet/model from the **Project Browser** window.

Merge into Current Model
The **Merge into Current Model** is used to merge the selected 3D model with the model opened in the Scene View.

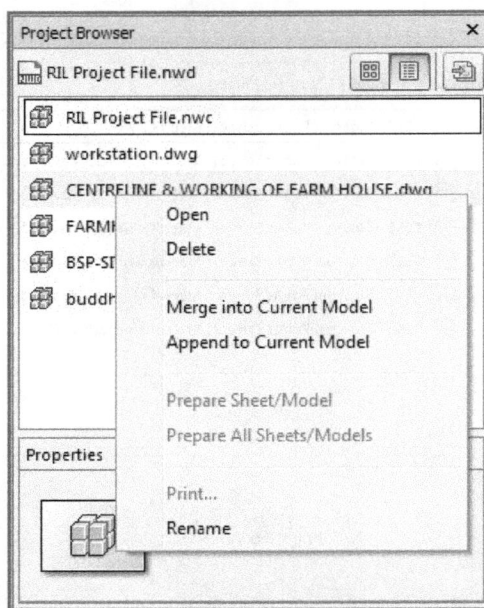

Figure 1-16 *The shortcut menu displayed in the **Sheet Browser** window*

Append to Current Model
The **Append to Current Model** option is used to add the selected 3D model to the model opened in the Scene View.

Prepare All Sheets/Models
The **Prepare All Sheets/Models** option is used to prepare all the added sheets/models to be used for Navisworks. Note that this option will be enabled only if the added sheets/models are not prepared for the use.

Print
The **Print** option is used to print the sheet/model opened in the Scene View.

Rename
The **Rename** option is used to change the name of selected sheet/model in the **Project Browser** window.

Autodesk Navisworks DIALOG BOXES
Certain options and commands, when invoked, display a dialog box. Figure 1-17 shows an example of a dialog box. The dialog box is a secondary window, which comprises elements such as title bar, tabs, area label, check boxes, radio buttons, edit boxes, drop-down list, and buttons. These elements are discussed next.

The title bar, as shown in Figure 1-17, displays the name of the dialog box. Tabs are the sections, which contain options related to the workflow. Area label is an area which contains options related to the chosen tab. Check boxes are toggle options which are used for making an option available or unavailable. Edit box is a box in which you can enter a value. Drop-down lists contain a list of options from which only one option can be selected at a time. There are more buttons in a dialog box which work as their name implies. The button [...] in a dialog box displays another dialog box. The **Help** button is used for accessing the help feature for the options available in the dialog box.

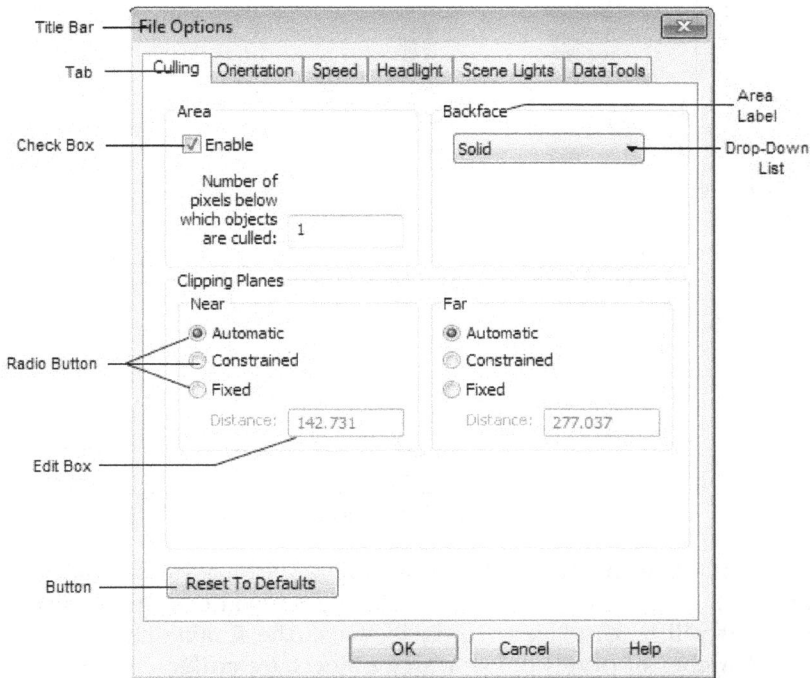

Figure 1-17 Components of a dialog box

CONFIGURING SETTINGS IN Navisworks

In Navisworks, you can configure the settings locally or globally by using the **File Options** dialog box and the **Options Editor** dialog box, respectively. Next, you will learn to configure the local and global settings using these dialog boxes.

Configuring the Local Settings

In Navisworks, you can save the model appearance, navigation speed, and any other viewing options with different file types such as NWD and NWF. Whenever a change is made in a file type, it will be saved and reloaded when you open the file. This can be done by using various file options available in the **File Options** dialog box, refer to Figure 1-18. To display the **File Options** dialog box, choose the **File Options** button from the **Project** panel of the **Home** tab; the **File Options** dialog box will be displayed. This dialog box contains various tabs and options, which are discussed next.

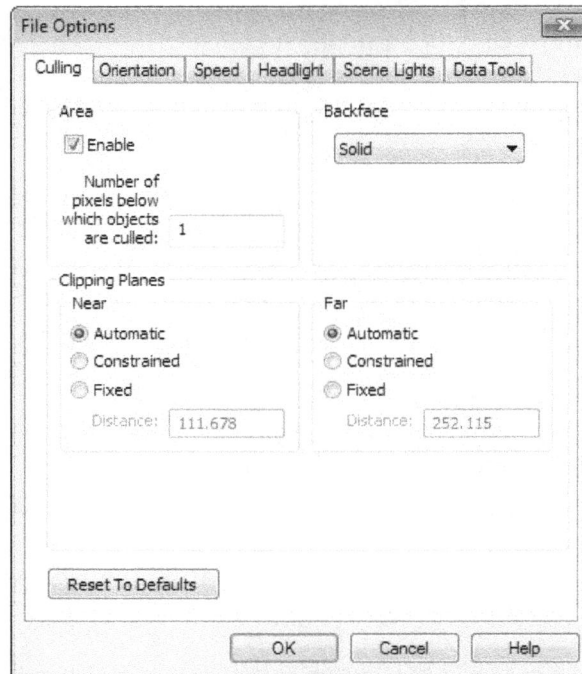

Figure 1-18 The File Options dialog box

In this dialog box, the **Culling** tab is chosen by default and has various options to adjust the geometry. This tab has four area labels. In the **Area** area, you can set the size of screen area in which the objects will be displayed. To do so, select the **Enable** check box to include the area culling; the **Number of pixels below which objects are culled** edit box will be activated. Specify a value for screen area in this edit box. For example, if you have specified 1000 in the edit box, then any object in the model drawn with a size lesser than the specified size will not be displayed. In the **Backface** area, select the required option from the drop-down list to enable the backface culling for all the elements in a project. In the **Clipping Planes** area, you can set the near clipping and far clipping planes. Both the clipping planes have three radio buttons: **Automatic**, **Constrained**, and **Fixed**. The **Automatic** radio button is used to automatically control the clipping plane position so that it can produce the best view of the model. The **Constrained** radio button is used to limit the clipping plane to the value specified in the **Distance** edit box. Note that the **Distance** edit box will be enabled only if you select the **Constrained** or **Fixed** radio button. The **Fixed** radio button is used to set the clipping planes to the value specified in the **Distance** edit box. In the **Orientation** tab, you can adjust the orientation of the model.

You can use various options in the **Speed** tab to adjust the frame rate speed to reduce the amount of drop-out during navigation. To adjust the frame rate speed, specify the desired value in the **Frame Rate** edit box. Note that higher the value specified, smoother the navigation produced.

In the **Headlight** and **Scene Lights** tabs, you can customize the lighting of the model in the Scene View.

In the **Data Tools** tab, you can link the current opened file to external databases such as Excel files. The **New** button is used to open the **New Link** dialog box where you can link the external files. The **Edit** button is used to edit the specified link. The **Delete** button is used to delete the added link. The **Import** button is used to load the previously saved data links. The **Export** button is used to save the added link as a data tool file.

Note

*The changes made in the **File Options** dialog box will be saved only for the current file in Navisworks.*

Configuring the Global Settings

Global options can be configured and saved for all the Autodesk Navisworks sessions. You can configure these options in the **Options Editor** dialog box. To display the **Options Editor** dialog box, choose **Options** from the Application Menu; the **Options Editor** dialog box will be displayed, refer to Figure 1-19. Alternatively, right-click in the Scene View; a shortcut menu will be displayed. Choose **Global Options** from the shortcut menu; the **Options Editor** dialog box will be displayed.

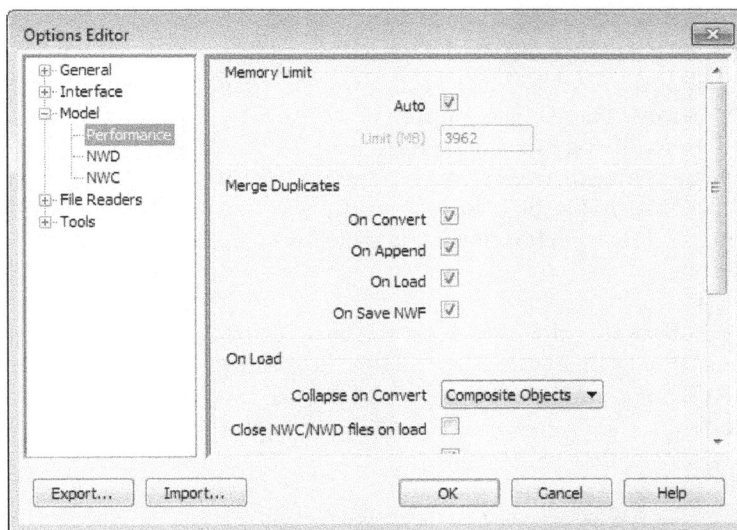

*Figure 1-19　The **Options Editor** dialog box*

The **Options Editor** dialog box is divided into left and right panes. The left pane contains several options which are grouped together and displayed in a hierarchy. The right pane displays the options selected from the left pane. There are five nodes in the left pane which have several sub-options. You can access any of the options by clicking on the plus node. These five nodes are discussed next.

In the **General** node, you can adjust the buffer size for undo/redo actions. You can specify the location where you will share the Navisworks settings such as workspaces, data tools, avatars, and so on. You can adjust the number of recent file shortcuts stored by Navisworks. You can also adjust the auto save options.

In the **Interface** node, you can use various options to customize the Navisworks interface. You can customize the units used, geometry selection method, the style and appearance of measuring lines, cursor snap settings, and so on.

The **Model** node contains the options for customizing Navisworks performance and file formats.

The **File Readers** node contains several file readers that are used while loading other CAD files.

In the **Tools** node, you can customize the settings of the options used for the **Clash Detective** window, **TimeLiner**, **Animator**, **Presenter**, and **Scripter**. These features will be discussed in the later chapters of this textbook.

> **Tip**
> *In the **Options Editor** dialog box, more options required for the advanced users can also be displayed. To do so, press and hold the SHIFT key and then choose the **Options** button from the Application Menu; the **Options Editor** dialog box will be displayed with more options.*

Exporting and Importing Global Options

You can share global options settings with other users by using the import and export feature. To export global options, choose the **Export** button in the **Options Editor** dialog box; the **Select options to export** dialog box will be displayed. In this dialog box, select the check boxes for the options which you want to export and choose the **OK** button; the **Save As** dialog box will be displayed. In this dialog box, browse to the location where you want to save the file. Specify a name for the file and choose the **Save** button; the file will be exported and saved at the specified location.

Similarly, you can import the global options settings. To do so, choose the **Import** button in the **Options Editor** dialog box; the **Open** dialog box will be displayed. In the **Open** dialog box, browse to the file location. Select the file and choose the **Open** button; the selected global options will be loaded in the **Options Editor** dialog box. Now you can use the imported global settings in the Navisworks session.

KEYBOARD SHORTCUTS

In Navisworks, keyboard shortcuts have been assigned to some of the frequently used tools to invoke them. The shortcut key corresponding to a tool is displayed when you hover the cursor over it. You can also enable the keytips by pressing the ALT key from the keyboard. As a result, various shortcut keys will be displayed in the Navisworks interface. Table 1-1 shows some of the frequently used shortcut keys in Navisworks.

Table 1-1 Various shortcut keys used in Navisworks

Keyboard Shortcut	Description
F1	Open the Help system
F2	Rename the selected Item
CTRL+N	Create a new file

CTRL+O	Display the **Open** dialog box
CTRL+A	Display the **Append** dialog box
CTRL+0	Turn on turntable mode
CTRL+1	Activate the **Select** tool
CTRL+2	Activate the **Walk** tool
CTRL+6	Activate the **Pan** tool
CTRL+S	Save the current file

FILE TYPES IN Navisworks

In Navisworks, you can access files originated from a variety of design and engineering applications. It has the capability to share and coordinate these file types and create a single Navisworks file with a project view of the entire model. This file contains all the geometry data and the data created by several teams, and it enables you to explore the model. Navisworks converts and compresses most of the files upto 80 percent of their original size, so that sharing and working with files may become easier. Autodesk Navisworks supports three types of file formats: **NWD File Format**, **NWC File Format**, and **NWF File Format**. These file formats are discussed next.

NWC File Format

In Navisworks, NWC files represent the cache files that are created whenever you open a CAD file such as AutoCAD and Revit file. The cache file is created with the name of its CAD file, but with .nwc extension. These files are smaller than the original CAD files which makes their access faster.

You can customize the parameters for .nwc files from the **NWC** page in the **Options Editor** dialog box. To do so, choose **Options** from the Application Menu; the **Options Editor** dialog box will be displayed. Expand the **Model** node in the left pane of the dialog box. Choose the **NWC** option from this node; the **NWC** page will be displayed in the right pane of the dialog box, as shown in Figure 1-20.

In the **Caching** area, select the **Read Cache** check box to read the cache files when you open a CAD file. Similarly, select the **Write** check box to save a cache file.

In the **Geometry Compression** area, select the **Enable** check box to reduce the size of *.nwc* files. This will create smaller .nwc files that will require less memory.

In the **Reduce Precision** area, select the **Coordinates** check box to reduce the precision of coordinates. On selecting this check box, the **Precision** edit box will be activated. In this edit box, you can specify the precision value of coordinates. Select the **Normals** check box to reduce the precision of normals. Select the **Colors** check box to reduce the precision of colors. Select the **Texture Coordinates** check box to reduce the precision of texture coordinates.

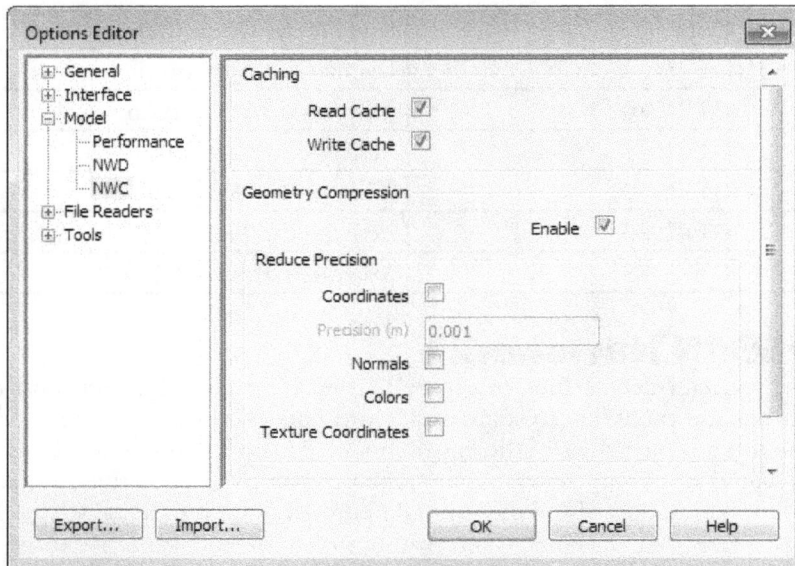

Figure 1-20 *The **NWC** page in the **Options Editor** dialog box*

NWD File Format

The NWD file stands for Navisworks Document File. The NWD file format is a basic file format, which contains all model geometry with relevant data such as review markup, clash tests, viewpoints, and so on. These files are very small as they compress the data upto 80 percent of their original size. Whenever you will publish a Navisworks file on the network, you need to publish it in the .nwd file format.

You can customize the parameters of .nwd files for saving and publishing. To do so, choose **Options** from the Application Menu; the **Options Editor** dialog box will be displayed. Expand the **Model** node in the left pane of the dialog box. Choose the **NWD** option from this node; the **NWD** page will be displayed in the right pane of the dialog box, as shown in Figure 1-21. The options in this page are discussed next.

In the **Geometry Compression** area, select the **Enable** check box to compress the model geometry. This results in less memory usage leading to smaller *.nwd* files.

In the **Reduce Precision** area, select the **Coordinates** check box to reduce the precision of coordinates. On selecting this check box, the **Precision** edit box will be enabled. In this edit box, you can specify the precision value of coordinates. Select the **Normals** check box to reduce the precision of normals. Select the **Colors** check box to reduce the precision of colors. Select the **Texture Coordinates** check box to reduce the precision of texture coordinates.

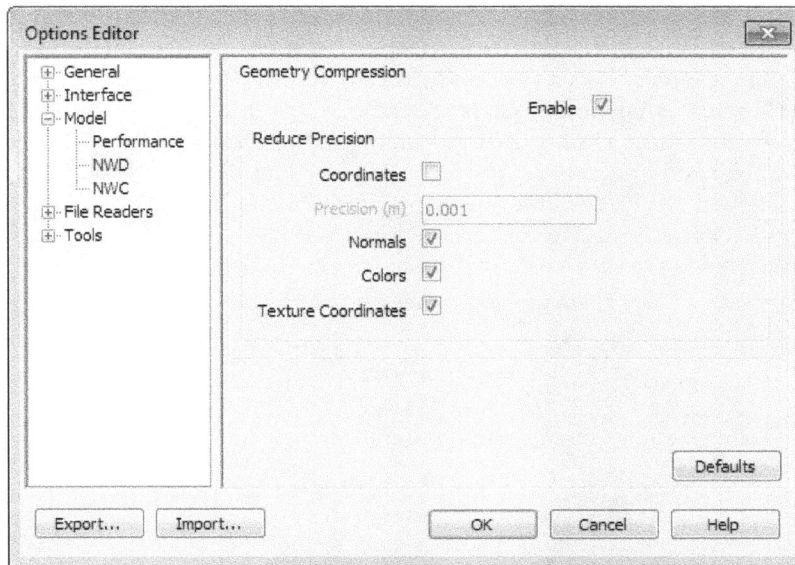

*Figure 1-21 The **NWD** page in the **Options Editor** dialog box*

NWF File Format

The *.nwf* format file contains link to the original Navisworks file along with specific data, graphics, viewpoints, search, timeliner, and information of the clash detection. The *.nwf* file does not store the model geometry, which makes it considerably smaller in size as compared to *.nwd* and *.nwc* file formats. You should always utilize *.nwf* file format while working on a project, so that the original source of file can be easily updated and re-cached.

Note
To avoid the external reference conflict while working with .nwf files make sure that you keep the .nwc and .nwf files in the same folder.

FILE READERS

Autodesk Navisworks supports a variety of additional file types such as CAD file format, laser scan formats, and other engineering design applications. The file readers provide support to these file types with several types of file formats available in Navisworks. You can modify the parameters of these file formats. To do so, expand the **File Readers** node in the right pane of the **Options Editor** dialog box. Next, choose the desired file format from the right pane under the **File Readers**; the related parameter will be displayed in the right pane of the dialog box. For example, choose the **Revit** option in the left pane of the dialog box; the related parameters will be displayed in the right pane of the window, refer to Figure 1-22. These parameters are discussed next.

The options in the **Convert element parameters** drop-down list are used to specify the parameters to be converted from Revit parameters. The **Convert element Ids** check box is used to export the Id numbers for each Revit element. The **Try and find missing materials** check box is used to find the missing material, if any. The options in the **Coordinates** drop-down list are used to specify whether to use internal or shared coordinates for file aggregation. The **Convert element properties** check box is used to convert the properties of each Revit element into Navisworks nwc properties. The **Convert URLs** check box is used to support the hyperlink in the converted file. Select the **Convert construction parts** check box to export the construction parts or clear this check box, if you want to export the original object. The **Convert linked files** check box is used to include the linked files in the exported NWC file. The **Divide File into Levels** check box is used to organize the Revit files in levels by file, category, family, and instance. The **Convert room geometry** check box is used to convert the room geometry into construction sub parts. The options in the **Convert** drop-down list are used to specify the viewing options for the Revit files.

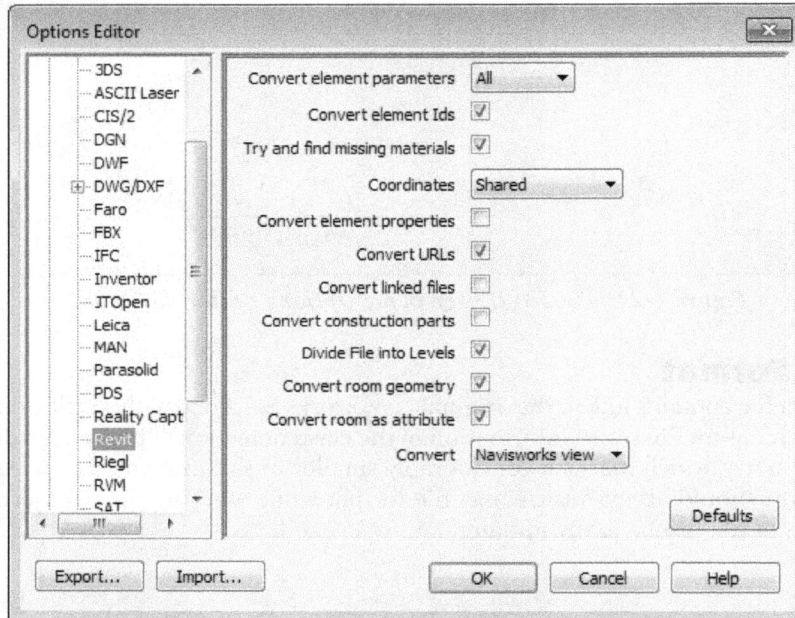

Figure 1-22   *Various options in the **Options Editor** dialog box*

MANAGING FILES IN Navisworks

In Navisworks, you can manage files in various ways. You can append geometry and data from multiple files, merge multiple files into single file, and save, publish, and email files. Various procedures and operations used while managing files are discussed next.

Opening Files

In Navisworks, you can open an existing Navisworks project or any supported design file. To open a file, choose **Open > Open** from the Application Menu; the **Open** dialog box will be displayed. In the dialog box, browse to the file location and select the appropriate file and choose the **Open** button; the file will be opened in Navisworks. In this method of file opening, Navisworks automatically uses the appropriate file reader according to the file format.

You can also open the .nwd files located on a web server. To do so, choose **Open > Open URL** from the Application Menu; the **Open URL** dialog box will be displayed, as shown in Figure 1-23. Next, specify the file address in the text box and choose the **OK** button; the file will be opened in Navisworks.

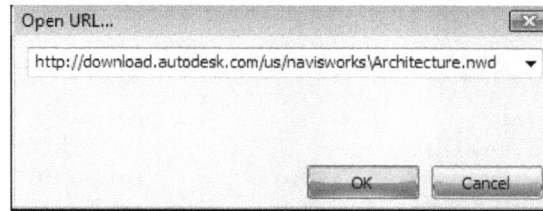

Figure 1-23 *The **Open URL** dialog box*

Appending Files

In Navisworks, you can create a composite model by adding geometry and data from multiple files to the currently open 3D model or 2D sheet. This process retains duplicate content such as geometry and mark ups. To add a file, choose **Open > Append** from the Application Menu; the **Append** dialog box will be displayed, as shown in Figure 1-24. In the dialog box, browse to the file location, select the appropriate file type, and then choose the **OK** button; the geometry and data from the file will be added to the currently opened file.

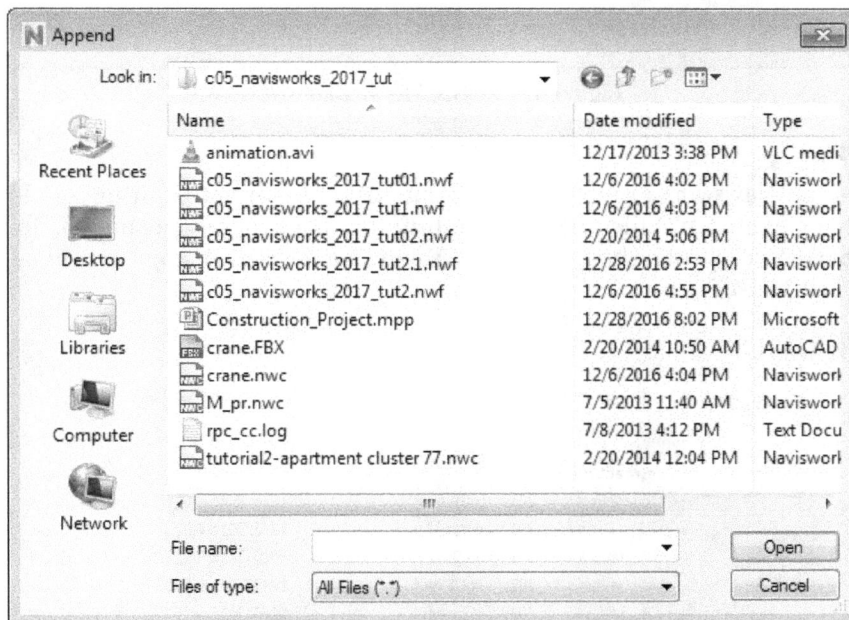

Figure 1-24 *The **Append** dialog box*

Merging Files

The users can review a model in different ways and their resultant models can be merged into a single Autodesk Navisworks file. Using the **Merge** feature, you can combine multiple copies of the same model without any duplication, but this will work only with NWF files. In NWD files, the **Merge** option will function like the **Append** option. To merge files, choose **Open > Merge** from the Application Menu; the **Merge** dialog box will be displayed, as shown in Figure 1-25. In the dialog box, browse to the file location, select the appropriate file type, and choose the **OK** button; the file will be merged with the currently opened file.

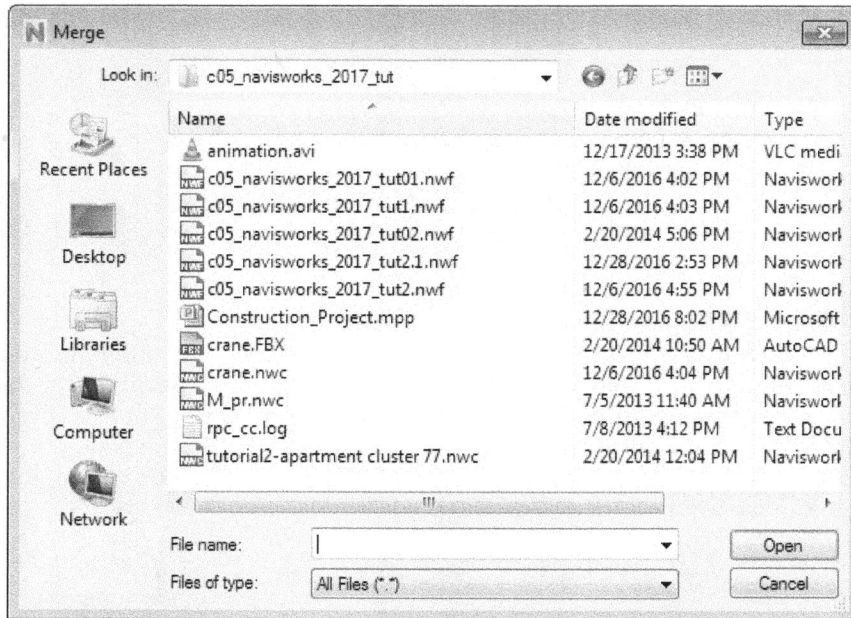

Figure 1-25 *The* **Merge** *dialog box*

Saving Files

You can save the Navisworks project either in an NWF file format by bringing all the model files together or in an NWD file format to capture the snapshot of your model. To save a file, choose the **Save As** tool from the Application Menu; the **Save As** dialog box will be displayed, as shown in Figure 1-26.

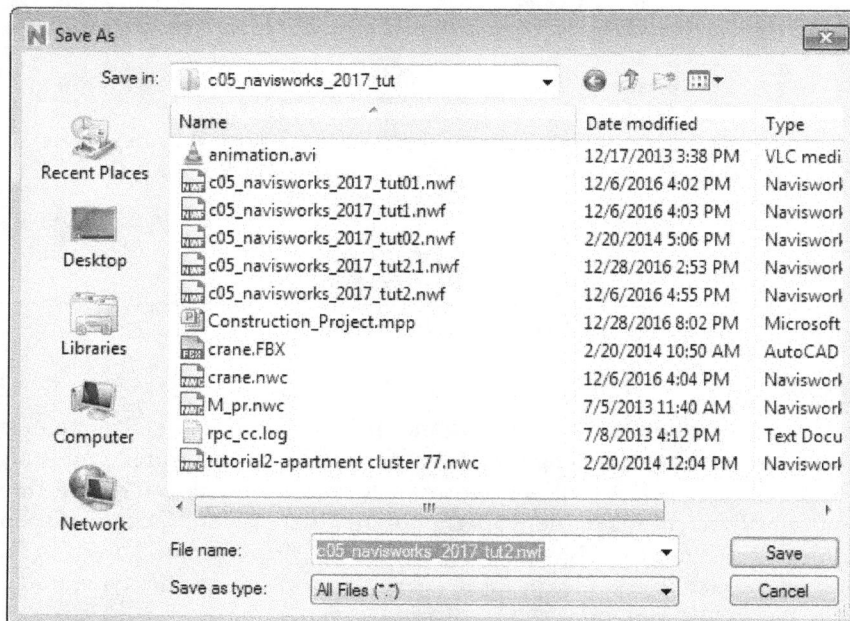

Figure 1-26 *The* **Save As** *dialog box*

In the **Save As** dialog box, browse to the file location, specify the file name, file type (*.nwd or .nwf*), and then choose the **Save** button; the file will be saved with the specified name at the specified location.

Publishing Files

When you want to share your files with others, you can do so by using the **Publish** option. Using this option, you can embed the additional document information such as title, subject, name of the author, publisher, and so on. This option provides an additional security feature file password and file expiration which allows the file to be accessed only with correct password, thus making it secure. To publish a file, choose the **Publish** option from the Application Menu; the **Publish** dialog box will be displayed as shown in Figure 1-27. Alternatively, to invoke the **Publish** dialog box, choose the **NWD** tool from the **Publish** panel in the **Output** tab.

Specify the parameters in the **Publish** dialog box and choose the **OK** button; the **Save As** dialog box will be displayed. In the **Save As** dialog box, browse to the file location, specify the file name, and choose the **Save** button; the file will be published at the specified location.

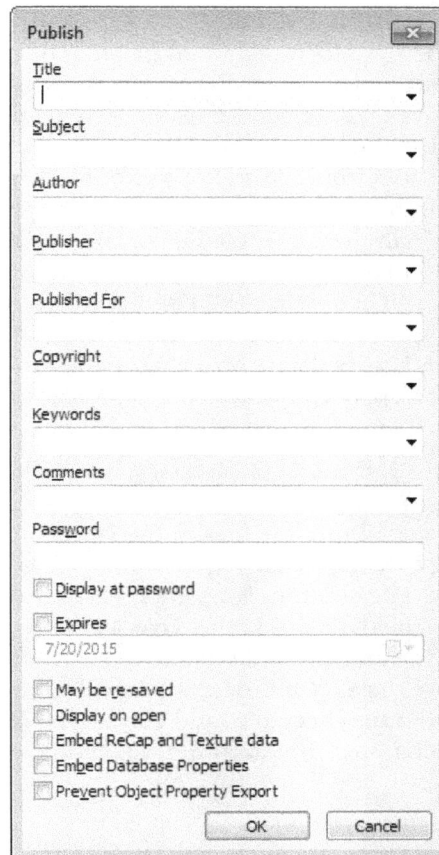

Figure 1-27 *The* **Publish** *dialog box*

Emailing Files

In Navisworks, because of the small size of the files, you can easily send and receive them through emails by using the **Send By Email** option. You can send both NWD and NWF files via email. To send an email, choose the **Send by Email** option from the Application Menu; an available mail software will open with a blank message. Your file will be attached with this message and ready to be dispatched. Alternatively, choose the **Send by Email** option from the **Send** panel in the **Output** tab.

WORKING WITH BATCH UTILITY

The **Batch Utility** allows you to append a batch of any supported format files into a single Navisworks file. You can convert multiple design files into a single NWD file. You can also generate a list of all the design files used in an integrated model in a text file. The **Batch Utility** is integrated with the **Windows Task Scheduler** option which allows you to schedule the conversions at specified times and intervals. The methods for combining and converting the files are discussed next.

Creating a List of Design Files

You can create a list of all the design files used in the current model. To do so, choose the **Batch Utility** tool from the **Tools** panel in the **Home** tab; the **Navisworks Batch Utility** dialog box will be displayed, as shown in Figure 1-28. In the left pane of the **Input** area, browse to the folder containing the design files; a list of design files will be displayed in the right pane of the **Input** area. Next, select the required files from the right pane of the **Input** area using the CTRL key and choose the **Add Files** button; files will be added to the file list box, refer to Figure 1-28. Next, choose the **Browse** button in the **As Single File** tab of the **Output** area in the dialog box; the **Save output as** dialog box will be displayed. In this dialog box, browse to the desired folder and enter the name for text file. Then, select the **File list (.*txt)** option from the **Save as type** drop-down list. Next, choose the **Save** button; the file will be saved at the specified location. Next, in the **Navisworks Batch Utility** dialog box, choose the **Run Command** button and then open the saved text file; it will display all the added design files and their paths.

Appending Multiple Files into a Single Navisworks File

You can append multiple design files into a single Navisworks file. To do so, browse to the file location and select the folder containing the files in the **Input** area; the files will be displayed in the right pane of the dialog box. Select the required files using the CTRL key. Next, add the selected files by using the **Add Files** button. Now, choose the **Browse** button in the **As Single File** tab of the **Output** area in the dialog box; the **Save As** dialog box will be displayed. Next, specify the name for the new file and select the required file format (NWD or NWF) from the **Save As** drop-down list. Choose the **Save** button; the file will be saved at the specified location. Next, select the **View file on output** check box and choose the **Run Command** button in the **Navisworks Batch Utility** dialog box; the saved file will be loaded in Navisworks with all the added files.

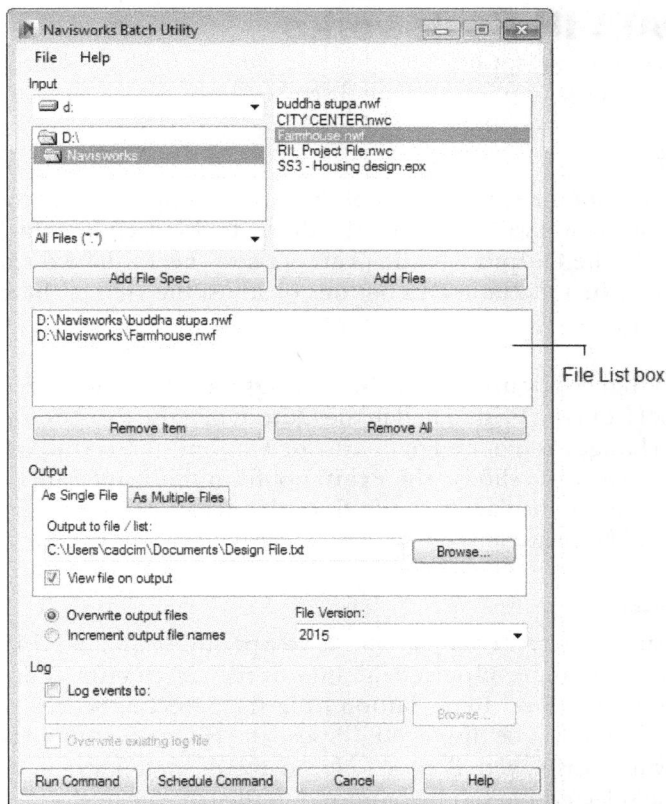

*Figure 1-28 The **Navisworks Batch Utility** dialog box*

You can also schedule the appending multiple design files into a single file. To do so, follow the same process as mentioned earlier, but instead of choosing the **Run Command**, choose the **Schedule Command** button; the **Save task file as** dialog box will be displayed. In this dialog box, browse to the desired location and choose the **Save** button; the **Schedule Task** dialog box will be displayed. In the **Schedule Task** dialog box, specify the task name if required and specify the user name and password, and then choose the **OK** button; the **Window Task Scheduler** dialog box will be displayed. In this dialog box, choose the **New** button in the **Schedule Tab**. Next, specify when and how often you want to run the task and then choose the **OK** button.

Converting Multiple Files into Single NWD File
To convert multiple design files into a single NWD file, follow the steps for selecting the files as discussed earlier. Next, select the **Output to directory** radio button in the **As Multiple Files** tab. Note that if you select the **Output to same directory as source files** option, then by default, files will be created at the same location of source files. Next, choose the **Browse** button; the **Browse For Folder** dialog box will be displayed. In this dialog box, select the desired folder and then choose the **Run Command**; the added files will be converted into the NWD files and will be saved at the specified location. You can also schedule the conversion of multiple design files into individual NWD file. To do so, instead of using the **Run Command**, choose the **Schedule Command** and for importing the files, follow the steps as discussed above.

SHARING DATA IN Navisworks

In Navisworks, there are three applications that you can use to share data: **Print** files, **Import** files, and **Export** files. These applications are discussed next.

Printing Files

In Navisworks, you can plot the existing view of the model using the **Print** tool. Before plotting the view, you need to see how it will appear on the sheet. To do so, choose the **Print Preview** tool from the **Print** panel in the **Output** tab; the print preview sheet will be displayed in the Scene View. Choose the **Zoom In** and **Zoom Out** buttons to adjust the view of the model. Choose the **Close** button to close the preview.

The print set up is configured by using the **Print Settings** tool. To change the print setup, choose the **Print Settings** tool from the **Print** panel in the **Output** tab; the **Print Setup** dialog box will be displayed. Make the changes as required in this dialog box, and choose the **OK** button to close it. Next, to print the current view, choose the **Print** tool from the **Print** panel in the **Output** tab; the **Print** dialog box will be displayed. Verify the printer settings as required and choose the **OK** button; the file will be printed.

Importing Files

In Navisworks, data such as current search criteria, viewpoints, search sets, PDS review data, PDS tags, and PDS display sets can be imported. To import the search criteria defined in other files into Navisworks, choose the **Find Items** button from the **Select & Search** panel in the **Home** tab; the **Find Items** window will be displayed. Choose the **Import** button in this dialog box; the **Import** dialog box will be displayed. Browse to locate the desired .xml file, select it and choose the **Open** button; the selected file will be imported to the current session.

You can also import the search sets that you have saved. To do so, choose **Manage Sets** from the **Home > Select & Search > Sets** drop-down; the **Sets** window will be displayed. In this window, choose the **Import Search Sets** tool from the **Import/Export** drop-down; the **Import** dialog box will be displayed. Browse to locate the desired .xml file, select it, and choose the **Open** button; the selected file will be imported as search sets in the **Sets** window.

The method of defining search criteria and saving the search sets is discussed in the later chapters.

Similarly, to import the PDS display sets, choose the **Import PDS Display Sets** option from the **Import/Export** drop-down; the **Import** dialog box will be displayed. Browse to the desired .xml file, select it, and choose the **Open** button; the file will be imported to the current session. To import a PDS tag file, expand the **Tag** panel in the **Review** tab; a drop-down will be displayed, as shown in Figure 1-29. Choose **Import PDS Tags** from this drop-down; the **Import** dialog box will be displayed. Browse to the desired *.tag file, select it and choose the **Open** button; the file will be imported in the current session.

Figure 1-29 The **Tag** *drop-down*

To import a viewpoint, right-click in the **Saved Viewpoints** window; a shortcut menu will be displayed. Select the **Import Viewpoints** option from this shortcut menu; the **Import** dialog box will be displayed. Browse to the desired .xml file, select it and choose the **Open** button; the viewpoint will be imported to the current session.

Exporting Files

You can export data into various file formats from Navisworks such as 3D DWF/ DWFx, Google Earth KML, FBX files, and so on. To export a 3D model as DWF/DWFx file format, choose the **3D DWF/DWFx** option from the **Export Scene** panel in the **Output** tab; the **Export** dialog box will be displayed. Browse to the folder, where you want to locate the file and enter the file name in the **File name** edit box. Choose the desired format to save the file from the options in the **Save as type** drop-down list and choose the **Save** button; the file will be exported and saved at the specified location.

To export the Google Earth KML files, choose the **Google Earth KML** tool from the **Export Scene** panel in the **Output** tab; the **KML Options** dialog box will be displayed, as shown in Figure 1-30. The options in the dialog box are discussed next.

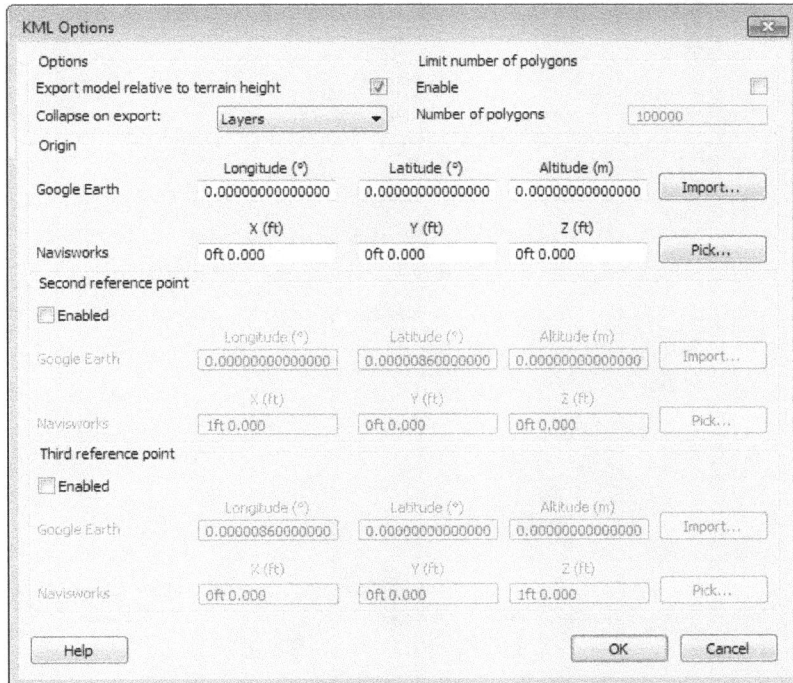

*Figure 1-30 The **KML Options** dialog box*

Select the **Export model relative to terrain height** check box in the **Options** area to put Google Earth in a mode where all heights are measured from ground. Clear this check box to measure the height from the sea level. To specify the model hierarchy in the exported file, select the required option from the **Collapse on export** drop-down list. To limit the amount of geometry exported into the output file, select the **Enable** check box in the **Limit number of polygons** area. The **Origin**, **Second reference point**, and **Third reference point** areas are used to show the reference point on Google Earth surface. To select the reference points from the Scene

View, choose the **Pick** button from the required area and click at a location in the Scene View. Next, choose the **OK** button; the **Export** dialog box will be displayed. Enter the file name and location, and choose the **Save** button; the file will be exported and saved at the specified location.

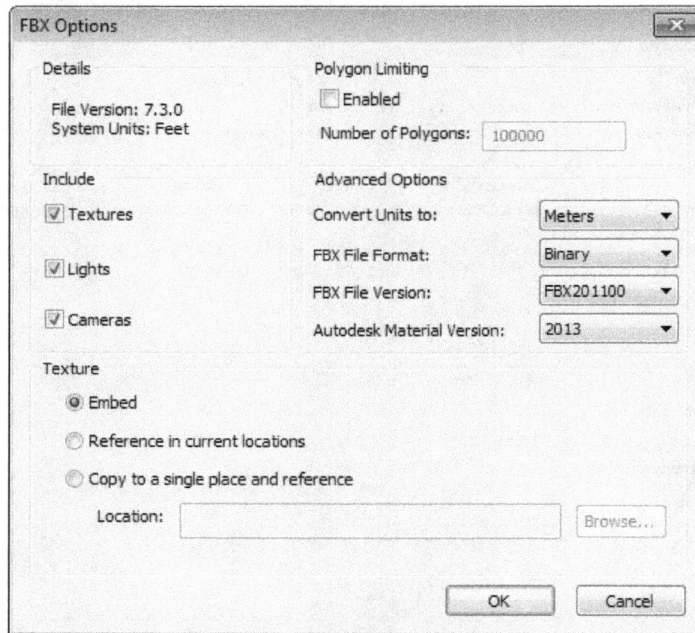

To export the FBX file, choose the **FBX** tool from the **Export Scene** panel in the **Output** tab; the **FBX Options** dialog box will be displayed, as shown in Figure 1-31. To limit the amount of geometry exported into the output file, select the **Enabled** check box in the **Polygon Limiting** area of the **FBX Options** dialog box and enter the number of polygons in the **Number of Polygons** edit box. To include **Textures**, **Lights**, and **Cameras**, select the corresponding check boxes in the **Include** area. Specify the units to be used in the exported FBX file by selecting an option from the **Convert Units to** drop-down list. Specify the format of exported file by selecting the option in the **FBX File Format** drop-down list. Select the version of exported file from the **FBX File Version** drop-down list. Choose the **OK** button to close the dialog box; the **Export** dialog box will be displayed. In this dialog box, enter the file name and location, and then choose the **Save** button; the file will be exported and saved at the specified location.

*Figure 1-31 The **FBX Options** dialog box*

Exporting Images and Animations

In Navisworks, you can export an image as JPEG or PNG file. You can also export the rendered image in different file formats. To export an image as JPEG image, choose the **Image** tool from the **Visuals** panel in the **Output** tab; the **Image Export** dialog box will be displayed, as shown in Figure 1-32. The options in this dialog box are discussed next.

In the **Output** area, select the format of the image from the **Format** drop-down list. For example, to export the image as JPEG, select the **JPEG** option from the list.

In the **Renderer** area, you can select a renderer from the **Renderer** drop-down list. Select the **Presenter** option to render the image with Presenter. Select the **Viewport** option to quickly render the image. Select the **Autodesk** option to render the image with Autodesk Rendering.

In the **Size** area, you can adjust the size of the image. In the **Options** area, you can smooth the edges of the exported image. Note that the **Options** area will be activated only if you have selected the **Viewport** option from the **Renderer** drop-down list in the **Renderer** area. After specifying all the parameters, choose the **OK** button; the **Save As** dialog box will be displayed. In this dialog box, specify the file name and location, and then choose the **Save** button; the file will be saved as an image at the specified location.

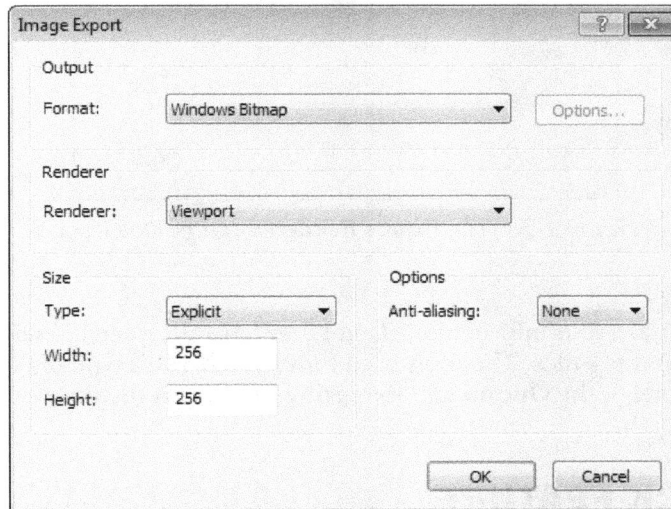

*Figure 1-32 The **Image Export** dialog box*

You can also export a rendered image in different file formats. To do so, choose the **Rendered Image** tool from the **Visuals** panel in the **Output** tab; the **Export Rendered Image** dialog box will be displayed, as shown in Figure 1-33. Select the type of image to be exported from the **Type** drop-down list. Choose the **Browse** button to specify the location. To set the size of file, specify the **Type**, **Width**, and **Height** parameters in the **Size** area. After specifying all the parameters, choose the **OK** button; the file will be saved as an image at the specified location.

Similarly, you can export a file as a Piranesi Epix file. To do so, choose the **Piranesi Epix** tool from the **Visuals** panel in the **Output** tab; the **Piranesi Epix** dialog box will be displayed. In this dialog box, choose the **Browse** button; the **Save As** dialog box will be displayed. In the **Save As** dialog box, specify the file name and location, and then choose the **Save** button. Next, specify the size in the **Size** area and choose the **OK** button; the file will be exported and saved at the specified location.

You can also export the animation created in Navisworks as an AVI file (Audio Video Interleave). To do so, choose the **Animation** tool from the **Visuals** panel in the **Output** tab; the **Animation Export** dialog box will be displayed. In the **Source** area of the dialog box, select the source of animation from the **Source** drop-down list. In the **Renderer** area, select a renderer from the **Renderer** drop-down list. Set the size of the animation by using the options available in the **Size** area. After specifying all the parameters, choose the **OK** button; the **Save As** dialog box will be displayed. In the **Save As** dialog box, specify the file name and location and choose the **Save** button; the file will be exported and saved at the specified location.

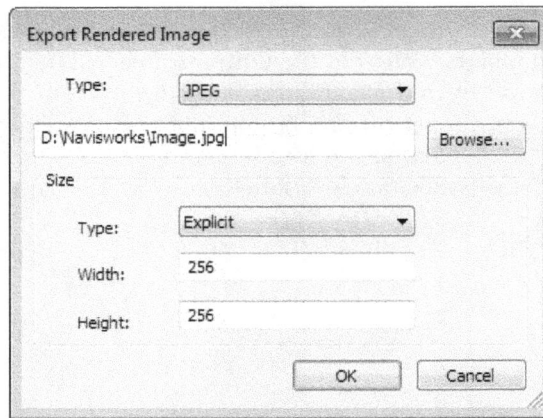

Figure 1-33 The Export Rendered Image dialog box

Export Data

You can also export clash tests information, TimeLiner schedule, search set files, viewpoint files, viewpoint reports, and tag files. The tools used for exporting this type of data are available in the **Export Data** panel of the **Output** tab. Exporting this data is discussed in detail in the later chapters.

SWITCHBACK FEATURE

In Navisworks, you can switchback to the native software by using the Switchback feature. This feature allows you to select an object in Navisworks and then locate, review, and modify that object in the native software where the model has been created. For example, to use the switchback feature with Revit, ensure that Revit is installed on the system. In the Revit software, choose the **Add-Ins** tab and select **External Tools > Navisworks Switchback**. Now return to Navisworks and select the required object in the Scene View; the **Item Tools** contextual tab will be displayed in the ribbon. Next, choose the **Switchback** tool from the **Switchback** panel in the **Item Tools** tab; the **Resolve** dialog box will be displayed. In this dialog box, you need to specify the location of the Revit file and then choose the **OK** button. The 3D view of the model will be loaded in the Revit. Next time when you select an object in the model in Navisworks and then use the **Switchback** feature, the model with the current view and the selected object will be loaded in Revit.

If the **Add-Ins** tab is not available in the Revit software, then go to **Control Panel > Programs and Features**. Double-click in the **Autodesk Navisworks Manage 2017 Exporter Plug-ins** displayed in the program list; the **Setup Initialization** window will be displayed. Choose the **Add or Remove Features** option; the description box will be displayed. In the description box, ensure that the **Revit 2017 Plugin** option has a green tick mark next to it. Next, choose the **Update** button and restart the computer; the **Add-Ins** tab will be added to the Revit software.

SETTING UNITS IN Navisworks

When you open a file in Navisworks, the units of the model elements in the file will be same as assigned in the original CAD application. You can also customize the display units. To do so, choose the **Options** button from the Application Menu; the **Options Editor** dialog box will be displayed. In the dialog box, expand the **Interface** node in the left pane of the dialog box and then select **Display Units**; various options will be displayed in the right pane of the dialog box, as shown in Figure 1-34. In the right pane of the dialog box, select the required linear units from the **Linear Units** drop-down list. Select the angular units from the **Angular Units** drop-down list. Specify the number of decimal places in the **Decimal Places** edit box. If you have selected the fractional unit instead of a decimal unit, you can also define the level of fraction to display the units by selecting options from the **Fractional Display Precision** drop-down list.

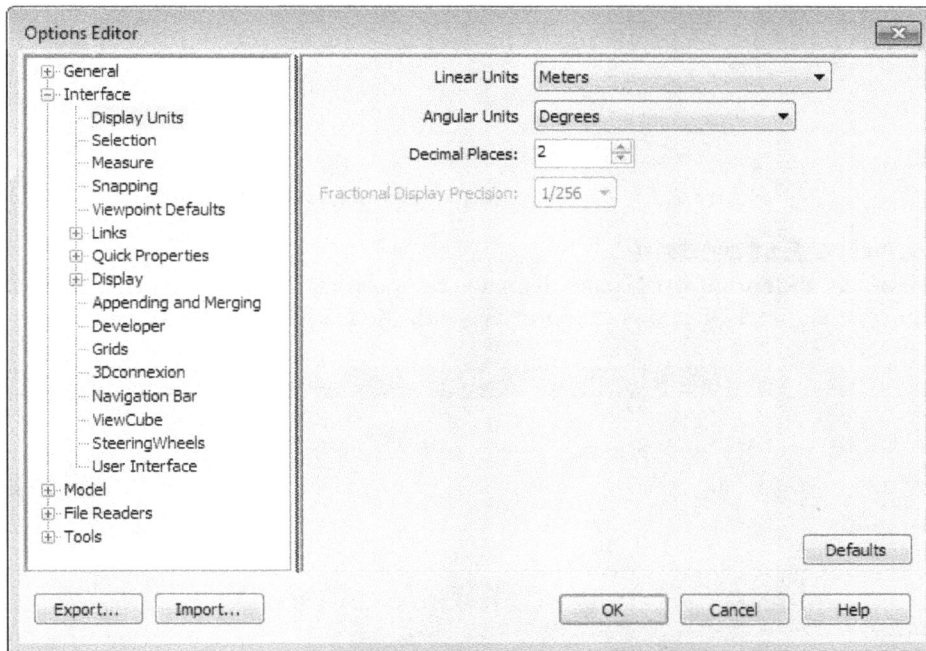

*Figure 1-34 Selecting the **Display Units** option in the **Options Editor** dialog box*

Navisworks WORKSPACES

In Navisworks, there are four pre-configured workspaces. You can customize these workspaces for the basic and advanced users. You can also set up a workspace as required. To use the pre-configured workspaces, select a workspace from the **View > Workspace > Load Workspace** drop-down. The options in the drop-down are: **Safe Mode**, **Navisworks Extended**, **Navisworks Minimal**, and **Navisworks Standard**. These workspaces are discussed next.

Safe Mode

The **Safe Mode** workspace displays the interface with minimal features, as shown in Figure 1-35.

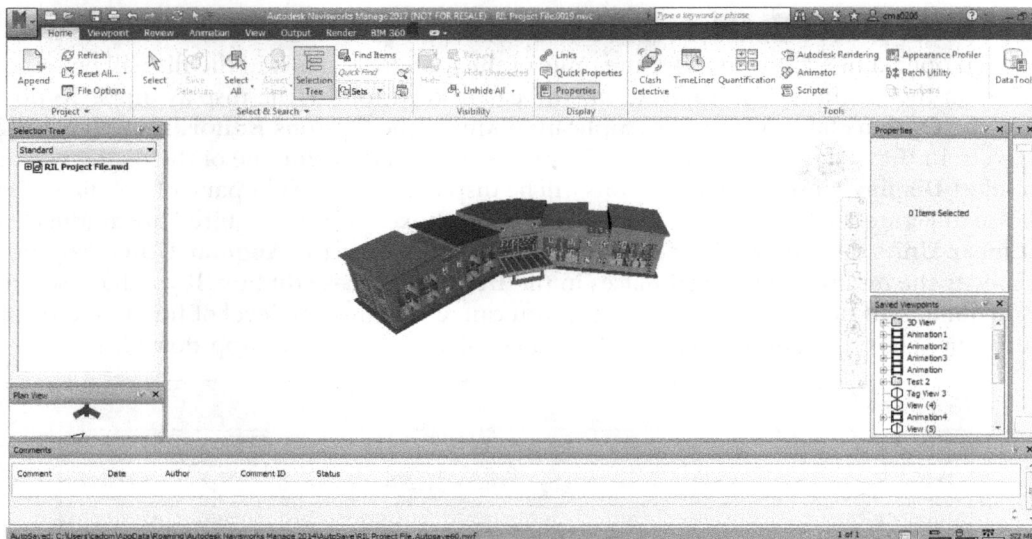

Figure 1-35 *The **Safe Mode** workspace in Navisworks*

Navisworks Extended

The **Navisworks Extended** workspace displays the interface with features recommended for advanced users. Figure 1-36 shows the interface with the **Navisworks Extended** workspace.

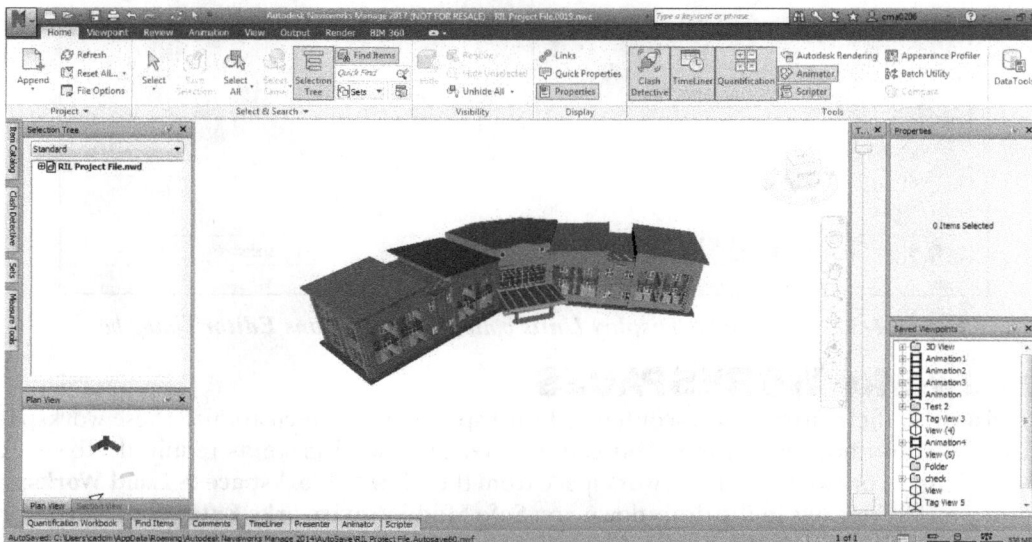

Figure 1-36 *The **Navisworks Extended** workspace*

Navisworks Standard

The **Navisworks Standard** workspace displays the interface with commonly used windows. These windows will be displayed as tabs in the Scene View. Figure 1-37 shows the interface with the **Navisworks Standard** workspace.

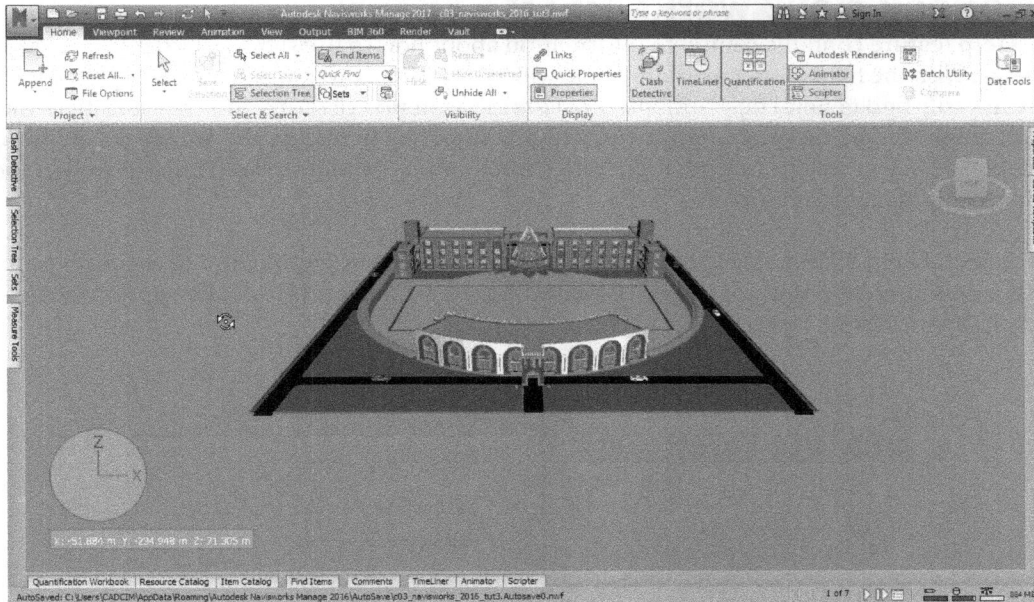

Figure 1-37 The **Navisworks Standard** *workspace*

Navisworks Minimal

The **Navisworks Minimal** is the default workspace that will be loaded when you start Navisworks. In this workspace, the dockable windows are not displayed in the Scene View and therefore you get the maximum area to work in.

You can save the current workspace configuration as a new workspace. To do so, choose the **Save Workspace** tool from the **Workspace** panel in the **View** tab; the **Save Current Workspace** dialog box will be displayed. In the dialog box, browse to the desired location and choose the **Save** button to save the workspace. The workspace will be saved as **Workspace files (*.xml)** file type.

Next, to load the saved workspace, select the **More Workspaces** option from the **View > Workspace > Load Workspace** drop-down list; the **Load Workspace** dialog box will be displayed. In the dialog box, browse to the file location, select the workspace and choose the **Open** button; the saved workspace will be loaded.

Autodesk Navisworks HELP SYSTEM

The **Help** feature contains complete information about how to use the Navisworks software. With the help of user assistance, you will learn to use Navisworks efficiently. You can easily find general descriptions, procedures, definition of terms, information about tools, details about the dialog boxes, and so on, with the use of Help feature. In Autodesk Navisworks Manage 2017, you can access online help documentation (Autodesk WikiHelp) as well as local (offline) help documentation.

To access the help feature, click, on the **Help** drop-down on the right of the **Favorites** button; a flyout containing help options will be displayed, as shown in Figure 1-38. The options to access the help are discussed next.

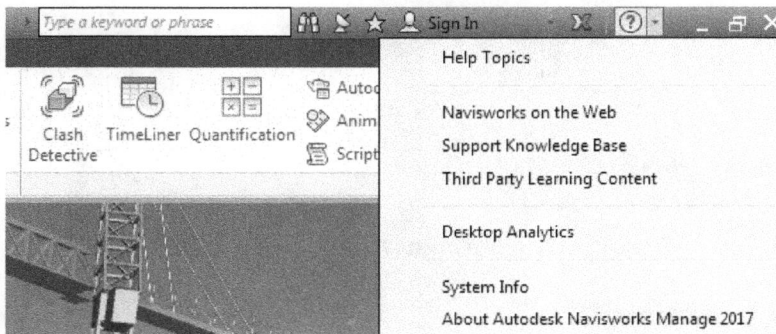

Figure 1-38 The flyout displayed

Using the Local Navisworks Manage 2017 Help Feature

To access the local Navisworks Manage 2017 Help, choose the **Help Topics** option from the **Help** menu; the Autodesk Navisworks **Help** page will be displayed, as shown in Figure 1-39. Note that if you want to access the Help page offline, then select the **Always use offline help** check box from **Options > Options Editor > General > Environment** from the Application Menu.

In the **Help** page, the left pane has various tabs that contain links of various help topics and different method to find the information. When you click on a link on this pane, the detailed information on that topic is displayed on the right. The first tab from the left is the **Contents** tab. This tab contains links of various topics available in the documentation. The next tab is the **Index** tab. On chosing this tab, an alphabetical list of keywords related to the various topics listed in the **Contents** tab will be displayed. With the help of the **Index** tab, you can access the information quickly. The tab right next to the **Index** tab is the **Search** tab. By using the **Search** tab, you can search all topics listed in the **Contents** tab. You can type the keywords in the **Search** box. The keywords are not case sensitive. Note that, you can only search for letters and numbers in the search box but not the punctuation marks. While searching for phrases, use double quotation marks to wrap the words so that they appear in defined sequence and display the correct result.

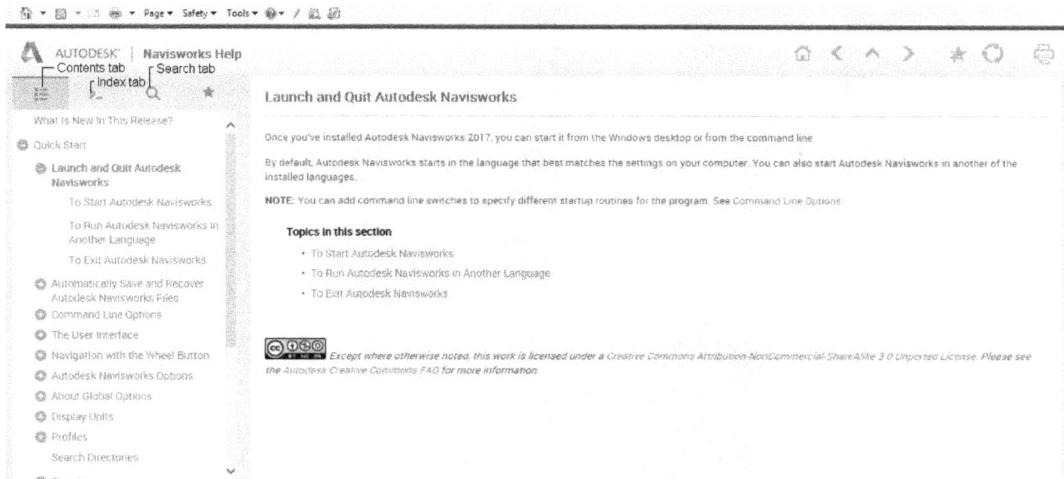

Figure 1-39 *Partial view of the Navisworks **Help** page*

In Navisworks **Help** page, there are various options to browse through the contents. You can use the the three arrow buttons displayed in the upper right corner of the page, refer to Figure 1-39, to browse next, up, and back. Also, you can return to the home page by choosing the **Home** button displayed ahead of the arrow buttons.

You can also print the topics in the **Help** window by using the **Print** option. To print a help topic, click on the topic you want to print and then right-click in the right side of the window; a shortcut menu will be displayed. Choose the **Print** option from the menu; the **Print** dialog box will be displayed. In this dialog box, choose the **Print** button.

Using the Autodesk Online Help

In Autodesk Navisworks Manage 2017, **Autodesk Online Help** has been introduced to access various help topics online. You can access the **Autodesk Online Help** for Navisworks Manage by choosing the **Help** tool from the **Info centre**. On doing so, the **Autodesk Help** page will be displayed, as shown in Figure 1-40. In the **Browse Help** area of this page, click on the required topic; the content related to the selected topic will be displayed in the help page.

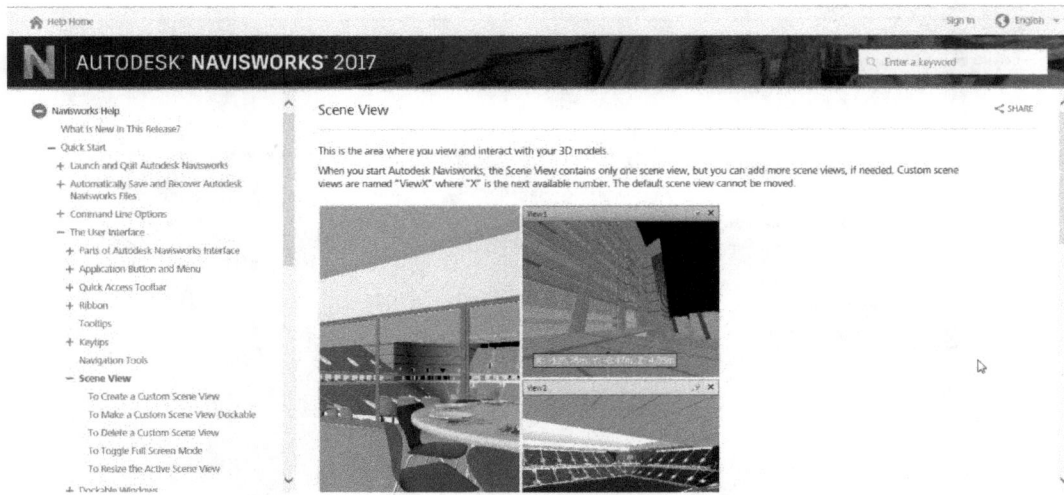

*Figure 1-40 The **Autodesk Online Help** page*

Using the Context Sensitive Help

If you need help regarding a selected tool or dialog box, Autodesk Navisworks Manage provides several options to access the relevant information. Many dialog boxes contain the **Help** button that can be used to view help for the options available in that dialog box. If the **Help** button is not available, you can press the F1 key to access related information. To inquire about a tool, place the cursor over it; the **Help** message box will be displayed. Now, you can press the F1 key to view more help topics.

Self-Evaluation Test

Answer the following questions and then compare them to those given at the end of this chapter:

1. The _____ window is used to import 2D sheets and 3D models.

2. The _____ feature is used to combine multiple models together without any duplication.

3. The _____ option is used to send Navisworks files to other users via email.

4. The _____ workspace displays the interface with minimum features.

5. The _____ comprises various navigation tools.

6. The NWC files are heavier than their original CAD applications. (T/F)

7. The NWD files contain model geometry with relevant data. (T/F)

8. Using the **Open URL** option, you can open the NWD files located on a web server. (T/F)

9. The **Batch Utility** option is used to print the Navisworks files. (T/F)

10. The NWF files does not contain any link to the original files. (T/F)

Review Questions

Answer the following questions:

1. Which of the following options is used to open a file in Navisworks?

   a) **Send by Email**   b) **Print**
   c) **Open**   d) **Export**

2. Which of the following options is used to display only tab titles in the Ribbon?

   a) **Minimize to Panel Titles**   b) **Minimize to Tabs**
   c) **Cycle through All**   d) **Open URL**

3. Which of the following options is used to split the Scene View horizontally?

   a) **Split Vertical**   b) **Split Horizontal**
   c) **Show Title Bars**   d) **Background**

4. Which of the following options is used to display the interface for advanced users?

   a) **Safe Mode**   b) **Navisworks Standard**
   c) **Navisworks Extended**   d) **Navisworks Minimal**

5. Which of the following options is not available in the Application Menu?

   a) **Open**   b) **Open URL**
   c) **Print**   d) **Load Workspace**

6. The File Options are used to save the changes within a particular file. (T/F)

7. In Navisworks, you cannot customize lighting in the Scene View. (T/F)

8. The Global Options are saved for all Navisworks sessions. (T/F)

9. The NWC file is a Navisworks Document File type. (T/F)

10. In Navisworks, you cannot export data in different file formats. (T/F)

Answers to Self-Evaluation Test

1. Project Browser, 2. Merge, 3. Send by Email, 4. Safe Mode, 5. Navigation Bar, 6. F, 7. T, 8. T, 9. F, 10. F

Chapter 2

Exploring the Navigation Tools in Navisworks

Learning Objectives

After completing this chapter, you will be able to:
- *Use Head-Up Display*
- *Set view orientations*
- *Use ViewCube*
- *Use navigation tools*
- *Use camera*
- *Use reference views*

INTRODUCTION

In Navisworks, you can explore an integrated project model using navigation tools. These tools provide a real-time navigation experience in a project. Using these navigation tools, you will be able to move around and explore the model in the Scene View. In this chapter, you will learn how to use various navigation tools and features in Navisworks.

USING THE Head-Up Display FEATURE

The **Head-Up Display** (**HUD**) feature in Navisworks helps in displaying information about the camera location and model orientation in 3D workspace. It consists of three display elements: XYZ triad, camera position, and grid position. You can display the XYZ triad by selecting the **XYZ Axes** option from the **HUD** drop-down list in the **Navigation Aids** panel of the **View** tab. On selecting this option, the information about the orientation of the model will be displayed, as shown in Figure 2-1.

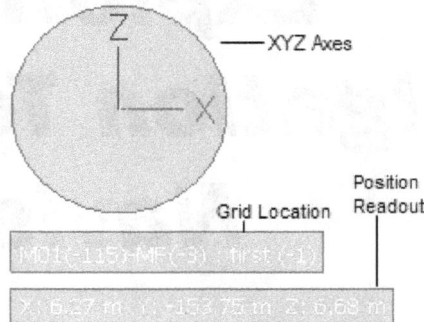

Figure 2-1 Displaying the camera location, position, and grid location using the Head-Up Display

Select the **Position Readout** option from the **HUD** drop-down list to view the absolute camera position, refer to Figure 2-1. Select the **Grid Location** option to view the grid and the location of the camera relative to the grid system currently used in the model.

SETTING VIEW ORIENTATIONS

In a project, you can set the view orientations by using various options. Although, Navisworks uses the cartesian coordinate system (X, Y, Z), the pointing direction of these axes is not fixed. By default, the Z axis points upward and the Y axis points toward north. However, you can change the up and north directions as per your requirement. To change the up direction, right-click in the Scene View; a shortcut menu will be displayed. Next, choose **Viewpoint > Set Viewpoint Up > Set Up** from the shortcut menu, refer to Figure 2-2.

For example, to set the view orientation of X axis upward, choose the **Set Up + X** option from the shortcut menu, as shown in Figure 2-2. On doing so, you will notice that the XYZ axis triad in the Scene View displays the X axis pointing upward. Also, the view of the model is changed accordingly. For example, if you are using the **Orbit** tool and you have selected the **Set Up+ Y** option from the shortcut menu, the axis of rotation will be Y axis.

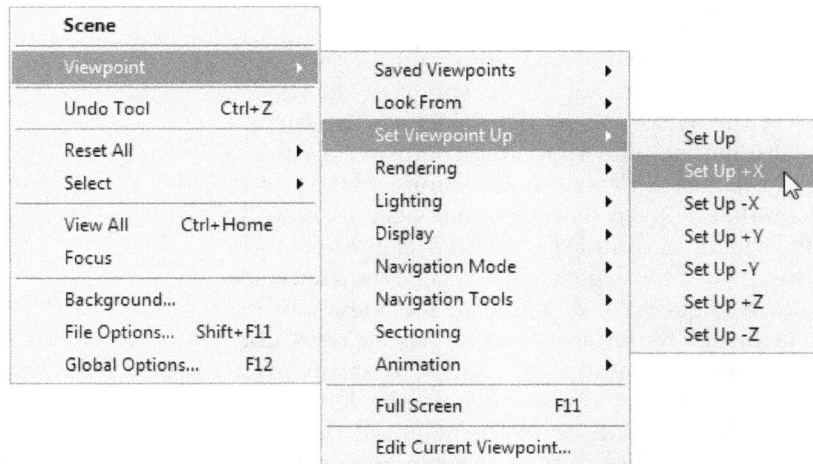

Figure 2-2 *The options to change the* **Up** *vectors*

You can also change the overall orientation of a view. To do so, choose the **File Options** option from the **Project** panel in the **Home** tab; the **File Options** dialog box will be displayed. Next, choose the **Orientation** tab in the dialog box; the options in this tab will be displayed, as shown in Figure 2-3. The options in the **Up** area are used to change the up direction for the view. For example, to orient the view with Z axis in the upward direction, specify **1** in the **Z** edit box and **0** in the **X** and **Y** edit boxes in the **Up** area. Similarly, in the **North** area, you can change the north direction for the view. For example, to orient the view with Y axis in north direction, specify **1** in the **Y** edit box and **0** in both the **X** and **Z** edit boxes in the **North** area. On doing so, the view will be oriented with Z axis upward and Y axis in north. You can notice the changes will be reflected in the ViewCube.

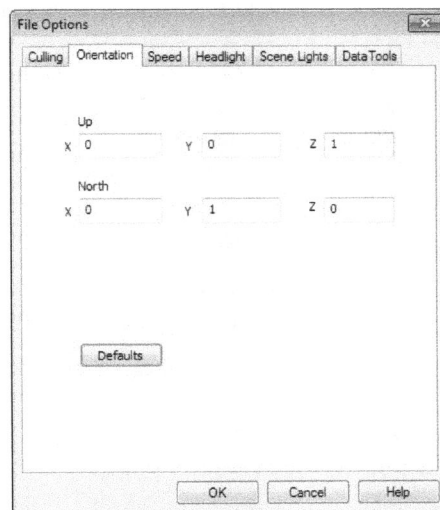

Figure 2-3 *The options displayed under the* **Orientation** *tab*

THE VIEWCUBE

Navisworks provides you with a navigation control called the
ViewCube that can be used to set the viewpoint of the model.
You can also use the ViewCube to visually assess the current
orientation of the model. The ViewCube comprises a cube, a
compass ring at the base with various directions marked on it,
and a home icon that helps to set the default view, as shown in
Figure 2-4. By default, it is always displayed at the top right
corner in the Scene View. If it is not visible, choose the **ViewCube**
button from the **Navigation Aids** panel in the **View** tab; the
ViewCube will be displayed in the Scene View. Using ViewCube,
you can rotate the model around a pivot point. To do so, place
the cursor on the ViewCube, press and hold the left mouse
button, and then drag the cursor in the required direction.

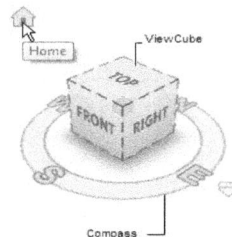

Figure 2-4 The ViewCube

Alternatively, click on any face of the ViewCube to get the corresponding view of the model.
The compass below the ViewCube indicates the north direction of the model. To rotate the
model using compass, place the cursor on the compass; it will be highlighted. Press and hold
the left mouse button and drag the cursor in the required direction. Alternatively, if you want
to display any particular view of the model, such as the east view, click on the key letter **E** on the
compass; the model will rotate in that direction.

You can configure various settings of the ViewCube such as its opacity level, size, display of
ViewCube, and the compass. To do so, invoke the **Options Editor** dialog box by choosing the
Options button from the **Application Menu**. In the left pane of the dialog box, expand the
Interface node and then select the **ViewCube** option displayed under this node; several options
will be displayed in the right pane of the **Options Editor** dialog box, as shown in Figure 2-5.

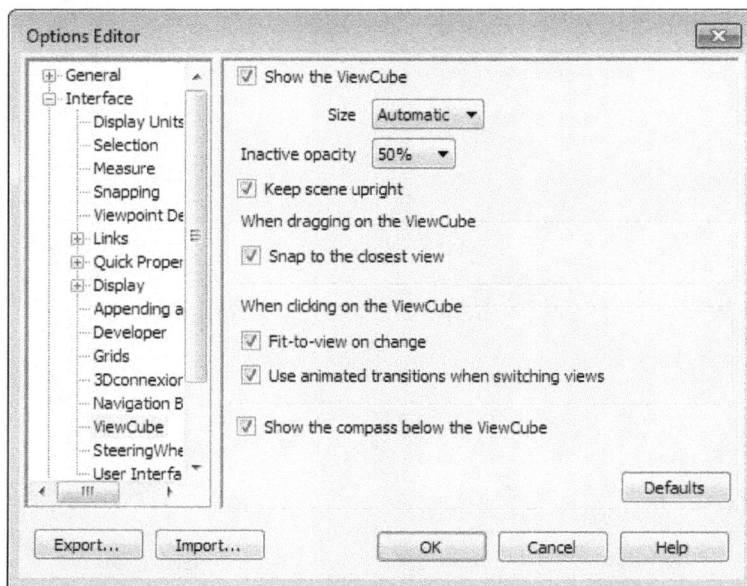

*Figure 2-5 The options displayed on selecting the **ViewCube** option*

You can toggle the display of the ViewCube by selecting or clearing the **Show the ViewCube** check box. Select the option from the **Size** drop-down list to configure the size of the ViewCube. By default, the **Automatic** option is selected in the **Size** drop-down list. To control the opacity level when the ViewCube is inactive, select the required value from the **Inactive opacity** drop-down list. Select the **Keep scene upright** check box to produce inverted orientations while dragging the ViewCube. In the **When dragging on the ViewCube** area, select the **Snap to the closest view** check box to snap the ViewCube snaps to one of the fixed views. Select the **Fit-to-view on change** check box in the **When clicking on the ViewCube** area to zoom in and out the Scene View while rotating the ViewCube. Select the **Use animated transitions when switching views** check box to display animated transitions. Select the **Show the compass below the ViewCube** check box to display the compass below the ViewCube.

In Navisworks, you can save a particular view of the model as the **Home** view. You can switch to this view by choosing the **Home** button in the ViewCube, refer to Figure 2-4. Alternatively, right-click on the ViewCube; a shortcut menu will be displayed, as shown in Figure 2-6. Now, choose the **Set Current View as Home** option from the menu; the current view will be saved as home view. You can also switch between the perspective and orthographic view projection modes of the model. To do so, choose the **Perspective** or **Orthographic** option from the shortcut menu, refer to Figure 2-6. You can also access the **ViewCube Options** page and the **Home** view using this shortcut menu.

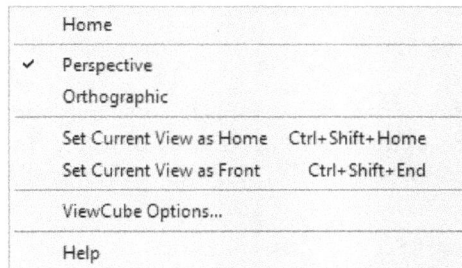

Figure 2-6 The shortcut menu displayed on right-clicking on ViewCube

THE NAVIGATION BAR

The Navigation Bar contains various tools such as **Pan**, **Orbit**, **Zoom**, **Look**, **Walk**, and **Fly**, refer to Figure 2-7. These navigation tools make it much comfortable and easier to navigate in the model. The Navigation Bar can be docked at the desired location in the Scene View. Also, the tools in Navigation Bar can be customized as per the requirement.

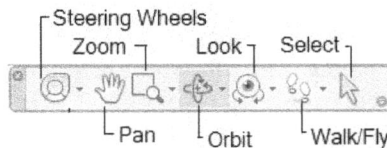

Figure 2-7 The Navigation Bar

By default, the Navigation Bar is displayed at the right corner of the Scene View. If it is not displayed, then choose the **Navigation Bar** button from the **Navigation Aids** panel in the **View** tab; the Navigation Bar will be displayed in the Scene View. If you want to hide this bar, click on the same button again. By default, the docking position of the Navigation Bar is linked with the ViewCube. To unlink them, click on the **Customize** arrow in the Navigation Bar; a shortcut menu will be displayed, as shown in Figure 2-8. Choose the **Docking Positions** option from the shortcut menu; a cascading menu will be displayed. Choose the **Link To ViewCube** option from the cascading menu; the ViewCube will be unlinked with the Navigation Bar. Now, you can move the Navigation Bar at any desired location in the Scene View without affecting the location of the ViewCube.

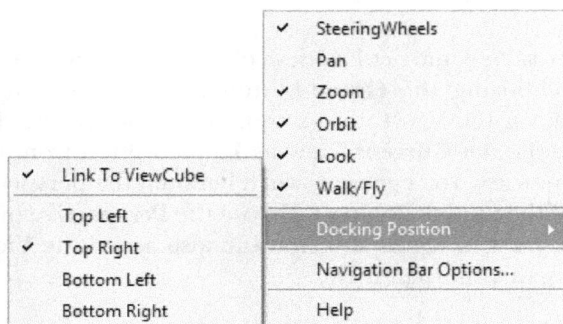

Figure 2-8 *The shortcut menu displayed*

You can also control the display of navigation tools in the Navigation Bar. To do so, choose the navigation tools that you want to display in the Navigation Bar from the shortcut menu, refer to Figure 2-8. In the next section, you will learn about various tools in the Navigation Bar.

Pan Tool

The **Pan** tool is used to move the model in the Scene View. During panning, the model moves along the direction of movement of the mouse. For example, if you drag the cursor in the upward direction, the model will move up; and if you drag the cursor in the downward direction, the model will move down. When you invoke this tool, a four sided arrow will be displayed in the Scene View. You can access this tool from the Ribbon as well. To do so, choose the **Pan** tool from the **Navigate** panel in the **Viewpoint** tab. You can also use the middle mouse button for panning while using the other tools such as **Orbit** and **Look**.

Zoom Tools

The Zoom tools are used to increase or decrease the view size of a model. You can increase or decrease the view size of any particular item or point in a model. To do so, place the zoom cursor on that particular point; a pivot point will be displayed. The pivot point can be changed by changing the location of the cursor. When you click on the **Zoom** drop-down in the Navigation Bar; a list of tools will be displayed. You can access these tools from the Ribbon as well. To do so, click on the **Zoom** drop-down in the **Navigate** panel of the **Viewpoint** tab; a list of tools will be displayed, as shown in Figure 2-9. The tools in the **Zoom** drop-down are discussed next.

The **Zoom Window** tool is used to zoom the specified area. To zoom an area using this tool, choose this tool and then draw a window around the area to be enlarged; the area within the rectangle window will be enlarged. The **Zoom** tool is used to zoom the view by clicking or dragging the cursor. When you choose this tool; a magnifying-glass shaped cursor is displayed. Press and hold the left mouse button and drag the cursor in upward or downward direction to zoom the selected area. Choose the **Zoom Selected** tool from the drop-down to zoom the selected object or group of selected objects to get an enlarged view of the respective region. If no object is selected in the model, then the **Zoom Selected** tool will work as the **Zoom All** tool and will zoom the model to the maximum possible magnification.

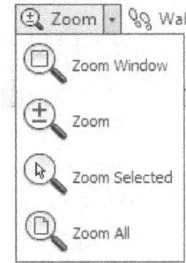

Figure 2-9 The Zoom drop-down

Orbit Tools

The Orbit tools allow you to visually maneuver around the 3D objects to obtain different views. While orbiting, the camera location changes whereas the view remains fixed. To orbit around an object, click on the **Orbit** drop-down in the Navigation bar; a list of tools will be displayed. You can use the desired tool from the list. You can access these tools from the Ribbon as well. To do so, click on the **Orbit** drop-down in the **Navigate** panel of the **Viewpoint** tab; a list of tools will be displayed, as shown in Figure 2-10. The tools in the **Orbit** drop-down are discussed next.

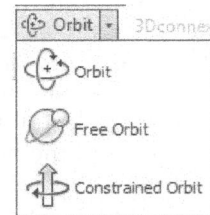

Figure 2-10 The Orbit tool drop-down

The **Orbit** tool is used to move the camera around the pivot point of the model. Choose the **Free Orbit** tool to rotate the model freely around the central point about any direction. Choose the **Constrained Orbit** tool to rotate the model around the central point. On choosing this tool, the model will rotate in the left and right directions but its upward and downward movement will be constrained. On invoking this tool, the model will rotate around a fixed axis.

You can also use the **Pan**, **Zoom**, and **Orbit** tools together. For example, if you are rotating a model by using the **Orbit** tool and you need to zoom in/out for a better view, you can roll the middle mouse button forward/backward.

Look Tools

The Look tools are used to adjust the view of a model. To use these tools, click on the **Look** drop-down in the Navigation Bar; a list of tools will be displayed. You can access these tools from the Ribbon as well. To do so, click on the **Look** drop-down in the **Navigate** panel of the **Viewpoint** tab; a list of tools will be displayed, as shown in Figure 2-11. The tools in the **Look** drop-down are discussed next.

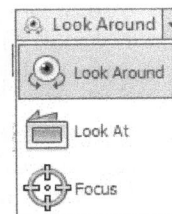

Figure 2-11 The Look drop-down

Choose the **Look Around** tool to look around the model from the current position. In this process, you can look around the scene without changing your location. Choose the **Look At** tool to look at a particular point or object in the model. When you choose this tool, the shape of the cursor is changed in the Scene View. Click on the specific object to analyze it. On doing so, the view will be adjusted. Note that in this case, only the camera position will be adjusted not the zoom factor. Choose the **Focus** tool to look at a particular point in the scene without changing the position of the camera. When you choose this tool, a cursor will be displayed in the Scene View. Place the cursor on the object you need to look at and click; the view will be adjusted according to the selected object.

Walk and Fly Tools

The **Walk** and **Fly** tools are used to walk/fly around the model to have a real-time experience during navigation. The **Walk** tool is used to walk through a model. Invoke the **Walk** tool from the Navigation Bar; a feet cursor will be displayed on the screen. You can also invoke this tool from the **Walk** drop-down in the **Navigate** panel of the **Viewpoint** tab. To start navigating in the model, press and hold the left mouse button, and drag in the desired direction. Move the mouse forward to go forward. Similarly, to take left and right turns, move the mouse accordingly. Instead of using mouse, you can also use the arrow keys (left, right, up, and down) on the keyboard to navigate inside the model. The **Fly** tool is used to take a birds-eye view of a model or project. This tool is used for large sites so that you can easily navigate around the outside area of your model. When you invoke the **Fly** tool, an aeroplane cursor will be displayed in the Scene View. You can also invoke this tool from the **Walk** drop-down in the **Navigate** panel of the **Viewpoint** tab. To fly around the model, invoke the **Fly** tool from the **Walk** drop-down in the Navigation Bar. Next, press and hold the left mouse button and click to fly straight. To change the elevation, move up and down. To zoom in and out, use the up and down arrow keys, and to revolve the camera, use the left and right arrow keys.

There are some additional options which will make the navigation easier and will give you a real experience. These options will become available in the **Walk** drop-down in the Navigation Bar after invoking the **Walk** tool. You can also invoke these options from the **Realism** drop-down list in the **Navigate** panel of the **Viewpoint** tab, as shown in Figure 2-12. These options are discussed next.

*Figure 2-12 The **Realism** drop-down list*

The **Collision** option is used to turn on collision detection. Turning this option on will prevent you from passing through the doors and columns while navigating. The **Gravity** option is used to follow the surface while walking and will prevent you from falling. This option is used with **Collision**. You can use **Collision** without **Gravity**, but **Gravity** cannot be used without **Collision**. This option works only with the **Walk** tool, and not with the **Fly** tool. The **Crouch** option is used to bend under the objects that are too low to walk under such as table. This option works only when **Collision** is activated.

In Navisworks, the **Third Person** option allows you to navigate a scene from a third person view. You can consider this third person as yourself navigating in the model and this third person is known as avatar, refer to Figure 2-13. When you select the **Third Person** option from the **Walk** drop-down in the **Navigation Bar**; an avatar will be displayed in the Scene View. You can also display the avatar by selecting the **Third Person** option from the **Realism** drop-down in the **Navigate** panel of the **Viewpoint** tab.

You can use the **Collision**, **Gravity** and **Crouch** options while working with the **Third Person** option. Using these options during the navigating, you can easily visualize how you will interact

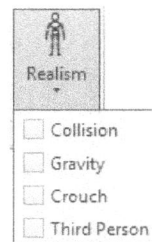

with the proposed model and can view the things inside the model in detail. There are several predefined avatars available in Navisworks and you can also create them.

Figure 2-13 *The third person view during the navigation in the model*

Customizing Navigation Parameters

In Navisworks, you are provided with the options which help you for controlling the walk speed during navigation. You can change the overall walk speed for navigation by using the **Edit Viewpoint** dialog box. To invoke this dialog box, choose the **Edit Current Viewpoint** button from the **Save**, **Load & Playback** panel in the **Viewpoint** tab; the **Edit Viewpoint - Current View** dialog box will be displayed, as shown in Figure 2-14.

You can change the linear speed by specifying the value in the **Linear Speed** edit box in the **Motion** area of this dialog box. To set the angular speed, specify the value in the **Angular Speed** edit box in the **Motion** area of the dialog box. You can also change the linear speed from the **Navigate** panel in the **Viewpoint** tab. For specifying speed, expand the **Navigate** panel; the options for setting speed will be displayed. Now you can specify values in the **Linear Speed** and **Angular Speed** edit boxes.

You can also customize various options to change the behavior of the **Walk** tool using the **Options Editor** dialog box. You can invoke this dialog box from the **Application Menu**. Next, expand the **Interface** node available in the left pane of the dialog box; several options will be displayed under this node. Select **Navigation Bar** from the left pane; various options for the **Orbit** and **Walk** tools will be displayed in the right pane of the dialog box, as shown in Figure 2-15.

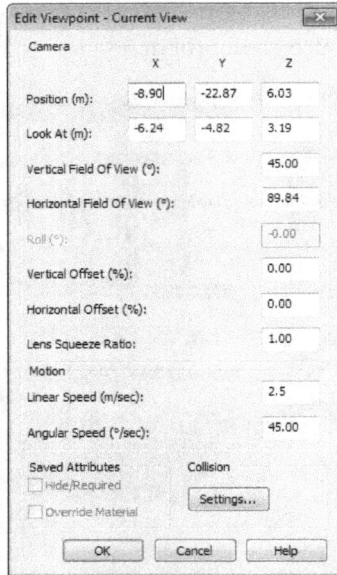

Figure 2-14 *The Edit Viewpoint-Current View dialog box*

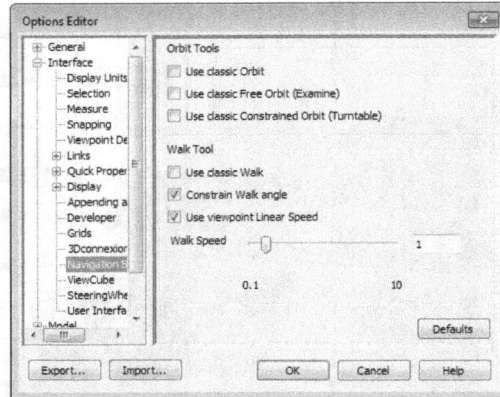

Figure 2-15 *The Navigation Bar options in the Options Editor dialog box*

Select the **Use classic Walk** check box in the **Walk Tool** area to switch to the classic Navisworks walk mode. Select the **Constrain Walk angle** check box to keep the camera vertical while navigating. Select the **Use viewpoint Linear Speed** check box to apply the viewpoint linear speed setting to the **Walk** tool. You can set the walking speed either by adjusting the **Walk Speed** slider or by entering the desired value in the corresponding edit box. This value ranges from 0.1 to 10.

Customizing Avatar Settings

The avatar settings such as avatar selection, dimensions, and positioning can be customized by using the options in the **Options Editor** dialog box. To do so, expand the **Interface** node in the left pane of the **Options Editor** dialog box; various options will be displayed under this node. Select the **Viewpoint Defaults** option; several options will be displayed in the right pane of the dialog box. Choose the **Settings** button in the **Collision** area of the dialog box; the **Default Collision** dialog box will be displayed, as shown in Figure 2-16. In this dialog box, select the **Enable** check box in the **Third Person** area; the options related to avatar position will be enabled. Select the desired option from the **Avatar** drop-down list and choose the **OK** button. You can also change the height of the avatar by specifying the value in the **Height** edit box in the **Viewer** area. Similarly, in the **Viewer** area, you can change the size of the avatar by specifying a value in the **Radius** edit box. You can change the location of the camera with respect to the top of the avatar by specifying a value in the **Eye Offset** edit box, refer to Figure 2-16.

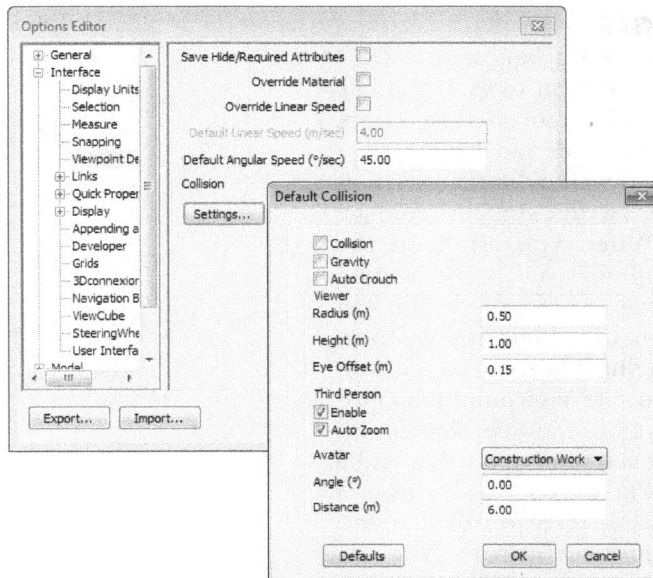

Figure 2-16 The **Default Collision** *dialog box*

You can also change the avatar for the current view by using the **Edit Viewpoint - Current View** dialog box. To do so, choose the **Edit Current Viewpoint** button from the **Save, Load & Playback** panel in the **Viewpoint** tab; the **Edit Viewpoint - Current View** dialog box will be displayed, as shown in Figure 2-17. Choose the **Settings** button in the **Collision** area; the **Collision** dialog box will be displayed, refer to Figure 2-16. You can now select a new avatar from the **Avatar** drop-down list. Next, choose the **OK** button to return to the **Edit Viewpoint - Current View** dialog box, and then choose the **OK** button to close the dialog box. The selected avatar will be displayed in the Scene View.

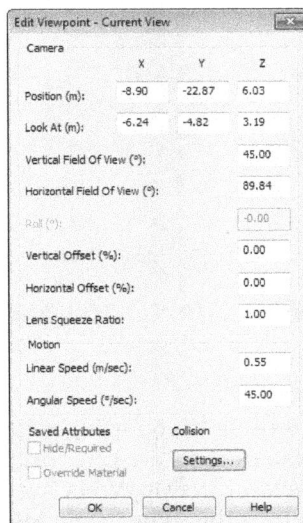

Figure 2-17 The **Edit Viewpoint - Current View** *dialog box*

SteeringWheels

The SteeringWheels are tracking menus that comprise multiple navigation tools in a single interface. These navigation tools are **Pan**, **Zoom**, **Walk**, **Look**, **Orbit**, **Rewind**, and **Up/Down**. A Steering Wheel is divided into different sections known as wedges. Each wedge contains a unique navigation tool. When you click on the **SteeringWheels** drop-down in the Navigation Bar; a list of tools will be displayed. You can access these tools from the ribbon as well. To do so, click on the **SteeringWheels** drop-down in the **Navigate** panel of the **Viewpoint** tab; a list of tools will be displayed, as shown in Figure 2-18. Select the wheel that you want to display. While using the SteeringWheels, you can activate a navigation tool available in it by pressing and holding the left mouse button over the corresponding wedge. Next, drag the tool over

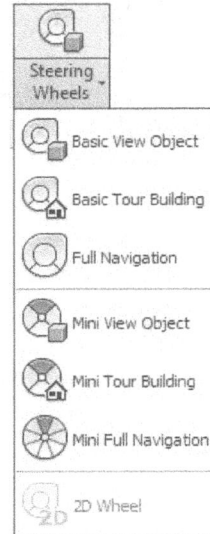

Figure 2-18   *The **SteeringWheels** drop-down*

the drawing area, you can use the selected navigation tool for reorienting your view. To exit the selected navigation tool, release the left mouse button. The SteeringWheels are categorized into two types and they are discussed next.

2D SteeringWheel

The **2D Wheel** SteeringWheel tool is used to work in the 2D view. Figure 2-19 shows a **2D Wheel**. The 2D navigation wheel has three navigation tools: **Zoom**, **Pan**, and **Rewind**. The **Zoom** tool is a common navigation tool used for enlarging or reducing the viewing scale of the model. You can use the **Pan** tool for traversing across the model view. The **Rewind** tool can be used to display the views of the previous zooming states which are saved temporarily.

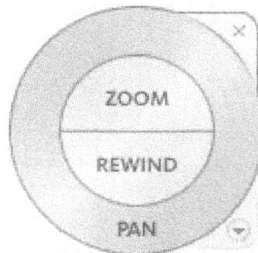

Figure 2-19   *The **2D Wheel***

3D SteeringWheels

The 3D SteeringWheels navigation tools help you navigate through 3D views. Based on the size and appearance, the 3D SteeringWheels tools are categorized into two groups: Mini Steering Wheels and Big SteeringWheels. The Mini SteeringWheels are further classified into three types: **Mini View Object** Wheel, **Mini Tour Building** Wheel, and **Mini Full Navigation** Wheel, as shown in Figure 2-20. The **Mini View Object** Wheel has four distinct navigation tools, **Pan**, **Zoom**,

Rewind, and **Orbit**. Similarly, the **Mini Tour Building** Wheel comprises four unique navigation tools: **Up/Down, Look, Walk**, and **Rewind**. The **Mini Full Navigation** Wheel comprises eight wedges with each wedge representing a unique navigation function. The **Mini Full Navigation** Wheel combines all the functions of the **Mini View Object** Wheel and **Mini Tour Building** Wheel.

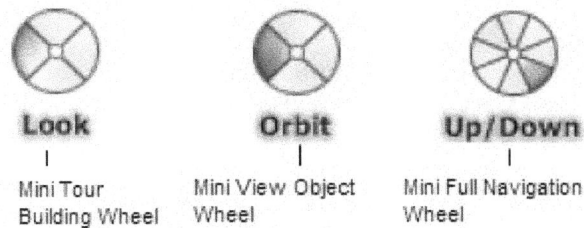

Figure 2-20 Different types of Mini SteeringWheels for 3D Views

The Big SteeringWheels contain same types of navigation tools as of the Mini SteeringWheels, it is clear from Figure 2-21. However, the appearance of these tools on screen is very different from the Mini SteeringWheels.

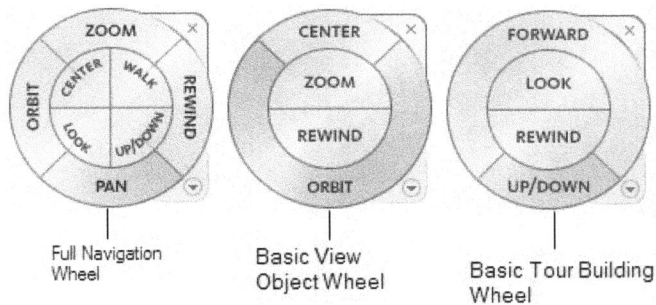

Figure 2-21 The big SteeringWheels

When you put the cursor over any of the navigation tools, the tooltips and messages are displayed, as shown in Figure 2-22. These tooltips give information about the tool on which the cursor is placed.

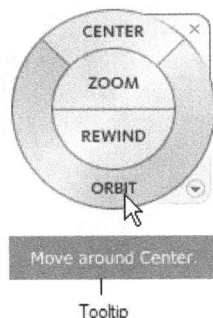

*Figure 2-22 The tooltip displayed on placing the cursor on the **Orbit** tool*

CAMERA

In Navisworks, you can specify various camera options such as camera projection, position and orientation during navigation. There are two types of cameras available in Navisworks: perspective and orthographic. In 3D workspace, you can use both the perspective and the orthographic cameras, but in 2D workspace, only orthographic camera can be used. To use any of the two cameras, select an option from the **Perspective** drop-down list in the **Camera** panel of the **Viewpoint** tab; the view will be adjusted according to the selected mode. The field of view is the area of scene that can be viewed through a camera. You can adjust the field of view by specifying a value in the **FOV** edit box in the **Camera** panel of the **Viewpoint** tab, as shown in Figure 2-23. You can also use the **FOV** slider to adjust the field of view angle.

Note
*When you move the **FOV** slider to the right, a wider field of the view will be displayed and on moving it to the left, a narrower field of the view will be displayed.*

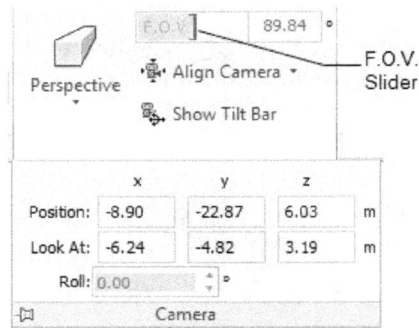

*Figure 2-23   Various options in the **Camera** panel*

The position of camera can be adjusted. To do so, slide the **Camera** panel in the **Viewpoint** tab. Next, specify the values in the **Position** edit boxes, refer to Figure 2-23. To change the focal point of the **Camera**, specify the values in the **Look At** edit boxes, refer to Figure 2-23. In the **Camera** panel, you can also adjust the rotation of the camera toward left or right by specifying the values in the **Roll** edit box. The **Roll** option is used to rotate the camera toward left or right. To rotate the **Camera** up or down, choose the **Show Tilt Bar** button in the **Camera** panel in the **Viewpoint** tab; the **Tilt Bar** will be displayed in the window. Drag the **Tilt bar** slider up or down to roll the camera.

You can align a camera to one of the axes which will change according to the camera's position and orientation in the Scene View. To do so, click on the **Align Camera** option; a drop-down list will be displayed. For example, if you select **Align X** from the list; camera position will align along the X axis and the model position and orientation will change accordingly.

REFERENCE VIEWS

Reference views are used to acquire an overall view of your position in the whole scene. It also helps you to quickly move the camera to a different location in a model. There are two types of reference views which can be used to view the model: **Section View** and **Plan View**. To access these options, click on the **Reference Views** option in the **Navigation Aids** panel of the **View** tab; a drop-down list will be displayed. Select the **Section View** option from the drop-down list to display the front view of the model; the view will be displayed in the **Section View** dockable window, as shown in Figure 2-24. You can also select the **Plan View** option from the drop-down list to display the top view of the model; the view will be displayed in the **Plan View** dockable window, as shown in Figure 2-25. In the **Section View** and **Plan View** window, a triangular marker will be displayed on invoking the **Walk** tool. This marker will represent the camera location and it will move in the window as you walk in the Scene View. You can also move the marker in the window by dragging it, the camera location will change accordingly in the Scene View.

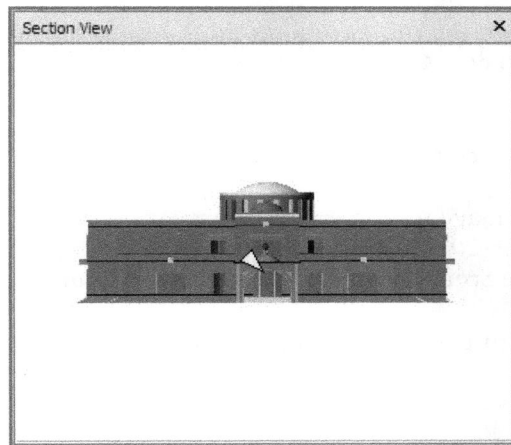

Figure 2-24 *The front view of the model in the* *Section View* *window*

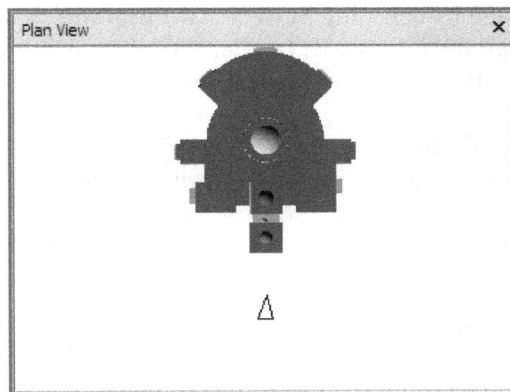

Figure 2-25 *The top view of the model in the* *Plan View* *window*

TUTORIALS
General instructions for downloading tutorial files:

1. Download the *c02_nws_2017_tut* zip file from *http://www.cadcim.com*. The path of the file is as follows: *Textbooks > Civil/GIS > Navisworks > Exploring Autodesk Navisworks 2017*.

2. Now, save and extract the downloaded folder at the following location:
   C:\ nws_2017

Note
*The default unit system used in the tutorials is metric. To change the units to imperial, select the required units from **Options Editor > Interface > Display Units**.*

Tutorial 1   Navigating Inside the Model

In this tutorial, you will open the *c02_navisworks_2017_tutorial.nwf* file and move around the model using the **Walk**, **Collision**, **Gravity**, **Third Person**, and other navigation tools.
(Expected time : 45min)

The following steps are required to complete this tutorial:

a.   Start the Navisworks Manage session.
b.   Open the *c02_navisworks_2017_tutorial.nwf* file.
c.   Orient the model in the Scene View using the **Orbit** and **Zoom** tools.
d.   Adjust the walk speed.
e.   Select Avatar and configure its dimension and position for the current viewpoint.
f.   Move around in the model.
g.   Move to the other floors of the building.
h.   Save the project.

Starting Autodesk Navisworks 2017
1.   Start **Navisworks Manage** by choosing **Start > All Programs > Autodesk > Navisworks Manage 2017 > Manage 2017** from the taskbar (Windows 7); the program is loaded and the user interface screen is displayed.

Opening the Existing Model
In this section, you will open the model created in Revit software.

1.   Choose the **Open** button from the Quick Access Toolbar; the **Open** dialog box is displayed.

Note
*You can also display the **Open** dialog box by choosing **Open > Open** from the **Application Menu**.*

2.   In this dialog box, browse to the following location:
     C:\nws_2017\c02_nws_2017_tut

3.   Select the **Navisworks File Set (*.nwf)** option from the **Files of type** drop-down list.

4. Next, select the *c02_navisworks_2017_tutorial.nwf* file; the file name is displayed in the **File name** edit box.

5. Choose the **Open** button on the right of the **File name** edit box; the model is displayed in the Scene view, as shown in Figure 2-26.

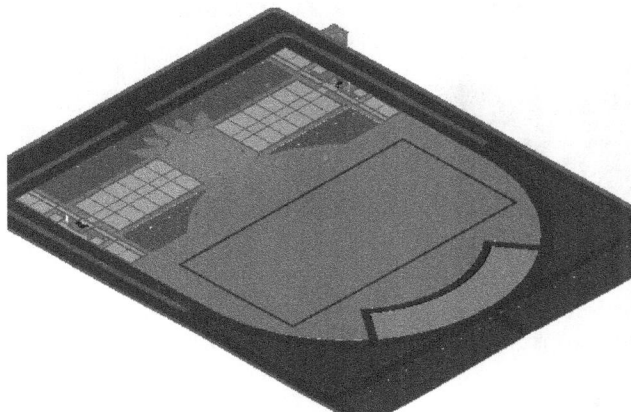

Figure 2-26 The model opened in Navisworks

Orienting the Model in the Scene View

In this section, you will adjust the model in the Scene View so that it fits properly in the workspace.

1. Invoke the **Orbit** tool from the **Orbit** drop-down in the **Navigate** panel of the **Viewpoint** tab; the Orbit cursor appears on the screen.

2. Press and hold the left mouse button, drag the mouse in the left direction, and then release it when the model looks similar to the one shown in Figure 2-27.

*Figure 2-27 The model after using the **Orbit** tool*

3. Next, to enlarge the viewing scale of the model, invoke the **Zoom** tool from the **Zoom** drop-down in the **Navigate** panel of the **Viewpoint** tab; the zoom cursor appears on the screen.

Note
*The **Orbit** and **Zoom** tools can also be invoked from the Navigation Bar.*

4.   Place the zoom cursor in the model, as shown in Figure 2-28.

*Figure 2-28 Placing the **Zoom** cursor on the model*

5.   Press and hold the left mouse button, drag the mouse in the forward direction, and release it when the model zooms in, as shown in Figure 2-29.

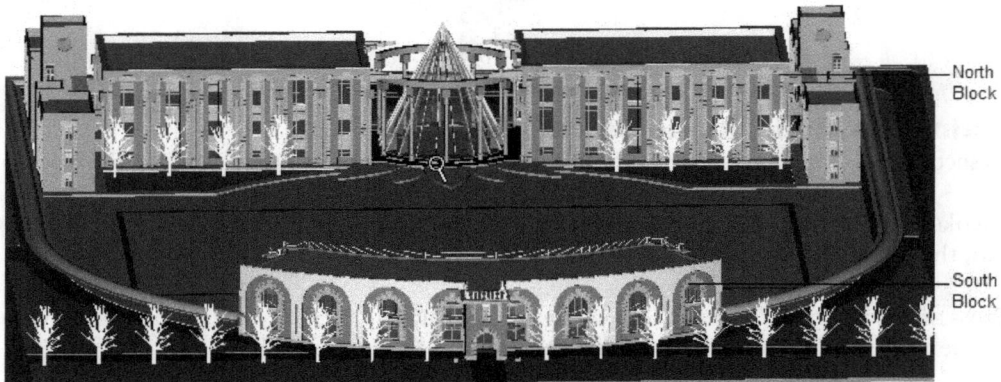

*Figure 2-29 The model after using the **Zoom** tool*

Specifying the Navigation Speed

In this section, you will adjust the walk speed while navigating inside a model.

1.   Choose the **Viewpoint** tab and then expand the **Navigate** panel by clicking on the down arrow next to it.

2.   Specify the walking speed **6.00** (**19ft 8.22**) in the **Linear Speed** edit box of this panel, as shown in Figure 2-30.

*Figure 2-30 Specifying the walk speed in the **Navigate** panel*

Note
*You can also specify the walk speed value in the **Linear Speed** edit box of the **Edit Current Viewpoint** dialog box.*

Selecting Avatars and Configuring its Parameters

In this section, you will select avatar and configure its parameters.

1. Choose the **Edit Current Viewpoint** button from the **Save, Load & Playback** panel in the **Viewpoint** tab; the **Edit Viewpoint-Current View** dialog box is displayed.

2. Choose the **Settings** button in the **Collision** area; the **Collision** dialog box is displayed, as shown in Figure 2-31.

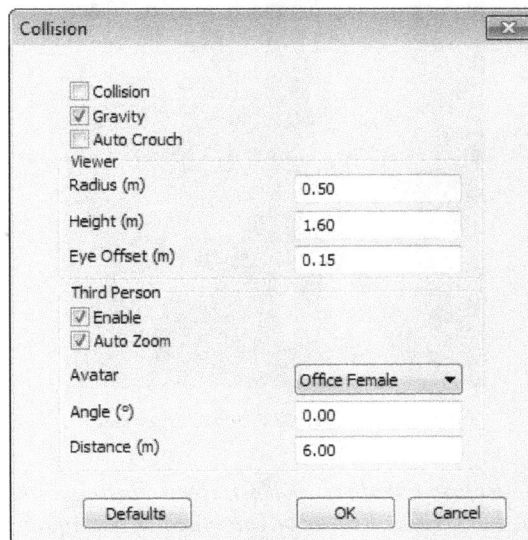

*Figure 2-31 The **Collision** dialog box*

3. In the **Collision** dialog box, select the **Enable** check box in the **Third Person** area. Ensure that the **Office Female** option is selected in the **Avatar** drop-down list.

4. Make sure that 0.00 degree is specified in the **Angle** edit box.

Note that on specifying 0.00 degree value in the **Angle** edit box, the camera will be placed directly behind the avatar.

5.   In the **Third Person** area, specify **6.00** (19 ft 8.16) in the **Distance** edit box.

6.   In the **Viewer** area, enter **0.50** (1ft 7.68) in the **Radius** edit box.

7.   Enter **1.60** (**5 ft 3.00**) in the **Height** edit box.

8.   Enter **0.15** (**0 ft 6.00**) in the **Eye Offset** edit box.

The **Eye Offset** value specifies the location of camera with respect to the top of the avatar.

9.   Next, choose the **OK** button; the **Collision** dialog box closes. Again choose **OK** to close the **Edit Viewpoint-Current View** dialog box.

10. Choose the **Walk** tool from the **Walk/Fly** drop-down in the **Navigate** panel of the **Viewpoint** tab; the **Walk** tool appears on the screen.

11. Press the left mouse button and click in the Scene View. Now, you can view the avatar, as shown in Figure 2-32.

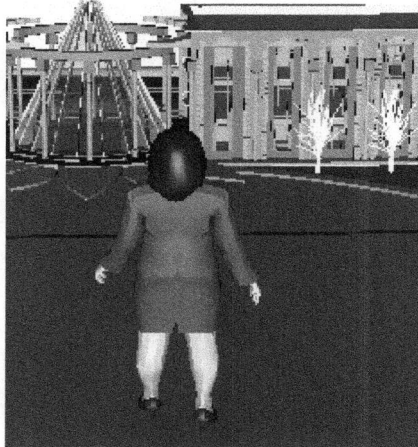

Figure 2-32 *Avatar near the model*

Note

*Make sure that the **Use classic Walk** check box is cleared to disable the classic walk tool icon. This check box is available in **Options Editor > Interface > Navigation Bar**. Also, if the avatar is displayed near the roof of the South block then drag it and place it near the North block.*

12. Select the **Gravity** check box from the **Realism** drop-down in the **Navigate** panel of the **Viewpoint** tab.

13. Next, press and hold the left mouse button, drag the cursor in the forward direction, and then release it when the avatar is placed on the ground, as shown in Figure 2-33. Here, avatar faces the North block.

Note

The position of avatar is not fixed, you can navigate in and around the model as per your choice.

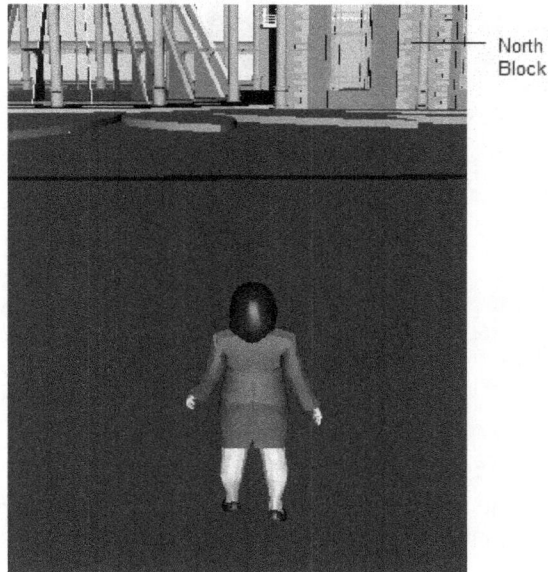

Figure 2-33 Avatar standing on the ground

Moving Around in the Model

In this section, you will walk around the model. Use the following steps to move around the model.

1. Press and hold the left mouse button, drag the mouse in the right direction until the arrow appears, refer to Figure 2-34.

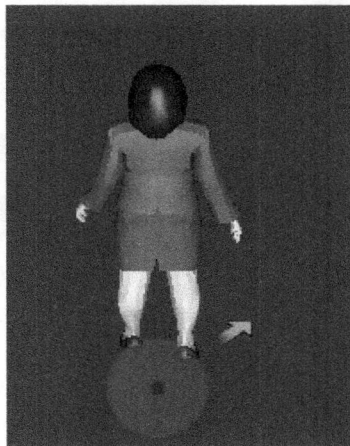

Figure 2-34 The cursor displayed in the model

2. Drag the cursor in the left direction and release it when the avatar faces the South block, as shown in Figure 2-35.

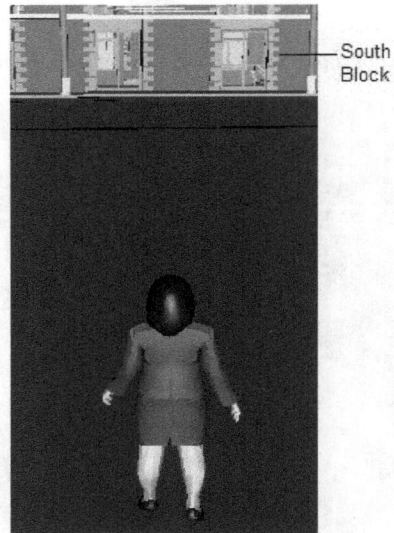

Figure 2-35 *The avatar facing the south block*

3. Drag the avatar in the forward direction and place it near the model, refer to Figure 2-36.

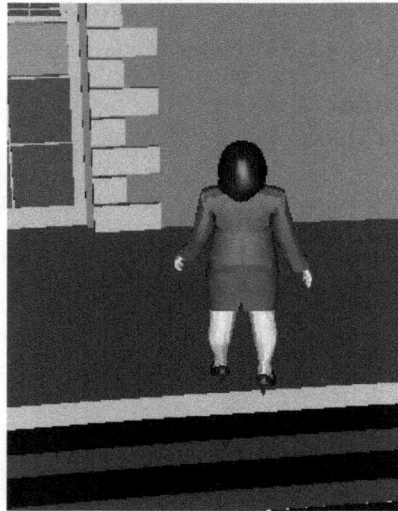

Figure 2-36 *Avatar moving around inside the model*

4. Drag the avatar in the left direction and place it near the fourth door from left in the model, refer to Figure 2-37.

Figure 2-37 *Avatar moving in the left direction*

5. Clear the **Collision** check box from the **Realism** drop-down in the **Navigate** panel of the **Viewpoint** tab.

6. Drag the avatar in the forward direction and release it when it passes through the doors, refer to Figure 2-38.

7. Drag avatar in the left direction and release it, refer to Figure 2-39.

Figure 2-38 *Avatar passing through the door*   *Figure 2-39* *Avatar moved in the left direction*

8. Select the **Gravity** check box from the **Realism** drop-down in the **Navigate** panel of the **Viewpoint** tab.

9. Move the avatar toward the stairs, as shown in Figure 2-40.

10. Drag the avatar upward to climb up the stairs, as shown in Figure 2-41.

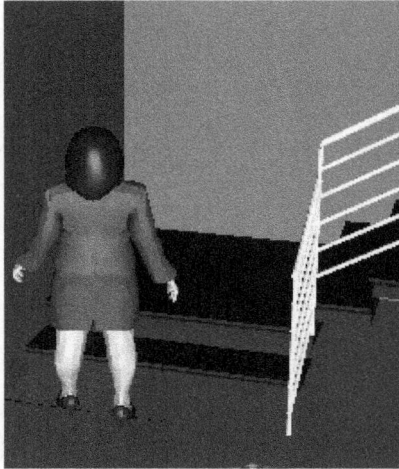

Figure 2-40 Avatar standing near the stairs

Figure 2-41 Avatar walking up the stairs

11. Drag the avatar to move up the stairs and place the avatar near the balcony door, as shown in Figure 2-42.

Figure 2-42 Avatar standing near the balcony

12. Clear the **Collision** check box from the **Realism** drop-down in the **Navigate** panel of the **Viewpoint** tab.

13. Next, drag the avatar to pass through the door and place it in the balcony area, as shown in Figure 2-43.

Figure 2-43 *Avatar standing in the balcony*

Navigating to the Other Levels of the Building

In this section, you will change the levels in the building using **Section View**.

1. To move to the other levels of the building, invoke the **Section View** window from the **Reference Views** drop-down list in the **Navigation Aids** panel of the **View** tab. The **Section View** of the building is displayed, as shown in Figure 2-44.

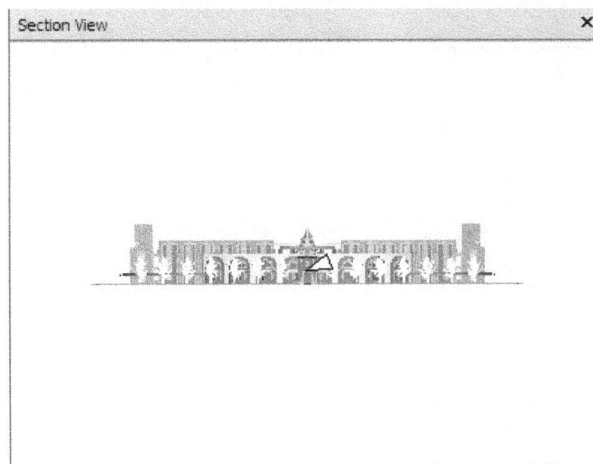

Figure 2-44 *The section view of the building in the **Section View** window*

2. In the **Section View** window, drag the triangular marker in such a way that the avatar moves to the roof of the building, as shown in Figure 2-45. Figure 2-46 shows the avatar standing on the roof. Next, close the **Section View** window.

Notice that as you move the triangular marker in the **Section View** window, the avatar moves to the roof in the Scene View.

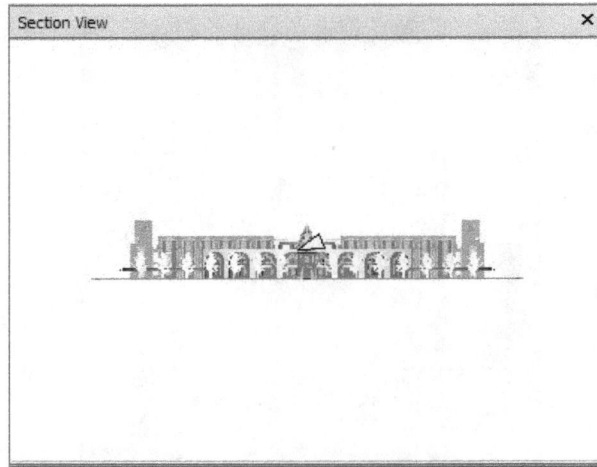

Figure 2-45 *Moving the triangular marker to the first floor of the building*

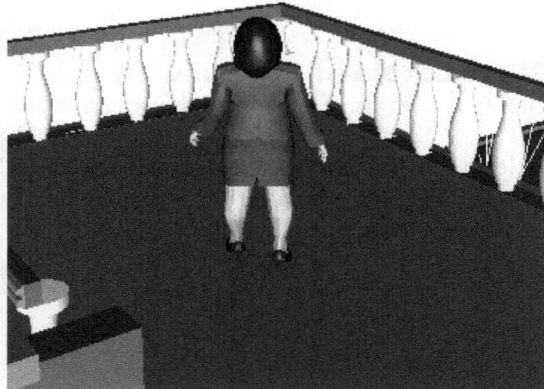

Figure 2-46 *Avatar standing on the roof of the building*

Saving the Project

In this section, you will save the project.

1. To save the project with the current view, choose **Save As** from the Application Menu; the **Save As** dialog box is displayed.

2. Browse to *nws_2017\c02_nws_2017_tut* folder and enter **c02_navisworks_2017_tut01** in the **File name** edit box. Next, choose the **Navisworks File Set (*.nwf)** file format from the **Save as type** drop-down list, and then choose the **Save** button.

Tutorial 2   Flying Around the Model

In this tutorial, you will open the *c02_navisworks_2017_tutorial.nwf* file and fly around the model using the **Fly** tool.   **(Expected time: 20 min)**

The following steps are required to complete this tutorial:

a. Start the Navisworks Manage session.
b. Open the *c02_navisworks_2017_tutorial* file.
c. Adjust the model in window using the **Orbit** and **Zoom** tools.
d. Fly around the model.
e. Save the project.

Starting Autodesk Navisworks 2017

1. Start **Navisworks Manage** by choosing **Start > All Programs > Autodesk > Navisworks Manage 2017 > Manage 2017** from the taskbar. The program is loaded and the user interface screen is displayed.

Opening the Existing File

In this section, you will open a file which is created in Revit.

1. To open the existing model, choose the **Open** button from the Quick Access Toolbar. Alternatively, choose **Open > Open** from the Application Menu; the **Open** file dialog box is displayed.

2. In this dialog box, browse to the following location:
    C:\nws_2017\c02_nws_2017_tut

3. Select the **Navisworks File Set (*.nwf)** file from the **Files of type** drop-down list.

4. Next, select the *c02_navisworks_2017_tutorial.nwf* file; the file name is displayed in the **File name** edit box.

5. Choose the **Open** button on the right of the **File name** edit box; the model is displayed in the Scene View, as shown in Figure 2-47.

Orienting the Model in the Scene View

In this section, you will adjust the model in the Scene View so that it fits into the Navisworks Scene View.

1. Invoke the **Orbit** tool from the **Orbit** drop-down in the **Navigate** panel of the **Viewpoint** tab; the Orbit tool cursor appears on the screen.

2. Place the cursor at the corner of the model, as shown in Figure 2-48.

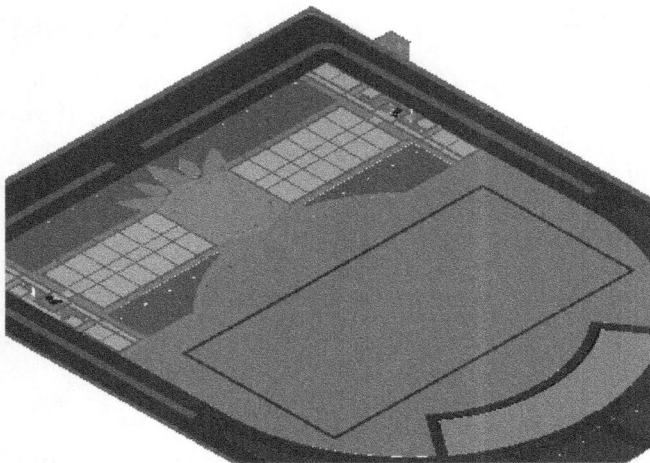

Figure 2-47 *The model displayed in the Scene View*

Figure 2-48 *Placing the cursor on the model*

3. Press and hold the left mouse button, drag the mouse in the downward direction, and then release it when the model is rotated, as shown in Figure 2-49.

4. Now drag the model in the left direction and release it when the model looks like, as shown in Figure 2-50.

Figure 2-49 *Model after rotating*

Figure 2-50 *The model after using the **Orbit** tool*

5. Invoke the **Zoom** tool from the **Zoom** drop-down in the **Navigate** panel of the **Viewpoint** tab; the Zoom tool cursor appears on the screen.

6. Press and hold the left mouse button, drag the mouse in the forward direction, and then release it when the model is zoomed, as shown in Figure 2-51.

Figure 2-51 *The model after using the **Zoom** tool*

Flying Around the Model

In this section, you will fly around the model using the **Fly** tool.

1. To fly around the model, invoke the **Fly** tool from the **Walk/Fly** drop-down list in the **Navigate** panel of the **Viewpoint** tab; the flight cursor appears on your screen.

2. Hold the left mouse button and drag the mouse in the right direction to fly around the model. Figure 2-52 shows a view in the fly mode.

Note
*Use the **Pan**, **Orbit**, and **Zoom** tools while flying around the model to adjust the model in such a way that you can clearly view the model and it becomes easier to fly around the model.*

Figure 2-52 Flying around the corner of the School building

Saving the Project

In this section, you will save the project.

1. To save the project with the current view, choose the **Save As** from the Application Menu; the **Save As** dialog box is displayed.

2. Browse to *nws_2017\c02_nws_2017_tut* folder and enter **c02_navisworks_2017_tut02** in the **File name** edit box. Next, choose the **Navisworks File Set (*.nwf)** file format from the **Save as type** drop-down list, and then choose the **Save** button.

Self-Evaluation Test

Answer the following questions and then compare them to those given at the end of this chapter:

1. The _____ tool is used to zoom a specified area.

2. You can link the position of Navigation Bar with the _____ .

3. The _____ tool is used to rotate the model freely around the pivot point in any direction.

4. The **Edit Viewpoint - Current View** dialog box can be displayed by choosing the _____ button.

5. The Section View can be displayed by selecting an option from the _____ drop-down list.

6. The docking position of the **Navigation** Bar cannot be changed. (T/F)

7. The **Orbit** tool is used to move the model to any position in the Scene View. (T/F)

8. The **Look At** tool is used to look around the model. (T/F)

9. In Navisworks, the walk speed cannot be changed while navigating inside a model. (T/F)

10. In Navisworks, the **Gravity** option can be used while flying around a model. (T/F)

Review Questions

Answer the following questions:

1. Which of the following options is used to pass the avatar through those objects that are too low to walk under?

    a) **Gravity**   b) **Crouch**
    c) **Collision**   d) **Fly**

2. Which of the following options is used for invoking the third person view?

    a) **Camera**   b) **Orbit**
    c) **Avatar**   d) **Zoom**

3. Which of the following tools is used to change the viewing scale of the model?

    a) **Orbit**   b) **Pan**
    c) **Fly**   d) **Zoom**

4. Which of the following options is used to display the absolute X, Y, Z position of the camera?

    a) **XYZ Axes**   b) **Grid Location**
    c) **Position Readout**   d) **Section View**

5. You can specify the radius of an avatar by entering values in the **Eye Offset** boxes. (T/F)

6. Clearing the **Collision** check box will enable the avatar to pass through the doors and columns. (T/F)

7. You cannot change the height of the avatar. (T/F)

8. Section View is used to display the top view of a model. (T/F)

9. In Navisworks, the **Walk** and **Fly** tools can be used simultaneously. (T/F)

10. The **Pan** tool is used to rotate a model. (T/F)

EXERCISE

Exercise 1   Navigating Inside the Residence Building

Download and open the *c02_navisworks_2017_ex1.nwf* file from *http://www.cadcim.com*. Navigate inside the Residence Building model, shown in Figure 2-53, by using various navigation tools.

(Expected time: 30min)

The steps required to complete this exercise are given next:

1. Open the *c02_navisworks_2017_ex1* file.
2. Orient the model using the **Orbit** and **Zoom** tool.
3. Adjust the walking speed.
4. Select avatar and configure its dimension and position for the current viewpoint.
5. Navigate inside the model and on the other floors of the building.
6. Save the file with the name *c02_navisworks_2017_ex01*.

Figure 2-53 The Residence Building

Answers to Self-Evaluation Test
1. Zoom Window, **2.** ViewCube, **3.** Free Orbit, **4.** Edit Current Viewpoint, **5.** Reference Views, **6.** F, **7.** F, **8.** F, **9.** F, **10.** F

Chapter 3

Selecting, Controlling, and Reviewing Objects

Learning Objectives

After completing this chapter, you will be able to:

- *Use direct selection tools*
- *Use different types of windows for selecting objects*
- *Save selected objects*
- *Analyze object properties*
- *Link databases to objects*
- *Control visibility of objects*
- *Control the appearance of model*
- *Control object attributes*
- *Use Measure and Redline tools*
- *Add tags and comments*
- *Use links*
- *Use the Appearance Profiler window*

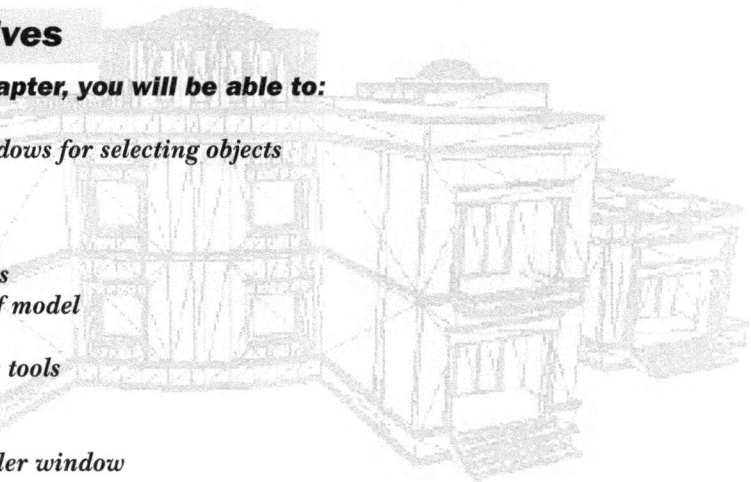

INTRODUCTION

In the previous chapter, you learned to use various navigation tools in a project. In this chapter, you will learn various methods and tools to select, control, and review objects. In Navisworks, there are several types of tools which are used for selecting objects in a model. After selecting objects, you can control their transformation and appearance. You can also use various measuring tools to measure distances, angle, and area. Using the **Link** tool, you can attach any file or geometry to the model. All these tools and methods are discussed in detail in this chapter. Various methods for selecting objects are discussed next.

OBJECT SELECTION METHODS

In order to work on an object or a group of objects, you need to select them. In Navisworks, you can do so by using the Direct Selection method. You can also select an object or group of objects by using the **Selection Tree** window and the **Find Items** window. Various methods for selecting objects in Navisworks are discussed next.

Direct Selection Method

In the Direct Selection method, you can select an object or group of objects directly from the Scene View by using the direct selection tools. The direct selection tools are available in the **Select & Search** panel of the **Home** tab. These tools are discussed next.

Select Tool

The **Select** tool is used to select an object(s) directly from the Scene View. To select an object, choose this tool from **Home > Select & Search > Select** drop-down, as shown in Figure 3-1. On doing so, a selection cursor will be displayed in the Scene View. Next, click on the desired object; the object will be selected and highlighted in blue color. To select more than one object at a time, press and hold the CTRL key while clicking on the objects.

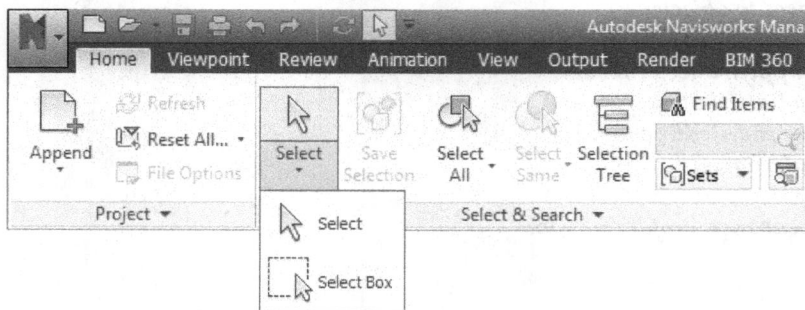

Figure 3-1 Invoking the Select tool

Note
*Sometimes, instead of selecting a single object, either the entire model or a particular layer will be selected in the Scene View. In this case, select the **Last Object** option from the **Resolution** drop-down list located at **Options Editor > Interface > Selection**.*

Select Box Tool

The **Select Box** tool is used to select object(s) by drawing a window in the Scene View. To select an object, invoke this tool from **Home > Select & Search > Select** drop-down, refer to Figure 3-1. On doing so, a selection cursor will be displayed in the Scene View. Next, click at a desired location in the Scene View and then drag the cursor to form a rectangle. You will notice that on dragging the cursor, a rectangle is formed in dashed lines and the objects that are completely inside it are selected.

Select All Tool

The **Select All** tool is used to select all the objects in the Scene View. To select objects, invoke this tool from **Home > Select & Search > Select All** drop-down, as shown in Figure 3-2. On doing so, all objects will be selected in the Scene View. Notice that all the selected objects will be highlighted in blue color.

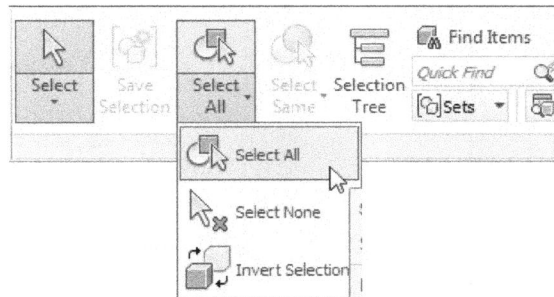

Figure 3-2 Invoking the Select All tool

Select None Tool

The **Select None** tool is used for deselecting the selected objects. Before using this tool, ensure that the objects are selected in the Scene View. Next, invoke this tool from **Home > Select & Search > Select All** drop-down, refer to Figure 3-2; all the selected objects will be deselected in the Scene View.

Invert Selection Tool

The **Invert Selection** tool is used for reversing the selection. To do so, first select an object in the Scene View. Next, invoke this tool from the **Home > Select & Search > Select All** drop-down, refer to Figure 3-2. On doing so, the selected objects will be deselected and the deselected objects will be selected in the Scene View.

In a project, you can also select multiple objects based on the their properties, type, name, and so on. To do so, first you need to select an object in the Scene View. After selecting the object, the **Select Same** drop-down will be enabled in the **Select & Search** panel. This drop-down contains various tools that are discussed next.

Same Name Tool

The **Same Name** tool is used to select multiple objects in a model with the same name as the currently selected objects. Before invoking this tool, ensure that an object is selected in the Scene View. To select the object with the same name, choose the **Same Name** tool from **Home > Select & Search > Select Same** drop-down, as shown in Figure 3-3; the objects with the same name as the currently selected object will be selected and highlighted in blue color. For example, if you have selected a wall named as Basic Wall in the model and you want to select all the walls named as Basic Wall in the model, then use the **Same Name** tool.

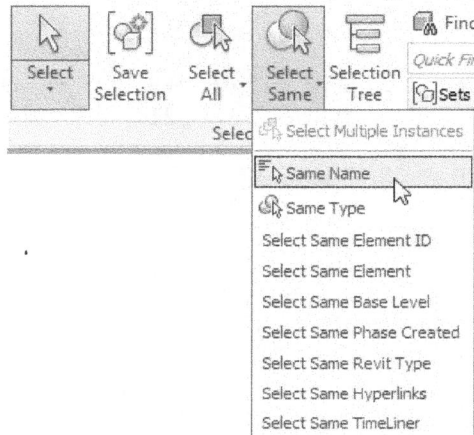

*Figure 3-3 Choosing the **Same Name** tool*

Same Type Tool

The **Same Type** tool is used to select objects of the same type as the currently selected object. Before invoking this tool, ensure that an object is selected in the Scene View. To select objects of the same type, choose the **Same Type** tool from **Home > Select & Search > Select Same** drop-down, refer to Figure 3-3. On doing so, the objects of the same type as the currently selected object will be selected and highlighted in blue color in the Scene View. For example, if you have selected an object of polygon geometry and you want to select all the objects with the polygon type, then you need to invoke the **Same Type** tool.

Same <Property> Tool

The **Same <Property>** tool is used to select objects whose properties such as element, element ID, and link are same as that of the currently selected object. To select the object, choose the **Same <Property>** tool from **Home > Select & Search > Select Same** drop-down, refer to Figure 3-3. On doing so, the objects that have the same properties as that of the current selected objects will be selected and highlighted in blue in the Scene View. For example, if you have selected an object in which glass material has been used and you want to select all the objects in which glass material is used, then you need to choose the **Select Same Material** tool.

The Selection Tree Window

In Navisworks, you can select the object(s) from the **Selection Tree** window. To use this window, choose the **Selection Tree** button from the **Select & Search** panel in the **Home** tab; the **Selection**

Tree window will be displayed, as shown in Figure 3-4. Alternatively, to display the **Selection Tree** window, select the **Selection Tree** check box from the **View > Workspace > Windows** drop-down. The **Selection Tree** window is a dockable window which contains a list of objects. These objects are arranged in a hierarchy. The arrangement of objects in the hierarchy is same as the sequence of object creation of the model in the native application. When you select an object from the **Selection Tree** window, the selected object will also be selected in the Scene View and highlighted in blue color, both in the Scene View as well as in the **Selection Tree** window.

You can control the display of hierarchy in the **Selection Tree** window by selecting the required option from the drop-down list displayed at the top in the **Selection Tree** window. The options available in the drop-down list are discussed next.

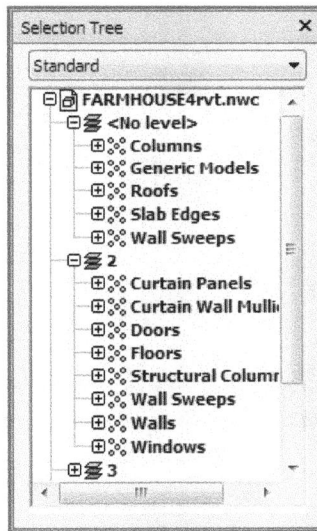

Figure 3-4 The Selection Tree window

Standard Option

In the **Selection Tree** window, select the **Standard** option from the drop-down list; the default hierarchy will be displayed which includes all the levels and parts of the model. You can arrange the displayed hierarchy in an alphabetical order. To do so, right-click on any of the items in the **Selection Tree**; a shortcut menu will be displayed. Choose **Scene > Sort** from the shortcut menu; the list will be arranged in an alphabetical order.

Compact Option

In the **Selection Tree** window, select the **Compact** option from the drop-down list; a simplified hierarchy will be displayed.

You can customize the level of details displayed on selecting the **Compact** option. To do so, choose the **Options** button in the **Application Menu**; the **Options Editor** dialog box will be displayed. Expand the **Interface** node in the left pane of the dialog box; several options will be displayed. Select the **Selection** option under the **Interface** node; various options related to the **Selection** option will be displayed in the right pane of the dialog box, as shown in Figure 3-5. In the right pane, select the required options from the **Compact Tree** drop-down list, refer to Figure 3-5. The options in the drop-down list are discussed next.

To display the model file only in the **Selection Tree** window, select the **Model** option from the drop-down list. To expand the hierarchy to layer level in the **Selection Tree** window, select the **Layers** option from the list. To expand the list to the objects level in the **Selection Tree** window, select the **Objects** option from the list.

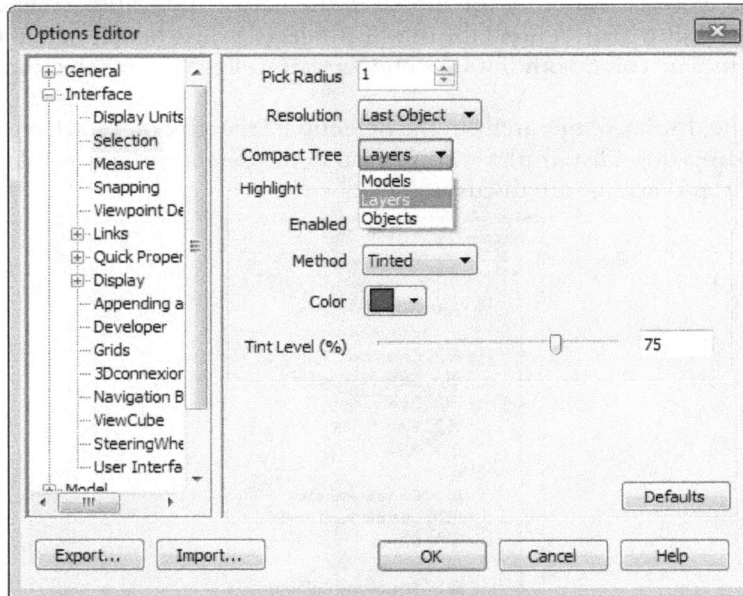

Figure 3-5 *Various options in the* ***Options Editor*** *dialog box*

Properties Option

In the **Selection Tree** window, select the **Properties** option from the drop-down list; a hierarchy based on the properties of items will be displayed. Now, in the **Selection Tree** window, you can select the objects based on their properties from the displayed hierarchy.

Sets Option

In the **Selection Tree** window, select the **Sets** option from the drop-down list; all the saved selections and search sets will be displayed in a hierarchy. Note that if no sets are defined, then the **Sets** option will not be available in the drop-down list of the **Selection Tree** window. The method of saving the selections and search sets is discussed later in this chapter.

Setting Selection Resolution

In Navisworks, at times, when you select an object in the Scene View, then either the entire model or a particular layer will be selected but not the single entity. In such a case, you can customize the selection resolution. To do so, choose the **Options** button in the **Application Menu**; the **Options Editor** dialog box will be displayed. In the left pane of the **Options Editor** dialog box, expand the **Interface** node and select the **Selection** option; several options will be displayed in the right pane of the dialog box, refer to Figure 3-5. In the right pane, select the required options from the **Resolution** drop-down list. The options in this drop-down list are used to define the level of selection. These options are discussed next.

To select all objects in the current file, select the **File** option from the **Resolution** drop-down list in the right pane of the **Options Editor** dialog box. Next, click on an object in the Scene View; all objects will be selected in the Scene View. To select objects within a layer, select the **Layer** option from the **Resolution** drop-down list. Next, click on the required object in the Scene View; all objects of the corresponding layer will be selected. Alternatively, you can select the first group of objects under the layer in a model. To do so, select the **First Object** option from the **Resolution** drop-down list and then click on an object in the Scene View; the object will be selected and the corresponding composite object will be highlighted in the **Selection Tree** window. To select the last branch of object in a model, select the **Last Object** option from the drop-down list. Next, click on the required object in a model; the object will be selected in the Scene View and the corresponding item will be highlighted in the **Selection Tree** window. By default, the **Last Object** option is selected in the **Resolution** drop-down list. Similarly, select the **Last Unique** and **Geometry** options from the **Resolution** drop-down list.

You can also access these options from the ribbon. To do so, expand the **Select & Search** panel by clicking on the down arrow. Next, select the required options from the **Selection Resolution** drop-down list.

Find Items Window

In a project, you may need to select or search an object on the basis of its property. To do so, choose the **Find Items** button from the **Select & Search** panel in the **Home** tab; the **Find Items** window will be displayed, as shown in Figure 3-6. Alternatively, to display the **Find Items** window, select the **Find Items** check box from the **View > Workspace > Windows** drop-down list. The **Find Items** window is a dockable window in which you can search for objects having common properties. These searched objects will be selected and highlighted in the Scene View and in the **Selection Tree** window.

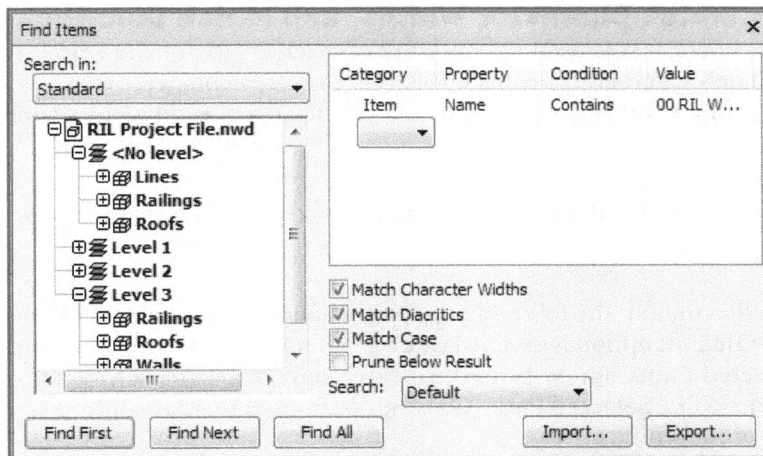

*Figure 3-6 The **Find Items** window*

The left pane of the **Find Items** window contains objects arranged in a hierarchy. In this pane, you can define the level of search by selecting the options from the drop-down list available at the top in the **Find Items** window. The options in this drop-down list are same as in the **Selection Tree** window. Using these options, you can control the levels of display of objects in the window.

Using the options available in the right pane of the **Find Items** window, you can specify the parameters for searching objects in the model. These options are discussed next.

Category

In the **Category** column, you can specify the category name for the object(s) to be searched. To do so, click in the **Category** column; a drop-down list will be displayed. This drop-down list contains the list of categories that are available in the model such as Item, Element, Material, and so on. You can select the required category name from the list.

Property

In the **Property** column, you can specify the property name for the object(s) to be searched. To do so, click in the **Property** column; a drop-down list will be displayed. This drop-down list contains the list of properties of the selected category. You can select the required property from the drop-down list.

Condition

In the **Condition** column, you can specify the condition operator for the objects such as **=**, **not equals**, **<**, **>**, and so on. To do so, click in the **Condition** column; a drop-down list will be displayed. You can select the required condition operator from it.

Value

In the **Value** column, you can specify the property value manually or select the required value from the drop-down list. To do so, click in the **Value** column; a drop-down list will be displayed. Select the required value from it. The values available in the drop-down list depend upon the selected category and property.

Match Case, Match Character Widths, and Match Diacritics

The **Match Case** check box is used to check the case sensitivity for the search statements. You can select the **Match Character Widths** check box to match character widths while searching. Select the **Match Diacritics** check box to match the diacritics terms while searching.

Prune Below Result

Select the **Prune Below Result** check box to stop the search when the first object is found.

Search

This drop-down list contains the following options: **Default, Below Selected Paths**, and **Selected Paths Only**. The **Default** option is used to search all items selected in the left pane of the window. The **Below Selected Paths** option is used to search only below the selected objects in the left pane of the window. The **Selected Paths Only** option is used to search from the objects selected in the left pane of the window.

The Find First Button

The **Find First** button is used to find the first object which will satisfy the search criteria. To find the first qualified object, choose the **Find First** button; the search results will be selected and highlighted in the Scene View as well as in the **Selection Tree** window.

The Find Next Button

The **Find Next** button is used to find the next object which will satisfy the search criteria. To find the next qualified object, choose the **Find Next** button; the search results will be selected and highlighted in the Scene View as well as in the **Selection Tree** window.

The Find All Button

The **Find All** button is used to find all objects that will satisfy the search criteria. To find all the qualifying objects, choose the **Find All** button; the search results will be selected and highlighted in the Scene View as well as in the **Selection Tree** window.

The Import Button

The **Import** button is used to import the previously saved search criteria. On choosing this button, the **Import** dialog box will be displayed. In this dialog box, browse and select the required file and then choose the **Open** button; the selected file will be imported.

The Export Button

The **Export** button is used to export the current search to an .xml file. On choosing this button, the **Export** dialog box will be displayed. In this dialog box, specify the file name and location and then choose the **Save** button; the file will be saved in the specified location.

Quick Find

The **Quick Find** text box is used to quickly find and select objects in a model. This text box is available in the **Select & Search** panel of the **Home** tab, as shown in Figure 3-7. To search an object, specify the related keywords in the **Quick Find** text box, and then choose the button in the text box; the objects matching the entered text will be selected and highlighted in the Scene View as well as in the **Selection Tree** window.

Figure 3-7 *The* **Quick Find** *tool in the* **Select & Search** *panel*

After selecting the required objects in a model, you can save the selection. The various methods of saving the selected objects are discussed next.

SAVING SELECTIONS

After selecting an object or a group of objects, you can save them as sets. They can be saved either by creating selection sets or by creating search sets. These two types of sets are discussed next.

Selection Sets

The selection sets are a group of objects selected either from the Scene View or from the **Selection Tree** window. To save the selected object, choose the **Save Selection** button from the **Select &**

Search panel in the **Home** tab; the **Sets** window will be displayed, as shown in Figure 3-8. The selected object will be saved as a selection set in this window. Alternatively, to display the **Sets** window, choose the **Manage Sets** option from the **Home > Select & Search > Sets** drop-down. This window displays a list of all the saved selections. The buttons located at the top of the **Sets** window are discussed next.

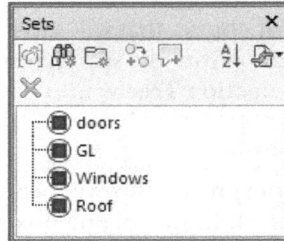

Figure 3-8 The Sets window

Save Selection

The **Save Selection** button is used to save the current selection. To save the current selected object, choose the **Save Selection** button; a text box with **Selection Set** written in it will be displayed in the **Sets** window. Specify a new name in the text box and press ENTER; the selection set with the new name will be displayed in the **Sets** window. The selection sets are represented by a ⬤ symbol, refer to Figure 3-8.

Save Search

The **Save Search** button is used to save the currently searched object. To save the currently searched object, choose the **Save Search** button; a text box with **Search Set** written in it will be displayed in the **Sets** window. Specify a new name in the text box and press ENTER; the search set with the new name will be displayed in the **Sets** window. Search sets are represented by a symbol, refer to Figure 3-9.

New Folder

The **New Folder** button is used to create a new folder. In the **Sets** window, you can create a new folder and group all the sets under it. To do so, choose the **New Folder** button; a folder with the text **New Folder** will be created in the **Sets** window. Specify a new name in the text box and press ENTER; the folder with the new name will be created. Next, drag and place the desired sets in this folder.

Duplicate

The **Duplicate** button is used to create a copy of the saved items. To do so, first select the desired set in the **Sets** window and then choose the **Duplicate** button; the duplicate set will be created with the same name and content.

Add Comment

The **Add Comment** button is used to add comments to the saved objects. To do so, select the desired set in the **Sets** window and then choose the **Add Comment** button; the **Add Comment** dialog box will be displayed. Specify a comment in the **Add Comment** area and choose the **OK** button; the comment will be added to the selected set.

Sort

The **Sort** button is used to arrange the saved sets in an alphabetical order. To do so, choose the **Sort** button; the content in the **Sets** window will be arranged alphabetically.

Import/Export

The **Import/Export** button is used to import and export the saved search sets. To export the saved search set, first select the desired set in the **Sets** window. Next, choose the **Import/ Export** button; a drop-down list will be displayed. Select the **Export Search Sets** option from the list; the **Export** dialog box will be displayed. In this dialog box, specify the file name and location of the exported file and then choose the **Save** button; the file will be exported and saved at the specified location. Similarly, to import the search sets, select the **Import Search Sets** option from the drop-down list; the **Import** dialog box will be displayed. In this dialog box, browse and select the required file and then choose the **Open** button; the file will be imported to the **Sets** window.

Delete

The **Delete** button is used to delete a selected set. To do so, select the set to be deleted and choose the **Delete** button; the selected set will be removed from the **Sets** window.

Search Sets

The search sets are a group of searched objects. When you run a search in the **Find Items** window, the searched objects will be selected and highlighted in blue color in the Scene View as well as in the **Selection Tree** window. To save these selected objects, choose the **Save Search** button in the **Sets** window; the objects will be saved as a search set, refer to Figure 3-9.

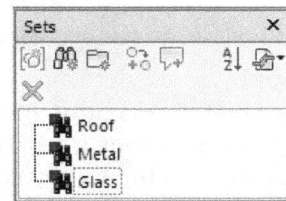

Figure 3-9 *The search sets in the* **Sets** *window*

The Selection Inspector Window

In a project, when you select an object or a group of objects, they will be displayed in the **Selection Inspector** window along with their properties. To invoke this window, choose the **Selection Inspector** button from the **Select & Search** panel in the **Home** tab; the **Selection Inspector** window will be displayed, as shown in Figure 3-10. Alternatively, to display the **Selection Inspector** window, select the **Selection Inspector** check box from the **View > Workspace > Windows** drop-down. The **Selection Inspector** is a dockable window. Select the object(s) from the Scene View or from the **Selection Tree** window; the selected items will be displayed in the **Selection Inspector** window, as shown in Figure 3-10. The buttons in this window are discussed next.

Show Item

The **Show Item** button is used to view the selected object in the Scene View. On choosing this button, the selected object will be zoomed in the Scene View.

Deselect

The **Deselect** button is used to remove an object from the **Selection Inspector** window. On choosing this button, the selected object will be removed from the **Selection Inspector** window.

Figure 3-10 The Selection Inspector window

Export

The **Export** button is used to export the selected object to a csv file. To do so, click on the **Export** button; a drop-down list will be displayed. Select the **Export CSV** option from the drop-down list; the **Export to CSV** dialog box will be displayed. In this dialog box, specify the file name and location of the file to be exported and then choose the **Save** button; the selected object will be saved at the specified location.

Save Selection

The **Save Selection** button is used to save the selected objects as a selection set. To do so, select the required object displayed in the **Selection Inspector** window. Next, choose the **Save Selection** button in the **Selection Inspector** window; the **Sets** window will be displayed and the selected object will be saved and added to the list of the **Sets** window. Specify the name of the saved object in the **Sets** window and press ENTER.

Quick Property Definitions

This button is used to add desired properties to the selected objects in the **Selection Inspector** window. To do so, select the desired object from the Scene View or from the **Selection Tree** window. Next, invoke the **Selection Inspector** window. Now, choose the **Quick Properties Definitions** button from this window; the **Options Editor** dialog box will be displayed with the **Quick Properties** options in the right pane. Specify the property definitions such as category and choose the **OK** button.

OBJECT PROPERTIES

In a project, you can analyze the properties of the selected object by using the **Properties** window. You can also customize the properties of selected object in the **Properties** window. The **Properties** window is discussed next.

Properties Window

After selecting an object in the Scene View, you need to invoke the **Properties** window for analyzing the object properties. To do so, choose the **Properties** button from the **Display** panel in the **Home** tab; the **Properties** window will be displayed, as shown in Figure 3-11. Alternatively, to display the **Properties** window, select the **Properties** check box from the **View > Workspace > Windows** drop-down. The **Properties** window is a dockable window. In this window, there are separate tabs for each property of the selected object. You can use these tabs to navigate between the available property categories. You can also create the custom tabs known as User Data tabs in this window. The method of creating a User Data tab is discussed next.

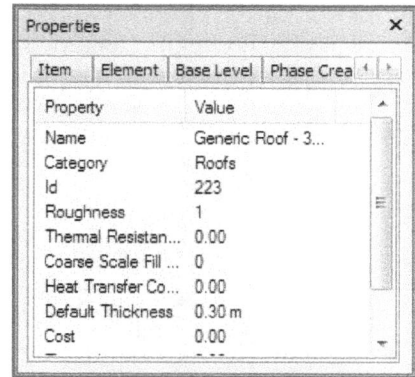

*Figure 3-11 The **Properties** window*

Creating the User Data Tab

In a project, you can also create custom tabs in which you can define the custom properties of a selected object in the Scene View. To create a custom tab, select an object from the Scene View and then right-click in the **Properties** window; a shortcut menu will be displayed, as shown in Figure 3-12. Choose the **Add New User Data Tab** option from the shortcut menu; the new **User Data** tab will be added to the **Properties** window. To rename this tab, right-click in the **Properties** window in the created **User Data** tab; a shortcut menu will be displayed. Choose the **Rename Tab** option from

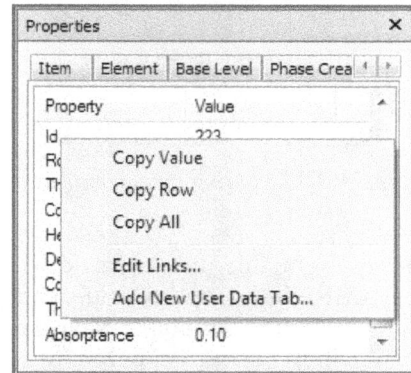

*Figure 3-12 Choosing the **Add New User Data Tab** from the shortcut menu*

the menu; the **Autodesk Navisworks Manage 2017** dialog box will be displayed. Specify a new name for the tab in the **Rename Tab** text box and then choose the **OK** button; the name of the tab will be changed. You can also delete the added **User Data** tab. To do so, select and right-click on the required tab; a shortcut menu will be displayed. Next, choose the **Delete User Data Tab** option from the shortcut menu; the selected tab will be deleted.

You can also add the properties to this **User Data** tab. The procedure to do so is discussed next.

Adding Properties in the User Data Tab

After creating the user data tab, you can add properties to it. To do so, right-click in the **User Data** tab in the **Properties** window; a shortcut menu will be displayed, as shown in Figure 3-13. Next, hover the cursor over the **Insert New Property** option; a flyout will be displayed, refer to Figure 3-13. You can choose the property type from this flyout. On doing so, a text box will be displayed in the tab. Specify a name for the newly created property in the text box and press ENTER. Next, you need to specify a value for the added property. To do so, double-click in the **Value** column corresponding to the required property in the **User Data** tab; the **Autodesk Navisworks Manage 2017** dialog box will be displayed. Specify the required value in the **Enter Property Value** edit box and choose the **OK** button. To rename an added property, right-click on the property to be renamed; a shortcut menu will be displayed, as shown in Figure 3-14. Choose the **Rename Property** option from the menu; a text box will be displayed. Specify a new name in the text box and press ENTER.

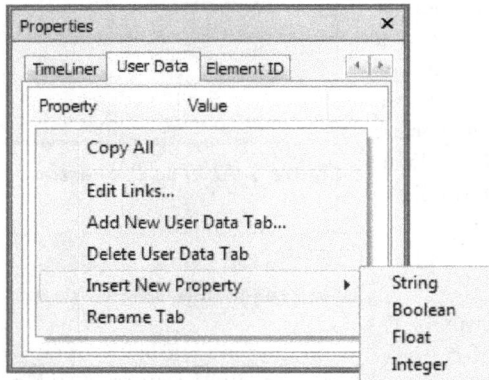

Figure 3-13 *Adding custom properties to the new tab*

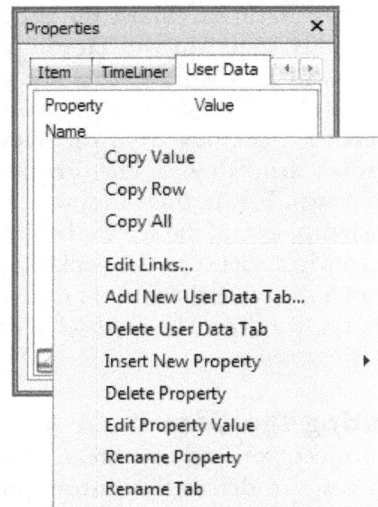

Figure 3-14 *The shortcut menu*

Similarly, from this menu, you can edit the property value and can also delete an added property. To edit a property value, choose the **Edit Property Value** option from the menu, refer to Figure 3-14; the **Autodesk Navisworks Manage 2017** dialog box will be displayed. Specify a new value in the **Enter Property Value** edit box and then choose the **OK** button; the value will be changed. To delete a property, choose the **Delete Property** option from the menu, refer to Figure 3-14; the property will be deleted. Similarly, you can delete the **User Data** tab by choosing the **Delete User Data Tab** option from the menu.

Note
If you have not selected an item in the Scene View then no property will be displayed in the ***Properties*** *window.*

You can also display the property information of an object in a tooltip style. To do so, choose the **Quick Properties** button from the **Display** panel in the **Home** tab. Next, move your cursor over the object; the property information related to the object will be displayed in a tooltip style in the Scene View. The properties to be displayed can be configured from the **Options Editor** dialog box.

LINKING EXTERNAL DATABASES TO OBJECTS

While working on a project, you may need to add some extra properties to the model object. These properties are displayed in the **Properties** window. These properties can be added by linking external databases such as an excel file to the model object. The database links can be added in a single Navisworks file (locally) and can also be added globally. Whenever you will load a particular file, the link will be established when you select an object. Also, the properties of the object will be displayed in the appropriate database tab in the **Properties** window. The process of adding a database link is discussed next.

Adding Database Link Locally

You can add external databases locally by using the **File Options** dialog box. You can invoke this dialog box by choosing the **File Options** tool from the **Project** panel of the **Home** tab. In the **File Options** dialog box, choose the **Data Tools** tab. In this tab, choose the **New** button; the **New Link** dialog box will be displayed. In this dialog box, specify a name in the **Name** edit box; the **Connection** area will be activated, refer to Figure 3-15.

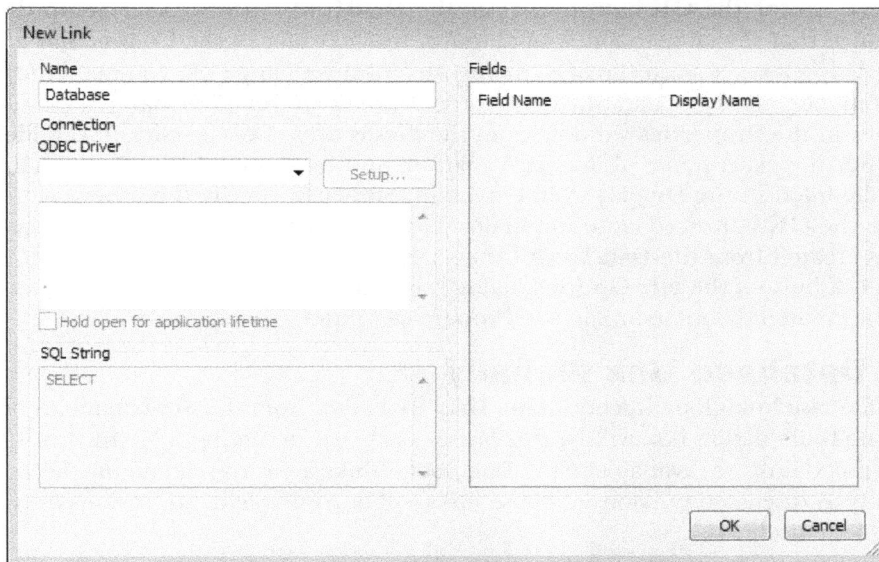

Figure 3-15 *The New Link dialog box*

Next, in the **Connection** area, select the driver connection type from the **ODBC Driver** drop-down list. For example, to link an excel file, select the **Microsoft Excel Driver** option from the list. Now choose the **Setup** button; the **ODBC Microsoft Excel Setup** dialog box will be displayed. Figure 3-16 shows the dialog box for the Microsoft Excel Driver.

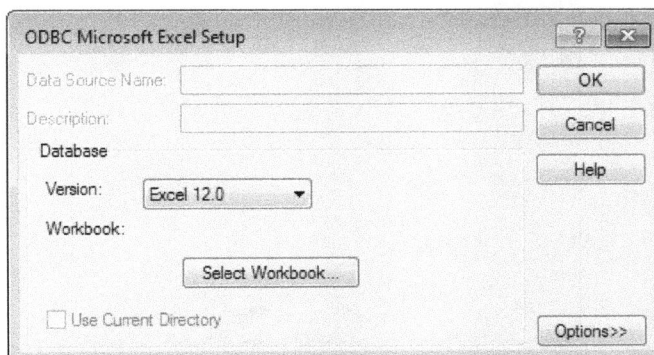

Figure 3-16 *The ODBC Microsoft Excel Setup dialog box*

Note
*The name of the ODBC setup dialog box and the options displayed in it depend on the driver connection type selected from the **ODBC Driver** drop-down list. For example, if you have selected **Microsoft Access Driver** option from the drop-down list, then the displayed dialog box will be **ODBC Microsoft Access Setup**.*

In this dialog box, you need to select the file to be linked and then choose the **OK** button to close the dialog box. For example, to link an excel file, select its version from the **Version** drop-down list and then choose the **Select Workbook** button; the **Select Workbook** dialog box will be displayed. In this dialog box, browse to the excel file, select it, and then choose the **OK** button. Next, choose the **OK** button to close the **ODBC Microsoft Excel Setup** dialog box. You will notice that the connection is updated in the text box in the **Connection** area of the **New Link** dialog box. Now, in the **SQL String** area of this dialog box, you need to specify the selection statement after the text SELECT. You can select the columns that you want to display as categories in the **Properties** window from the **Fields** area. Double-click in the **Field Name** column, enter the exact name of database column, and then press ENTER; the name will be automatically filled in the **Display Name** column. Similarly, specify the other column names and choose the **OK** button to close the dialog box. Now, select the check box corresponding to the links created from the **DataTools Links** area in the **File Options** dialog box, and then choose the **OK** button; the **File Options** dialog box will be closed. You can now select an object and view the created database link in the **Properties** window.

Adding Database Link Globally

To add a database link globally, choose the **DataTools** tool from the **Tools** panel of the **Home** tab; the **DataTools** dialog box will be displayed, as shown in Figure 3-17. In this dialog box, some predefined links are available in the **DataTools Links** area. You can define the new links in the same way as discussed previously. These links will be available in all sessions of Navisworks and can also be modified.

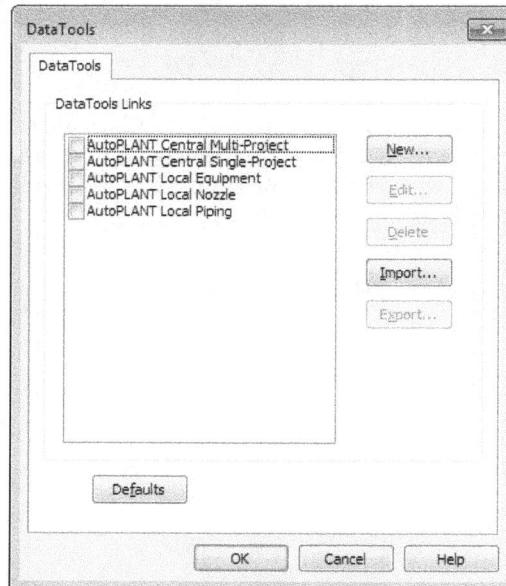

*Figure 3-17 The **DataTools** dialog box*

There are few buttons in the **DataTools** tab of the **File Options** dialog box which are also available in the **DataTools** dialog box. The usage of these buttons is discussed next.

To modify a database link, select the check box corresponding to that link from the **DataTools Links** area and then choose the **Edit** button; the **Edit Link** dialog box will be displayed. In this dialog box, you can make the changes as required. You can remove a link from the **DataTools Links** area by first selecting the corresponding check box and then choosing the **Delete** button. To export a database link, select the corresponding check box and then choose the **Export** button; the **Save As** dialog box will be displayed. In this dialog box, browse to the desired folder and save the file with a distinct name. Similarly, to import a database link, choose the **Import** button; the **Open** dialog box will be displayed. In this dialog box, browse to the required file and choose the **Open** button; the link will be added in the **DataTools** tab of the **File Options** dialog box.

MEASURE TOOLS

In a project, you can perform several functions such as calculating distances, angle, and area. To perform these functions, you can use measure tools. Before measuring an object, you need to adjust the cursor snapping. To do so, invoke the **Options Editor** dialog box by choosing the **Options** button from the **Application Menu**. Expand the **Interface** node in the left pane of this dialog box and select the **Snapping** option; several options will be displayed in the right pane of the dialog box, as shown in Figure 3-18.

In the **Picking** area, the **Snap to Vertex** check box is selected by default. As a result, the cursor will snap to the nearest vertex while measuring distances. Select the **Snap to Edge** check box to snap the cursor to the nearest triangular edge. The **Snap to Line Vertex** check box is selected by default. As a result, the cursor will snap to the nearest line end. To define the snapping margin, specify a value in the **Tolerance** edit box. By default, the value in the **Tolerance** edit box is **5**. In the **Rotation** area, specify the snapping angle value in the **Angles** edit box. By default, the value displayed in the **Angles** edit box is **45.00**. To define the snapping tolerance, enter the value in the **Angle Sensitivity** edit box. The default value displayed in the **Angle Sensitivity** edit box is **5.00**.

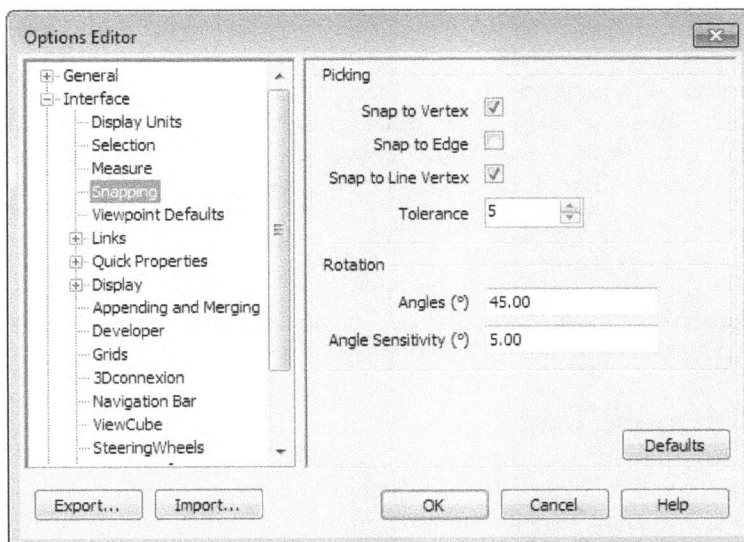

*Figure 3-18 The **Snapping** page in the **Options Editor** dialog box*

After defining all the parameters in the **Options Editor** dialog box, close it by choosing the **OK** button. Next, invoke the measuring tools from the **Review > Measure > Measure** drop-down. The measuring tools in this drop-down are discussed next.

Point to Point Tool

This tool is used to measure the distance between two points in a model. To do so, choose the **Point to Point** tool from the **Measure** drop-down; the measuring cursor will appear on the screen. Click on the required points between which the distance is to be measured; the measured distance will be displayed in the Scene View.

Point to Multiple Points Tool

This tool is used to measure the distance from a point to several other points in a model. To measure the distance between multiple points, choose the **Point to Multiple Points** tool from the **Measure** drop-down. Click on the start point and the first end point; the measured distance will be displayed in the Scene View. Next, click on the end points to be measured.

Point Line Tool

This tool is used to measure the total distance along the route. To measure the total distance along the route, choose the **Point Line** tool from the **Measure** drop-down. Next, click on the start point and then click on the points along the route to be measured; the measured distance will be displayed in the Scene View.

Accumulate Tool

This tool is used to calculate the total distance between several point-to-point measurements. To do so, choose the **Accumulate** tool from the **Measure** drop-down. Then, click on the start and end points of the first distance to be measured. Next, click on the start and end points of the next distance to be measured; the sum of all point to point distances will be displayed in the Scene View.

Angle Tool

This tool is used to measure an angle between two points. To do so, choose the **Angle** tool from the **Measure** drop-down. Click on the start point of the first line. Next, click on the intersection point of the first and the second line. Click on the end point of the second line; the measured angle will be displayed in the Scene View.

Area Tool

This tool is used to calculate the area of a plane. To do so, choose the **Area** tool from the **Measure** drop-down. Click on the desired points to define the periphery of the area to be calculated; the calculated area will be displayed in the Scene View.

Shortest Distance Tool

This tool is used to measure the shortest distance between two objects in a model. To do so, select the required objects in the Scene View. Choose the **Shortest Distance** tool from the **Measure** panel in the **Review** tab; the shortest distance will be displayed in the Scene View.

Clear Tool

The **Clear** tool is used to delete the measurements. To do so, choose the **Clear** tool from the **Measure** panel in the **Review** tab; the measurements made will be deleted.

Convert to Redline Tool

The **Convert to Redline** tool is used for converting the measurements displayed in the Scene View into redlines. To do so, first take the measurements using any of the **Measure** tools. Next, choose the **Convert to Redline** tool from the **Measure** panel in the **Review** tab; the measurements will be converted into redlines, and saved as a viewpoint in the **Saved Viewpoints** window. The thickness and color of these lines can be modified. To change the color of redline, select the color from the **Color** drop-down list of the **Redline** panel in the **Review** tab. To change the thickness of line, enter the thickness value in the **Thickness** edit box.

When you perform measurements in the Scene View, you can view the results of the measurement such as the distance, coordinates of the start and end points, and the difference in the co-ordinates in the **Measure Tools** dockable window, as shown in Figure 3-19. To invoke this window, choose the **Measure Options** button from the **Measure** panel of the **Review** tab.

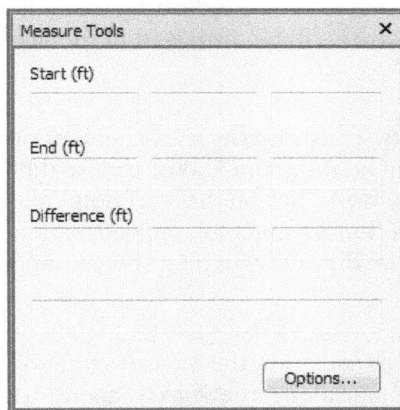

*Figure 3-19 The **Measure Tools** dockable window*

In the **Measure Tools** window, the **Start** and **End** edit boxes will display the coordinates of the start and end points of the measured distance, respectively. The **Difference** and **Distance** edit boxes in this window will display the difference of coordinates between the start and end points and the distance between them. If we calculate area and angle, the **Area** and **Angle** edit boxes will be displayed in this window.

You can also control the display of measurements in the Scene View. To do so, choose the **Options** button from the **Measure Tools** window; the **Options Editor** dialog box will be displayed. In this dialog box various measuring options are displayed in the right pane. Note that you can also access these options by expanding the **Interface** node in the **Options Editor** dialog box and then clicking on the **Measure** sub-node.

In the **Options Editor** dialog box, you can specify the thickness of measurement lines in the **Line Thickness** edit box. You can change the color of measurement lines by selecting an appropriate color from the **Color** flyout. To draw the measurement lines in 3D, select the **In 3D** check box. To clear the measurement values in the Scene View, clear the **Show measurement values in Scene**

View check box, which is selected by default. The **Use centre lines** check box is also selected by default. As a result, the shortest distance measurement will snap to the centre lines of the parametric object. Additionally, the **Auto - zoom when measuring shortest distances** check box is selected by default. As a result, on calculating the shortest distance between two objects, the measured distance will zoom in automatically in the Scene View.

DIRECTION CONSTRAINTS FOR MEASUREMENTS

In a project, you can constrain the direction of measuring objects. The constraints can be applied in X, Y, and Z directions and they can also be applied in a direction parallel or perpendicular to the surface of an object. To apply directional constraints, first you need to invoke any of the measuring tools from **Review > Measure > Measure** drop-down. The tools used for constraining the direction are discussed next.

X Axis Tool

The **X Axis** tool is used to constrain the measurement of an object in the X direction. To constrain the measurement along the X axis, choose the **X Axis** tool from the **Review > Measure > Lock** drop-down. Click on the start point of the distance to be measured; the measuring line will be displayed in red color indicating that it is locked in the X direction. Next, click on the end point; the distance will be displayed in the Scene View.

Y Axis Tool

The **Y Axis** tool is used to constrain the measurement of an object in the Y direction. To constrain the measurement along the Y axis, choose the **Y Axis** tool from the **Review > Measure > Lock** drop-down. Click on the start point of the distance to be measured; the measuring line will be displayed in green color indicating that it is locked in the Y direction. Next, click on the end point; the distance will be displayed in the Scene View.

Z Axis Tool

The **Z Axis** tool is used to constrain the measurement of an object in the Z direction. To constrain the measurement along the Z axis, choose the **Z Axis** tool from **Review > Measure > Lock** drop-down. Click on the start point of the distance to be measured; the measuring line will be displayed in blue color indicating that it is locked in the Z direction. Next, click on the end point; the distance will be displayed in the Scene View.

Perpendicular Tool

The **Perpendicular** tool is used to constrain the measurement of an object in a direction perpendicular to the surface of the start point. To use the perpendicular lock, choose the **Perpendicular** tool from the **Review > Measure > Lock** drop-down. Click on the start point of the distance to be measured; the measuring line will be displayed in yellow color indicating that it is locked in the perpendicular direction. Next, click on the end point; the distance will be displayed in the Scene View.

Parallel Tool

The **Parallel** tool is used to constrain the measurement of an object in a direction parallel to the surface of the start point. To use the parallel lock, choose the **Parallel** tool from the **Review > Measure > Lock** drop-down. Click on the start point of the distance to be measured; the measuring line will be displayed in magenta color indicating that it is locked in the parallel direction. Next, click on the end point; the distance will be displayed in the Scene View.

REDLINE TOOLS

In a project, you can mark up the viewpoints and clash results by adding texts and drawing revision clouds. These mark ups are added by using the redline tools. The redline tools are discussed next.

Adding Text to Redline

In a project, you can mark or annotate a saved viewpoint and clash results by adding text. To add text, choose the **Text** tool from the **Redline** panel in the **Review** tab; a pencil cursor will be displayed on the screen. Click at the required location in the Scene View; the **Autodesk Navisworks Manage 2017** dialog box will be displayed. Specify the text in the **Enter Redline Text** text box and choose the **OK** button; the text will be added at the specified location. To move the text, right-click on it; a shortcut menu will be displayed. Choose the **Move** option from the menu, and click at the desired location; the text will be placed at the specified location. To edit the text, choose the **Edit** option from the shortcut menu; the **Autodesk Navisworks Manage 2017** dialog box will be displayed. Make the required changes and choose the **OK** button to apply the changes. To delete the text, choose the **Delete Redline** option from the shortcut menu.

> **Note**
> *By default, the redline text is written in a single line. To write it in multiple lines, specify \P after the text from where you want to start the next line. To left align the text, leave no space between the \P and the succeding text.*

Drawing Cloud

In a project, you can mark a saved viewpoint and clash result by drawing cloud, line, ellipse, and so on. For example, to draw a cloud, first you need to choose the desired saved viewpoint. Next, choose the **Cloud** tool from the **Review > Redline > Draw** drop-down; the pencil cursor will appear in the Scene View. Now, click in the required direction to create a cloud. Similarly, you can use the other tools from the **Draw** drop-down to mark the saved viewpoints.

You can delete redline clouds and texts as per your requirement. To do so, choose the **Erase** tool from the **Redline** panel in the **Review** tab; the erase cursor will appear in the Scene View. Now, draw a box around the text or line to be deleted; the redlines will be erased.

TAGS AND COMMENTS

In Navisworks, you can tag anything in the Scene View which allows you to add comments or notes to the viewpoints. The comments added to the saved viewpoints and animations can be used for reference later. In the next section, you will learn about adding tags.

Adding Tags

In a project, you can tag anything in the Scene View. To do so, choose the **Add Tag** option from the **Tags** panel in the **Review** tab; a pencil cursor will be displayed on the screen. In the Scene View, first click on the object you want to tag and then click at the point where you want to label the tag; the **Add Comment** dialog box will be displayed. Alternatively, you can click thrice at the point where you want to create a tag. Specify the text in the **Add Comment** area. To set the tag status, select the required options from the **Status** drop-down list and choose the **OK** button. The tag will be added to that particular object, and it will be saved as a viewpoint in the **Saved**

Viewpoints window. The comment which you have added in the **Add Comment** dialog box will be displayed in the **Comments** window. This window is discussed next.

The Comments Window

In a project, when you add comments to viewpoints or results, you can view them in the **Comments** window. In this window, you can add all the comments related to the model and can obtain any information quickly. The data will be displayed in a separate window rather than the Scene View. To display the window, choose the **View Comments** button from the **Comments** panel in the **Review** tab; the **Comments** window will be displayed, as shown in Figure 3-20.

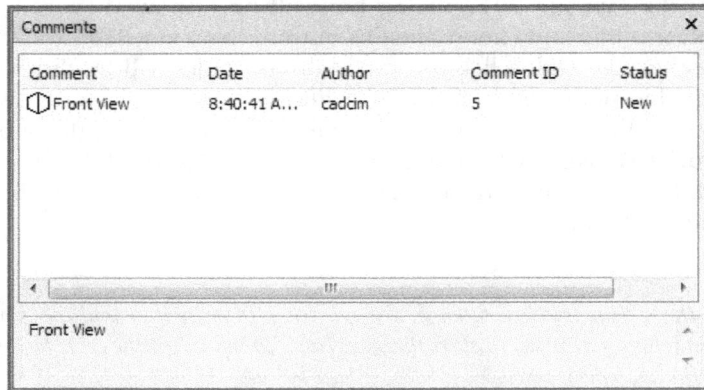

Figure 3-20 The Comments window

In the **Comments** window, you can view the information related to comments such as the author of comments, date on which comment was added, the content of comment, and so on. These informations will be displayed in different columns. You can add comments to selection and search sets, Viewpoints, Animation, Timeliner, and Clash results. The method of adding comment is discussed next.

Adding Comment

In a project, you can add comments to a selection and search set, viewpoints, clash results, animation, and so on. To add a comment to a selection/search set, select the required set from the **Sets** window. Choose the **Add Comment** button; the **Add Comment** dialog box will be displayed. Specify a comment and choose the **OK** button; the comment will be added to the set. Similarly, you can add comments to a particular viewpoint. To do so, choose the **Saved Viewpoints Dialog Launcher** button from the **Save, Load & Playback** panel in the **Viewpoint** tab; the **Saved Viewpoints** window will be displayed, as shown in Figure 3-21. Now, select the desired viewpoint and right-click on it; a shortcut menu will be displayed. Choose the **Add Comment** option from the menu; the **Add Comment** dialog box will be displayed. Enter the comment in the dialog box, assign the status from the **Status** drop-down list, and choose the **OK** button; the comment will be added to the window. Similarly, you can add comments to clash results and timeliner which will be discussed in later chapters. To view the added comments,

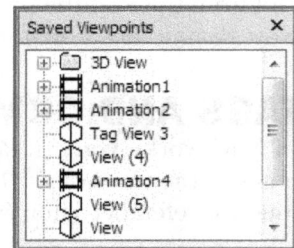

Figure 3-21 The Saved Viewpoints window

click on the source item to which you have added the comment; the comments will be displayed in the **Comments** window.

You can modify the added comments. The procedure of modifying comments is discussed next.

Modifying Comment
In a project, after adding comments, you can edit them directly in the **Comments** window. To do so, right-click on the comment in the **Comment** window; a shortcut menu will be displayed. Choose the **Edit Comment** option from the menu; the **Edit Comment** dialog box will be displayed. Next, type the desired comment and choose the **OK** button; the comment will be edited. To delete a comment, choose the **Delete Comment** option from the shortcut menu; the comment will be deleted.

In a project, you may need to search for comments based on data, date, and source. To do so, the **Find Comments** window is used, which is discussed next.

The Find Comments Window
The **Find Comments** window is a dockable window. Using the options in this window, you can search for the added comments. To invoke the **Find Comments** window, choose the **Find Comments** button from the **Comments** panel in the **Review** tab; the **Find Comments** window will be displayed, as shown in Figure 3-22.

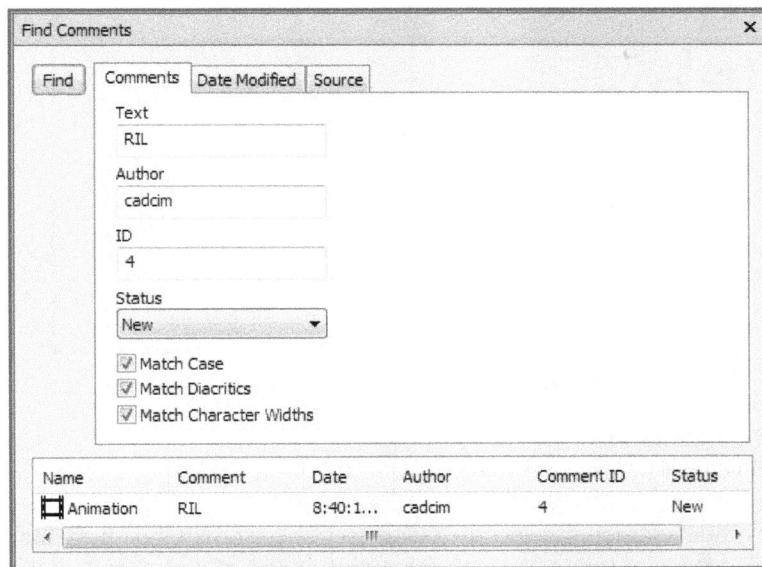

*Figure 3-22 The **Find Comments** window*

This window comprises of three tabs: **Comments**, **Date Modified**, and **Source**. To search for a particular **Text, Author, ID** or **Status**, choose the **Comments** tab. To search within a certain timeframe, choose the **Date Modified** tab. To search for the comments attached to the selected source, choose the **Source** tab. After specifying all the search criteria, choose the **Find** button; the results will be displayed at the bottom of the window.

You can also use the **Quick Find Comments** text box from the **Comments** panel in the **Review** tab to quickly find the comments. To do so, type the keywords in the **Quick Find Comments** edit box in the **Comments** panel of the **Review** tab. Next, choose the button next to the text box; the **Find Comments** dialog box will be displayed with a list of all comments that match the text.

COMPARING MODELS

While working on a project, sometimes you may need to find the differences between two elements in a model. Alternatively, you may need to compare two models of different versions. In Navisworks, you can find out the differences or compare two models by using the **Compare** tool. To compare two elements in a model, first select the required elements using the CTRL key. Next, choose the **Compare** tool from the **Home > Tools** panel; the **Compare** dialog box will be displayed, as shown in Figure 3-23. To compare two different versions of the same model, first load the first file that you want to compare. Next, append the second file and then select both the files in the **Selection Tree** window using the CTRL key. Next, invoke the **Compare** tool; the **Compare** dialog box will be displayed. The options in this dialog box are discussed next.

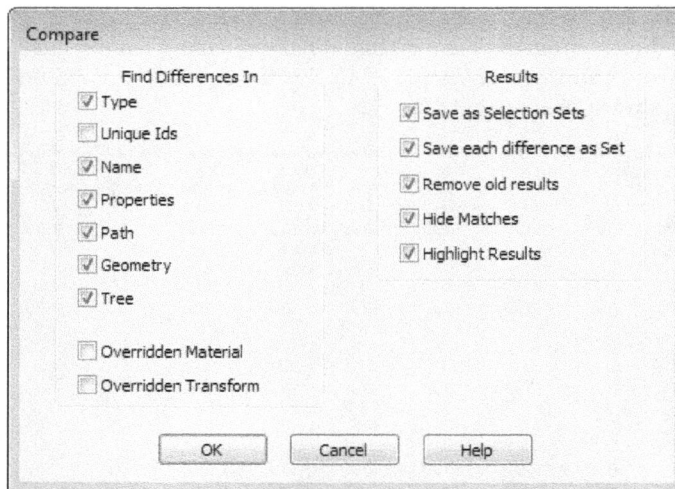

Figure 3-23 The **Compare** dialog box

Find Differences In

In the **Find Differences In** area of the dialog box, select the check boxes corresponding to the categories in which you want to find the differences. For example, to find differences in the geometry of two model elements, select the **Geometry** check box. If you have made any changes in the model appearance or made any transformation in the model, then you need to select the **Overridden Material** and **Overridden Transformation** check boxes to compare the elements with material and location overrides.

Results

In the **Results** area, there are various options that allow you to specify how you want to display the comparison results. The options in this area are discussed next.

Save as Selection Sets

Select this check box to save the comparison results of matched and unmatched objects as a separate selection set in the **Sets** window. The comments related to the comparison result of the matched and unmatched objects will be available in the **Comments** window.

Save each difference as Set

Select this check box to save the differences that are found while comparing the elements as a selection set. The selection set will also contain comments related to the comparison results. You can view these comments in the **Comments** window.

Remove old results

Select this check box to delete the results of old comparison from the **Sets** window.

Hide Matches

Select this check box to hide the matching objects after the comparison finishes. This will allow you to view the differences clearly.

Highlight Results

Select this check box to highlight all the objects that have differences in different color overrides.

After the comparison finishes, the results will be displayed in the Scene View. The matched and unmatched items, and the differences in the items will be displayed in different color codes. The matching elements will be highlighted in white color. The elements with differences will be highlighted in red color. Items found in the first element but not in the second element will be highlighted in yellow, and the one found in the second element but not in the first element will be highlighted in cyan.

CONTROLLING THE VISIBILITY OF OBJECTS

In a project, you may need to control the visibility of some objects in the Scene View. In Navisworks, you have several options to control the visibility of objects in the model. You can hide and display an object or a group of objects in the model. To hide an object in the model, first select the desired object in the model. Next, choose the **Hide** button from the **Visibility** panel in the **Home** tab; the selected object will be hidden. Notice that the hidden object will appear grey in the **Selection Tree** window. To display the

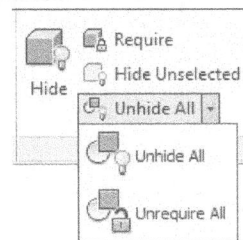

*Figure 3-24 The **Unhide All** drop-down*

hidden object(s), choose the **Hide** button again. To hide all the unselected objects in the model, select the object which you want to display and then choose the **Hide Unselected** button from the **Visibility** panel in the **Home** tab; the unselected objects will be hidden. To display all the hidden objects, choose **Unhide All** from the **Unhide All** drop-down, refer to Figure 3-24. To make the object visible during navigation, regardless of the performance setting, select the item in the **Selection Tree** window. Next, choose the **Require** button from the **Visibility** panel; the required objects will appear red in the **Selection Tree** window. To remove all objects from selection in the window, select the **Unrequire All** option from the drop-down list in the **Visibility** panel, refer to Figure 3-24.

CONTROLLING THE APPEARANCE OF MODEL IN THE SCENE VIEW

You can control the appearance of a model in the Scene View. To control the display of your model in the Scene View, use the tools available in the **Render Style** panel of the **Viewpoint** tab. There are four render modes available to control the delivery of items in the **Scene View**. To apply these modes, click on the **Mode** option in the **Render Style** panel of the **Viewpoint** tab; a drop-down list will be displayed. Select any of the four options from the drop-down list. Select the **Full Render** option to render the model with smooth shading including the materials used. Select the **Shaded** mode to deliver the model with smooth shading but without textures. Select the **Wireframe** mode to deliver the model in wireframe. Select the **Hidden Line** mode to deliver the model in wireframe but in this case only outline edges will be displayed, as shown in Figure 3-25.

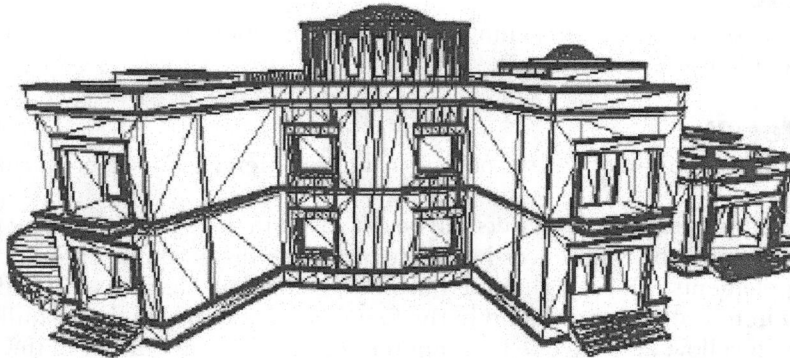

*Figure 3-25 The Wireframe view of the model on using the **Hidden Line** mode*

In Navisworks, the 3D scene illumination can be controlled by making it brighter or lighter by using four lighting modes available in the **Lighting** mode drop-down list. To use these options, click on the **Lighting** option in the **Render Style** panel of the **Viewpoint** tab; a drop-down list will be displayed. You can select any of the following four options: **Full Lights**, **Scene Lights**, **Head Light**, and **No Lights**.

You can also change the background color and its effects to be used in the Scene View as per your choice. To configure the background effects, invoke the **Background** tool from the **Scene View** panel in the **View** tab; the **Background Settings** dialog box will be displayed, as shown in Figure 3-26.

Select the type of background effect from the **Mode** drop-down list, and then choose the color from the **Color** drop-down list. Select the **Plain** mode to fill the background of the scene with the selected color. Select the **Graduated** mode to fill the background of the scene with two colors. Select the **Horizon** mode to split the scene across the horizontal plane which gives the effect of sky and ground.

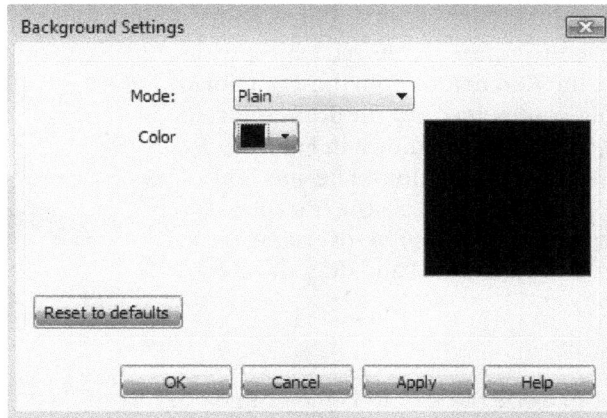

Figure 3-26 *The **Background Settings** dialog box*

CONTROLLING OBJECT ATTRIBUTES

In a project, you can control the object transformation and its appearance. All these changes are displayed in the Scene View. You can alter the object's position, size, and rotation using the visual manipulation tools (Gizmos). You can also make these changes numerically. The method of controlling object attributes using various tools will be discussed individually.

Controlling Object Attributes Using Gizmo

You can move, rotate, and resize the objects using gizmo. To move an object to the required position in the Scene View, select the required object; you will notice that the **Item Tools** contextual tab appears in the ribbon. This tab contains all the transformation tools which are also called gizmos. These gizmos are discussed next.

Move Gizmo

The **Move** gizmo is used to adjust the position of the selected object. To do so, choose the **Move** tool from the **Transform** panel in the **Item Tools** contextual tab; the **Move** gizmo will be displayed in the Scene View, as shown in Figure 3-27. To move the selected object in the X, Y, and Z direction, place the mouse over the gizmo; the cursor will change into a hand symbol. Next, place the mouse over the required axis; the axis will be highlighted. Now, drag the gizmo in the required direction. You can also change the position of the gizmo in the Scene View. To do so, press and hold the CTRL key. Place the cursor on the center ball of the gizmo. Next, drag the gizmo and place it at the desired location.

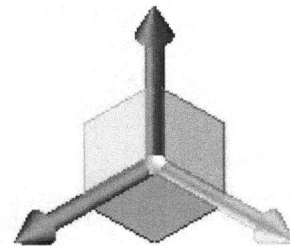

Figure 3-27 *The **Move** gizmo*

Tip
*You can also move an object by specifying exact values. To do so, select the object and right-click to display a shortcut menu. Next, choose the **Override Item > Override Transform** option from the menu; the **Override Transform** dialog box will be displayed. Specify the coordinate value in the dialog box and choose the **OK** button; the selected object will be moved to the specified location.*

Rotate Gizmo

The **Rotate** gizmo is used to rotate a selected object with gizmo. To do so, choose the **Rotate** tool from the **Transform** panel in the **Item Tools** contextual tab; the **Rotate** gizmo will be displayed in the Scene View, as shown in Figure 3-28. Now, adjust the center point of rotation. The method of adjusting the center point is same as the method for adjusting the **Move** gizmo. Place the mouse over one of the curves in the middle, refer to Figure 3-24 and drag the mouse to rotate the selected object.

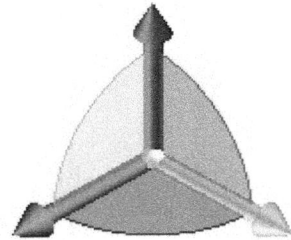

*Figure 3-28 The **Rotate** gizmo*

Scale Gizmo

The **Scale** gizmo is used to resize the selected object. To do so, choose the **Scale** tool from the **Transform** panel in the **Item Tools** contextual tab; the **Scale** gizmo will be displayed in the Scene View, as shown in Figure 3-29. Now, to resize the object along a single axis, drag any of the three axes. To resize the object across two axes, drag the colored triangles in the middle of the two axes. To resize the object across all the three axes at the same time, use the ball in the center of the gizmo.

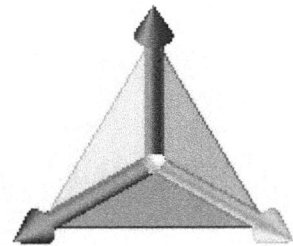

*Figure 3-29 The **Scale** gizmo*

Controlling Object Attributes Numerically

You can also move, rotate, and scale the selected objects numerically. To do so, first invoke any of the gizmos, and then click on the down arrow in the **Transform** panel; various edit boxes such as **Position**, **Rotation**, **Scale**, and **Transformation Centre** will be displayed. Now, specify the values in the required edit boxes.

You can also change the appearance such as color and transparency of the selected objects. To change the color, click on the **Color** drop-down in the **Appearance** panel of the **Item Tools** tab; a list of colors will be displayed. Select the desired color; the color of the object will be changed. To adjust the transparency, drag the **Transparency** slider in the **Appearance** panel of the **Item Tools** tab.

LINKS

Navisworks uses several sources of links such as the links converted from CAD files, links added by the users, and links which are automatically generated by the program, like viewpoint links. To display these links, choose the **Links** button from the **Display** panel in the **Home** tab; all links will be displayed as icons in the Scene View.

Types of Links

There are two types of links: Standard and User Defined. Standard links are hyperlink, label, viewpoints, and clash detection. User defined links are those in which a user can create different categories other than the standard ones. You can control the display of both the links. To control the display of the Standard links, choose the **Options** button from the **Application Menu**; the **Options Editor** dialog box will be displayed. Expand the **Interface** node in the left pane of the dialog box. Next, expand the **Links** sub node and select **Standard Categories**, as shown in Figure 3-30.

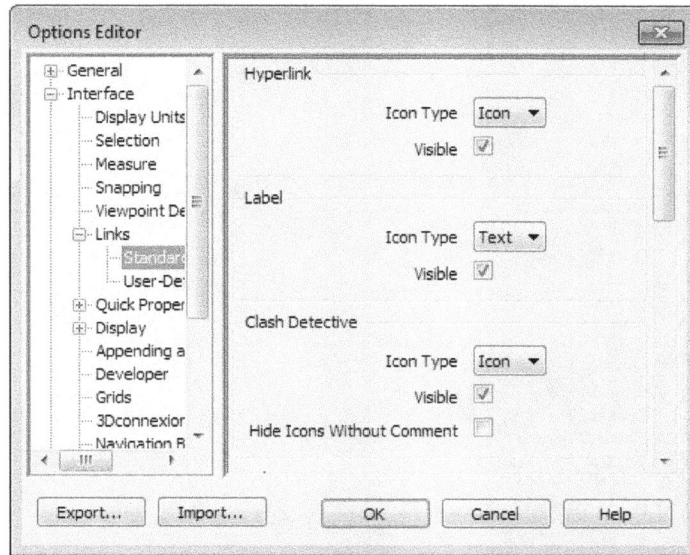

Figure 3-30 *The* **Standard Categories** *option selected in the* **Options Editor** *dialog box*

In the right pane of the dialog box, select the corresponding **Visible** check box to display the links and to hide them clear the check box. To change the appearance of links, select the required option from the **Icon Type** drop-down list for the corresponding link category, and choose the **OK** button to close the dialog box.

Similarly, to control the display of user defined links, select the **User Defined Categories** option from the **Links** sub node under the **Interface** node; several options will be displayed in the right pane of the dialog box. In this pane, select the **Visible** check box to display the corresponding links and clear the check box to hide them. To change the appearance of links, select the required option from the **Icon Type** drop-down list. Next, choose the **OK** button to close the dialog box. Note that the user defined options will not be available if you have not added the user defined categories. You can also search for links by using the **Find Items** window.

Customizing Appearance of Links in the Scene View

To avoid the confusion arising due to the large number of links displayed in the Scene View, you can customize the appearance of links displayed in the Scene View by using options in the **Options Editor** dialog box. To do so, expand the **Links** sub node under the **Interface** node in the left pane of the **Options Editor** dialog box, as shown in Figure 3-31; several options will be displayed in the right pane of the dialog box. Specify the value in the **Max Icons** edit box to control the display of icons in the Scene View. Select the **Hide Colliding Icons** check box to hide any overlapping link icon in the Scene View. To specify the distance between camera and the drawn links, enter the value in the **Cull Radius** edit box.

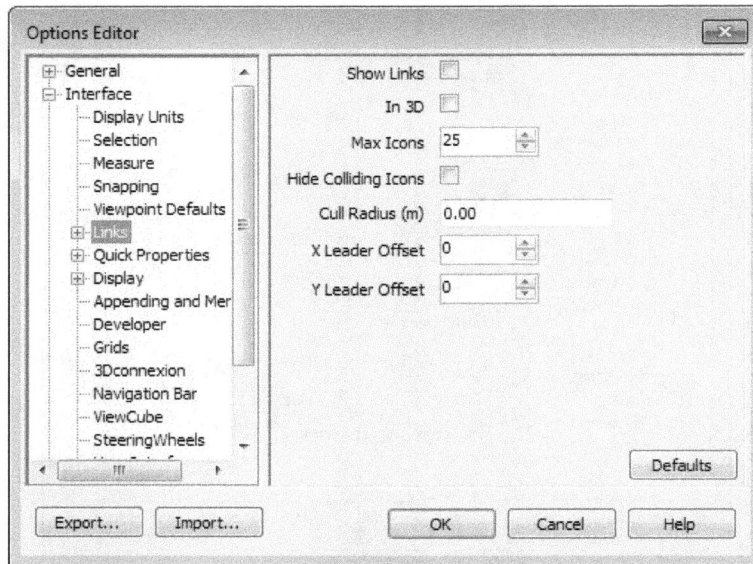

*Figure 3-31 The **Links** options in the **Options Editor** dialog box*

Sometimes links can be hidden in the Scene View. To avoid this, select the **In 3D** check box; the links will float in front of their attachment points. To show the leader lines, specify the distance in the **X Leader Offset** and **Y Leader Offset** edit boxes; links will be displayed with the leader lines pointing to the attached items.

Adding Links

The links can be appended from various data sources such as spreadsheets, webpages, audio, and video files. You can attach multiple links to a single object. To add a link to an object, select the desired object in the model; the **Item Tools** contextual tab will be displayed in the ribbon. Choose the **Add Link** tool from the **Links** panel in the **Item Tools** tab; the **Add Link** dialog box will be displayed, as shown in Figure 3-32. Alternatively, right-click on the selected object; a shortcut menu will be displayed. Choose the **Links** option from the shortcut menu; a flyout will be displayed. Choose the **Add Link** option; the **Add Link** dialog box will be displayed. Specify a desired name for the link in the **Name** edit box. To specify the file to be linked, choose the Browse

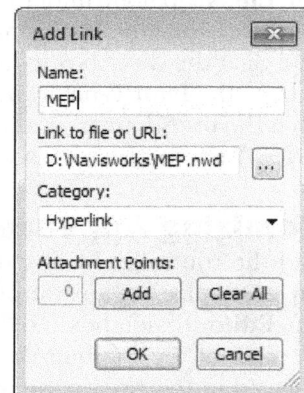

*Figure 3-32 The **Add Link** dialog box*

button displayed to the right of the **Link to file or URL** edit box; the **Choose Link** dialog box will be displayed. Browse to the folder location and select the required file. Next, choose the **Open** button from the **Choose Link** dialog box; the path of the selected file will be displayed in the **Link to file or URL** edit box. Choose the link category from the **Category** drop-down list. To attach the link to a specified point on the selected item, choose the **Add** button under the **Attachment Points** area; a pick cursor will be displayed in the Scene View. Click on the required location on the selected object in the Scene View. Next, choose the **OK** button; the dialog box is closed and a link will be added. The link icon will be displayed at the specified location in the Scene View. To delete the attached points, choose the **Clear All** button in the **Attachment Points** area and then choose the **OK** button.

You can add several links to an object by repeating the above mentioned steps.

Managing Links

In Navisworks, you can edit, reset, and delete all the links except those which are automatically generated such as links which point to the clash results, viewpoints, TimeLiner tasks, and so on. To edit a link, right-click on the required link; a shortcut menu will be displayed. Choose the **Edit Link** option from the menu; the **Edit Links** dialog box will be displayed, as shown in Figure 3-33.

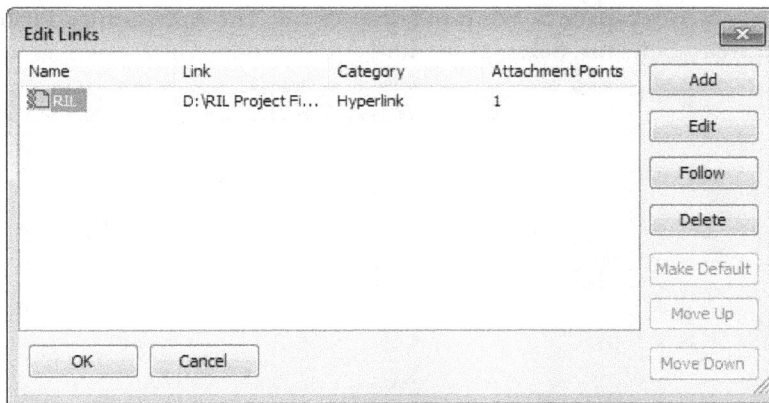

Figure 3-33 The **Edit Links** dialog box

Now, in the dialog box, select the link that you want to edit and then choose the **Edit** button; the **Edit Link** dialog box will be displayed, as shown in Figure 3-34. Enter the link details in the **Name**, **Link to file or URL**, and **Category** edit boxes. Next, specify the desired **Attachment Points** to change the location of links in the Scene View, and choose the **OK** button to close the dialog box.

Similarly, to delete a link, select the required link from the **Edit Links** dialog box, refer to Figure 3-34. Choose the **Delete** button and then the **OK** button; the link will be deleted. To reset all links for an object, select the required object. Next, choose the **Reset Links** tool from the **Links** panel in the **Item Tools** contextual tab;

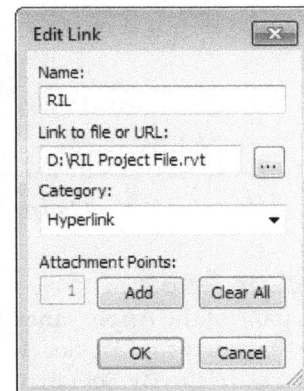

Figure 3-34 The **Edit Link** dialog box

all links will be readjusted to their original state. Similarly, you can reset all links in the Scene View to their original state. To do so, select the **Links** option from the **Reset All** drop-down list in the **Project** panel of the **Home** tab; all links will be readjusted.

Tip
*You can also display the **Edit Links** dialog box by choosing the **Edit Links** tool from the*
***Links** panel of the **Item Tools** tab, refer to Figure 3-29.*

THE APPEARANCE PROFILER WINDOW

In a project, you can customize the appearance of a model by defining the custom display settings based on the property data or selection/search sets using the **Appearance Profiler** window. The **Appearance Profiler** window is a dockable window. You can use this window to differentiate between the objects by color coding them and adding transparency in the model. To invoke the **Appearance Profiler** window, choose the **Appearance Profiler** button from the **Tools** panel in the **Home** tab; the **Appearance Profiler** window will be displayed, as shown in Figure 3-35. Alternatively, to display the **Appearance Profiler** window, select the **Appearance Profiler** check box from the **View > Workspace > Windows** drop-down. The **Appearance Profiler** window is divided in three areas: **Selector**, **Selector** list, and **Appearance**. These areas are discussed next.

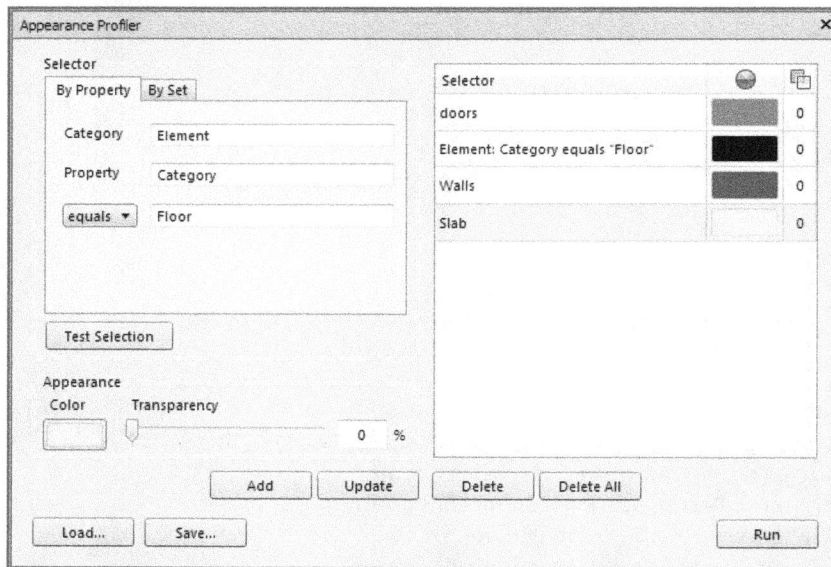

*Figure 3-35 The **Appearance Profiler** window*

Selector Area

The left pane of the **Appearance Profiler** window is the **Selector** area. In this area, there are two tabs: **By Property** and **By Set**. In the **By Property** tab, you can define the search criteria based on property. In the **By Set** tab, you can define the search criteria based on selection and search sets. The **Test Selection** button is used for checking the defined search criteria. On choosing this button, the objects satisfying the search criteria will be selected in the model in the Scene View.

To define the selector by the property data, choose the **By Property** tab from the **Appearance Profiler** window. Enter the property category and property type in the **Category** and **Property** edit boxes. Select the appropriate condition operator from the **equals/does not equals** drop-down list. After defining the search criteria, choose the **Test Selection** button; all objects meeting the search criteria will be selected in the Scene View.

Similarly, you can define the selector by using the selection/search set. To do so, first ensure that a search/selection set is defined in the project. Next, choose the **By Set** tab in the **Selector** area. Then select the set from the list. Next, choose the **Test Selection** button; all objects meeting the search criteria will be selected in the Scene View.

After defining the selector using any of the above discussed methods, you can configure its appearance in the **Appearance** area which is discussed next.

Appearance Area

The **Appearance** area is located at the right corner in the **Appearance Profiler** window. In this area, you can configure the appearance of the defined selector. The **Color** button in this area is used for selecting colors. On choosing this button, a **Color** palette will be displayed. Select the desired color from it and choose the **OK** button; the **Color** palette will be closed and the color will be selected. Specify the transparency level value in the **Transparency** edit box. After specifying the color and transparency, choose the **Add** button; the selector will be added in the **Selector** list area. Next, choose the **Run** button; the objects in the model will be color coded with the new color.

If you make any changes in the added selector, then choose the **Update** button for saving those changes. To remove a selector from the list, choose the **Delete** button. The **Delete All** button is used to clear the **Selector** list. The **Load** button is used to open an existing appearance profile. On choosing this button, the **Open** dialog box will be displayed. In the dialog box, browse to the desired file, and choose the **Open** button; the file will be loaded in the **Appearance Profiler** window. The **Save** button is used for saving the current appearance profile. On choosing this button, the **Save As** dialog box will be displayed. Specify the name for the file in the **File name** edit box. Choose the file type from the **Save as type** drop-down list, and choose the **Save** button; the file will be saved as a data file (.DAT file type).

Selector List Area

The right pane of the **Appearance Profiler** window is the **Selector** list. This area contains the list of all added selectors.

TUTORIALS

General instructions for downloading tutorial files:

1. Download the *c03_tutorial* zip file for the tutorial from *http://www.cadcim.com*. The path of the file is as follows: *Textbooks > Civil/GIS > Navisworks > Exploring Autodesk Navisworks 2017*.

2. Now, save and extract the downloaded folder at the following location:
   C:\ Navisworks_2017

Note
*The default unit system used in the tutorials is metric. To change the units to imperial, select the required units from **Options Editor > Interface > Display Units.***

Tutorial 1   Taking Measurements

In this tutorial, you will open the file *c03_navisworks_2017_tut1.nwf*. Next, calculate the distance between mullions in the door by using various measuring tools such as the **Point to Point** tool, the **Point Line** tool, and the **Accumulate** tool.

(Expected time: 30 min)

The following steps are required to complete this tutorial:

a.   Open the file *c03_navisworks_2017_tut1.nwf*.
b.   Display the saved views.
c.   Measure distance using the **Point to Point** tool.
d.   Measure distance using the **Point Line** tool.
e.   Measure distance using the **Accumulate** tool.
f.   Save the project.

Opening the File

In this section, you will open the file in Navisworks.

1.   Choose the **Open** button from the Quick Access Toolbar; the **Open** dialog box is displayed.

2.   In this dialog box, browse to the following location:
      C:\nws_2017\c03_nws_2017_tut.

3.   Select **Navisworks File Set (*.nwf)** from the **Files of type** drop-down list.

4.   Select the file *c03_navisworks_2017_tut1.nwf* from the displayed list of files; the file name is displayed in the **File name** edit box.

5.   Choose the **Open** button from the dialog box; the model is displayed in the Scene View, as shown in Figure 3-36.

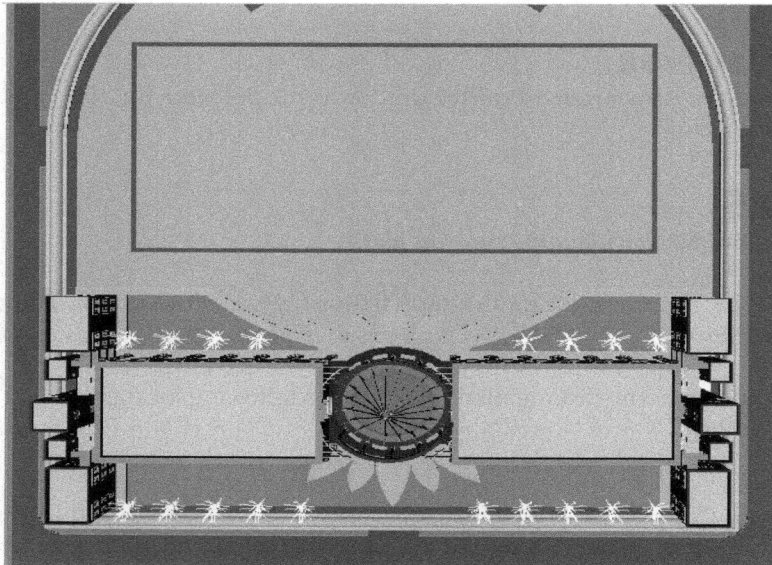

Figure 3-36 Model opened in Navisworks

Displaying the Saved Views

In this section, you will display the saved views by using the **Saved Viewpoints** window.

1. Choose the **Saved Viewpoints Dialog Launcher** button from the **Save, Load & Playback** panel of the **Viewpoint** tab; the **Saved Viewpoints** window is displayed, as shown in Figure 3-37.

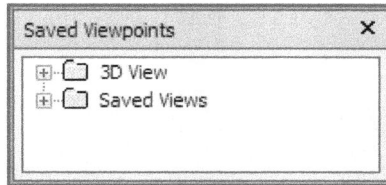

Figure 3-37 The Saved Viewpoints window

2. In the **Saved Viewpoints** window, click on the **+** icon corresponding to the **Saved Views** viewpoint folder; the folder is expanded, as shown in Figure 3-38.

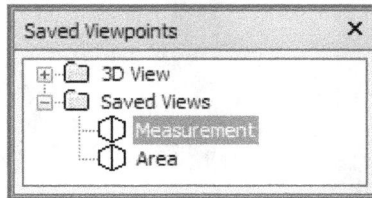

Figure 3-38 Expanded view of the Saved Views folder

3. Select the **Measurement** option from the **Saved Views** folder, refer to Figure 3-38; the view is displayed in the Scene View, as shown in Figure 3-39. Close the **Saved Viewpoints** window.

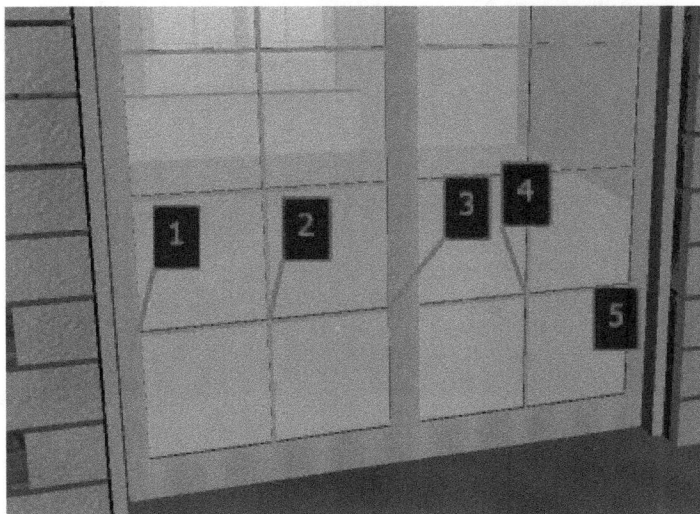

Figure 3-39 The Measurement view in the Scene View

Measuring Distance Using the Point to Point Tool

In this section, you will measure distance using the **Point to Point** tool.

1. Invoke the **Point to Point** tool from **Review > Measure > Measure** drop-down; the cursor is displayed on the screen.

2. Move and place the cursor at point **1** and click when the snap to vertex cursor appears, as shown in Figure 3-40.

Figure 3-40   *The cursor placed at point 1*

3. Next, place the cursor on point **2** and click when the snap to vertex cursor appears; the calculated distance is displayed in the Scene View, as shown in Figure 3-41.

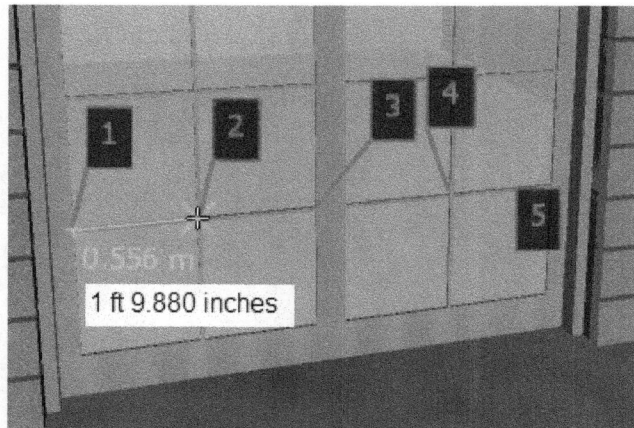

Figure 3-41   *Distance calculated between point 1 and 2*

Measuring Distances Using the Point Line Tool

In this section, you will measure distance using the **Point Line** tool.

1. Invoke the **Point Line** tool from the **Review > Measure > Measure** drop-down; the cursor appears on the screen and the previously calculated distance disappears.

2. Place the cursor on point **1** and click when the snap to vertex cursor appears, as shown in Figure 3-42.

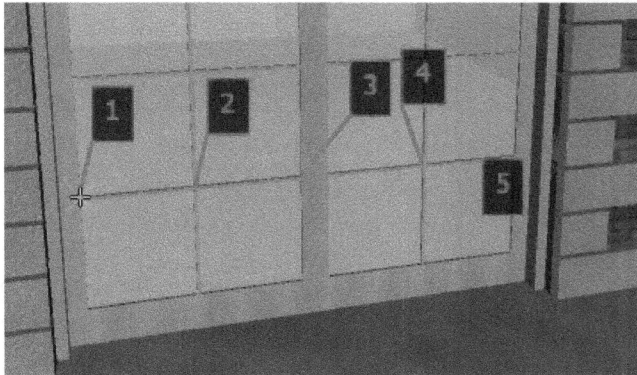

Figure 3-42 The cursor placed at point **1**

3. Next, place the cursor on point **2** and click when the snap to vertex cursor appears; the calculated distance is displayed in the Scene View, as shown in Figure 3-43.

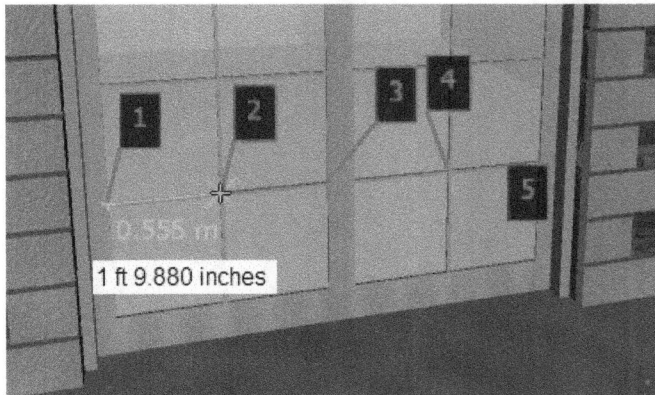

Figure 3-43 Distance calculated between point **1** and **2**

4. Repeat the procedure followed in steps 1 through 3 to calculate the total distance between **1** and **5**; the calculated distance is displayed in the Scene View, as shown in Figure 3-44.

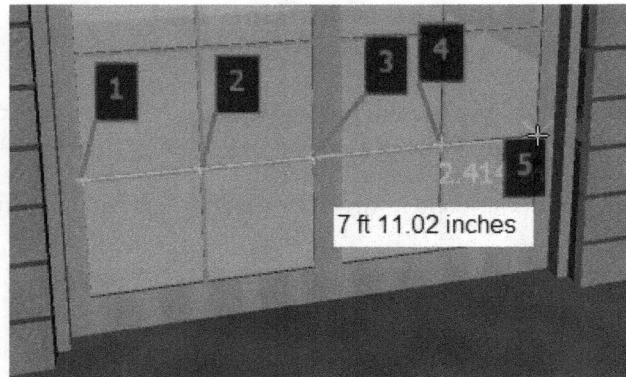

Figure 3-44 *The total distance calculated between point 1 and 5*

Measuring Distances Using the Accumulate Tool

In this section, you will measure distances using the **Accumulate** tool.

1. Invoke the **Accumulate** tool from **Review > Measure > Measure** drop-down; the cursor appears on the screen and the previously calculated distance disappears.

2. Place the cursor on point **1** and click when the snap to vertex cursor appears, as shown in Figure 3-45.

Figure 3-45 *The cursor placed at point 1*

3. Next, place the cursor on point **2** and click when the snap to vertex cursor appears; the calculated distance is displayed in the Scene View, as shown in Figure 3-46.

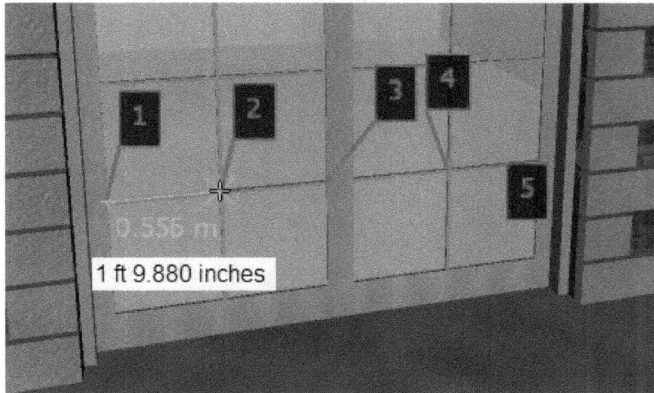

Figure 3-46 *Distance calculated between point **1** and **2***

4. Again, click on the point **2** and then click on point **3**; the calculated distance is displayed in the Scene View, as shown in Figure 3-47.

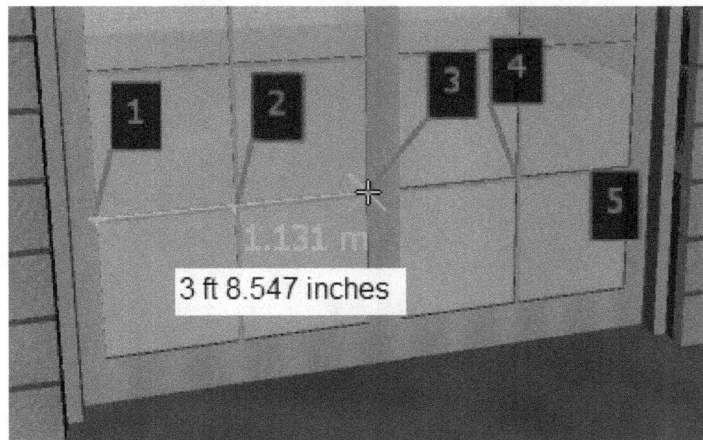

Figure 3-47 *Distance calculated between point **1** and **3***

5. Repeat the procedure followed in step 3 and calculate the distance between point **1** and **4,** and point **1** and **5**, refer to Figure 3-48 and Figure 3-49.

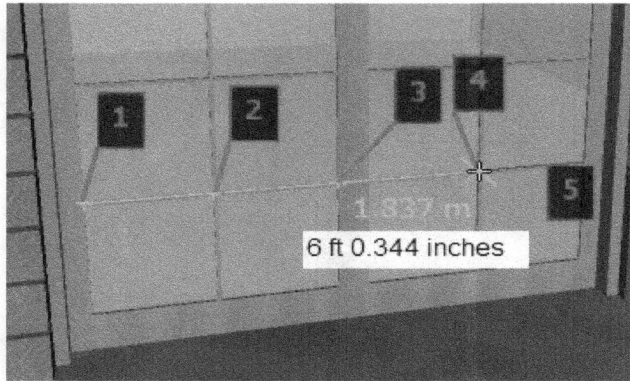

Figure 3-48 *Distance calculated between point 1 and 4*

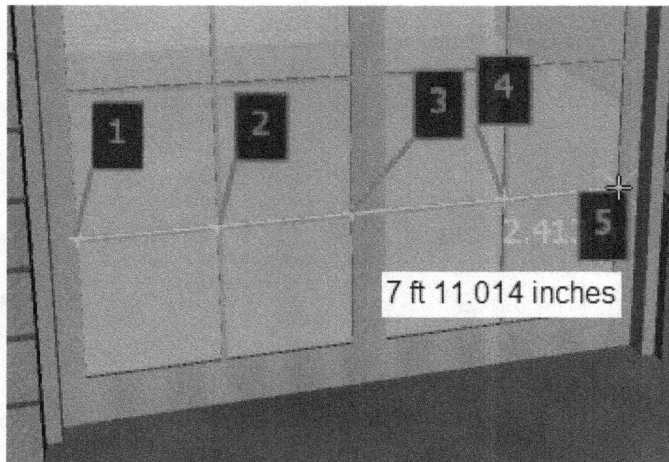

Figure 3-49 *Distance calculated between point 1 and 5*

Note

*To save the dimensions, choose the **Convert to Redline** tool from the **Measure** panel in the **Review** tab; the view will be saved in the **Saved Viewpoints** window.*

Saving the Project

In this section, you will save the project.

1. Choose **Save As** from the Application Menu; the **Save As** dialog box is displayed.

2. Browse to *Navisworks_2017* folder and enter **c03_navisworks_2017_tut01** in the **File name** edit box.

3. Next, select the **Navisworks File Set (*.nwf)** file format from the **Save as type** drop-down list and then choose the **Save** button; the project is saved.

Tutorial 2   Creating Tags and Calculating Area

In this tutorial, you will open the file *c03_navisworks_2017_tut2.nwf*. Next, you will add tags to four corners of the roof and calculate the area of roof using the **Add Tag** and **Area** tools.

(Expected time: 45 min)

The following steps are required to complete this tutorial:

a. Open the file *c03_navisworks_2017_tut2.nwf*.
b. Display the saved views.
c. Add tags.
d. Calculate area.
e. Add text.
f. Save the project.

Opening the File

In this section, you will open the file in Navisworks.

1. Choose the **Open** button from the Quick Access Toolbar; the **Open** dialog box is displayed.

2. In this dialog box, browse to the following location:
   C:\nws_2017\c03_nws_2017_tut.

3. Select **Navisworks File Set (*.nwf)** from the **Files of type** drop-down list.

4. Select the file *c03_navisworks_2017_tut2* from the displayed list of files; the file name is displayed in the **File name** edit box.

5. Choose the **Open** button from the dialog box; the model is displayed in the Scene View, refer to Figure 3-36 (same as Tutorial 1).

Displaying the Saved Views

In this section, you will display the saved views by using the **Saved Viewpoints** window.

1. Choose the **Saved Viewpoints Dialog Launcher** button from the **Save, Load & Playback** panel of the **Viewpoint** tab; the **Saved Viewpoints** window is displayed, refer to Figure 3-37.

2. In the **Saved Viewpoints** window, click on the **+** icon corresponding to **Saved Views** viewpoint folder; the folder expands, refer to Figure 3-38.

3. Select the **Area** option from the **Saved Views** folder, as shown in Figure 3-50; the view is displayed in the Scene View, as shown in Figure 3-51. Now, close the **Saved Viewpoints** window.

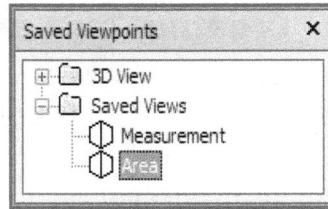

*Figure 3-50 Expanded view of the **Saved Views** folder*

Figure 3-51 The Area view in the Scene View

Adding Tags

In this section, you will add tags to the four corners of the roof using the **Add Tag** tool.

1. Choose the **Add Tag** tool from the **Tags** panel in the **Review** tab; the shape of the cursor is changed to pencil.

2. Place the cursor at the corner of **Wing A**, as shown in Figure 3-52.

3. Double-click on the specified location; the tag 6 is created and the **Add Comment** dialog box is displayed.

4. In the dialog box, add the comment **Corner point 1** in the text box and choose the **OK** button; the dialog box is closed and the tag is created, as shown in Figure 3-53.

*Figure 3-52 Placing the **Tag** cursor*

Figure 3-53 Tag 1 created at the corner

5. Repeat the procedure followed in steps 1 through 4 and add tags on both the wings, as shown in Figure 3-54.

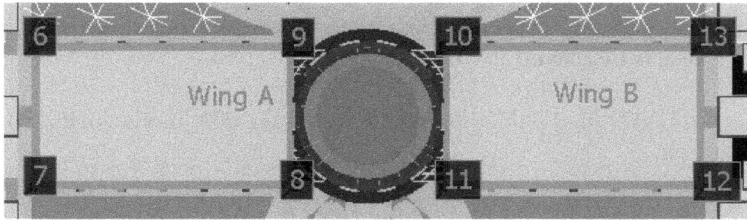

Figure 3-54 Tags created on both the wings

Note
*Enter the comment as **Corner point 2**, **Corner point 3**, and so on in the **Add Comment** dialog box.*

Calculating Area

In this section, you will calculate area of both the roofs.

1. Choose the **Area** tool from **Review > Measure > Measure** drop-down.

2. Place the cursor on the corner **6** in **Wing A** and click when the snap to vertex cursor appears.

3. Similarly, click on the corners **7**, **8**, and **9**; the calculated area is displayed in the Scene View, as shown in Figure 3-55.

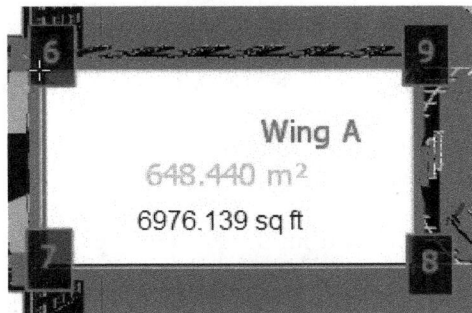

Figure 3-55 Area of roof WingA

Note
Right-click to change the location of cursor while measuring the area.

4. Similarly, click on the four corners **10**, **11**, **12**, and **13**; the calculated area is displayed, which is almost equal to the area of **Wing A**.

Saving the Project

In this section, you will save the project.

1.   To save the project with the current view, choose **Save As** from the Application Menu; the **Save As** dialog box is displayed.

2.   Browse to *nws_2017\c03_nws_2017_tut* folder and enter **c03_navisworks_2017_tut02** in the **File name** edit box.

3.   Next, select **Navisworks File Set (*.nwf)** file format from the **Save as type** drop-down list, and choose the **Save** button; the project is saved.

Tutorial 3   Finding and Saving Items as Search Sets

In this tutorial, you will open the file *c03_navisworks_2017_tut3.nwf* and then find items based on their types and properties. After finding items, you will save them as search sets.

(Expected time: 20 min)

The following steps are required to complete this tutorial:

a.   Open the file *c03_navisworks_2017_tut3.nwf*.
b.   Display the **Find Items** and **Sets** windows.
c.   Specify the search criteria to find the items and save them as search set.
d.   Save the project.

Opening the File

In this section, you will open the file in Navisworks.

1.   Choose the **Open** button from the Quick Access Toolbar; the **Open** dialog box is displayed.

2.   In this dialog box, browse to the following location:
     C:\nws_2017\c03_nws_2017_tut.

3.   Select **Navisworks File Set (*.nwf)** from the **Files of type** drop-down list.

4.   Select the file *c03_navisworks_2017_tut3* from the displayed list of files; the file name is displayed in the **File name** edit box.

5.   Choose the **Open** button from the dialog box; the model is displayed in the Scene View.

Displaying the Dockable Windows

In this section, you will invoke the **Find Items** and **Sets** windows and dock them on the left side of the interface.

1.   Choose the **Find Items** tool from **Home > Select & Search** panel; the **Find Items** window is displayed.

2.   Place the cursor on the title bar of the window and double-click; the window is docked at the left side of the interface.

3. Choose the **Manage Sets** option from **Home > Select & Search > Sets** drop-down list; the **Sets** window is displayed.

4. Place the cursor on the Title Bar in the **Sets** window, press and hold the left mouse button, drag the **Sets** window and over the Title Bar of the **Properties** window; the two windows are grouped together, refer to Figure 3-56.

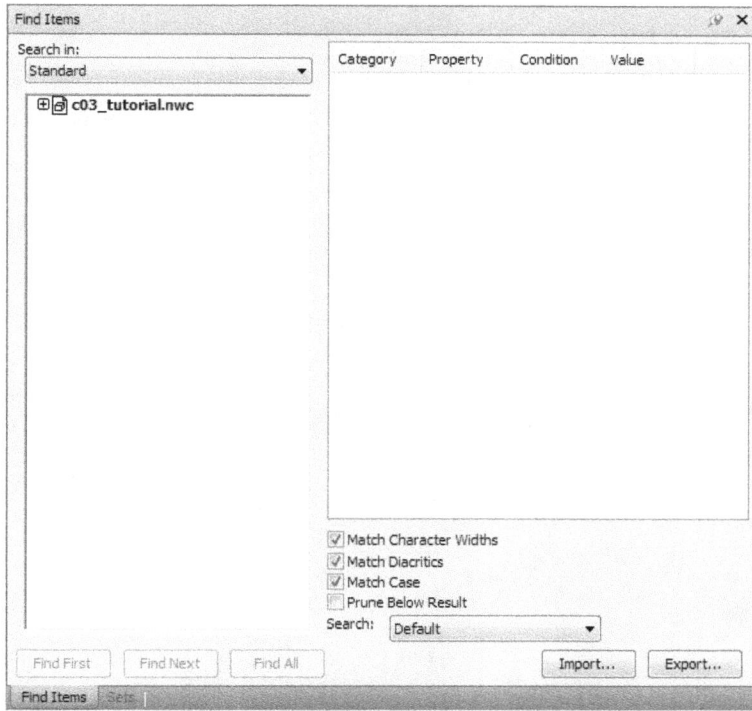

Figure 3-56 Two windows grouped together

Finding Items and Saving Items

In this section, you will specify the search criteria to find the items and then save them as search sets.

1. In the **Find Items** window, click in the **Category** column; a drop-down list is displayed. Select **Item** from the drop-down list. Similarly, select **Type**, **Contains**, and **Columns** from the **Property**, **Condition**, and **Value** columns, respectively, refer to Figure 3-57.

2. Choose the **Find All** button; the columns are selected and highlighted in the Scene View.

3. Choose the **Hide Unselected** tool from **Home > Visibility** panel; all the selected columns are displayed, refer to Figure 3-58.

4. Choose the **Unhide All** tool from **Home > Visibility** panel; the whole model will be displayed.

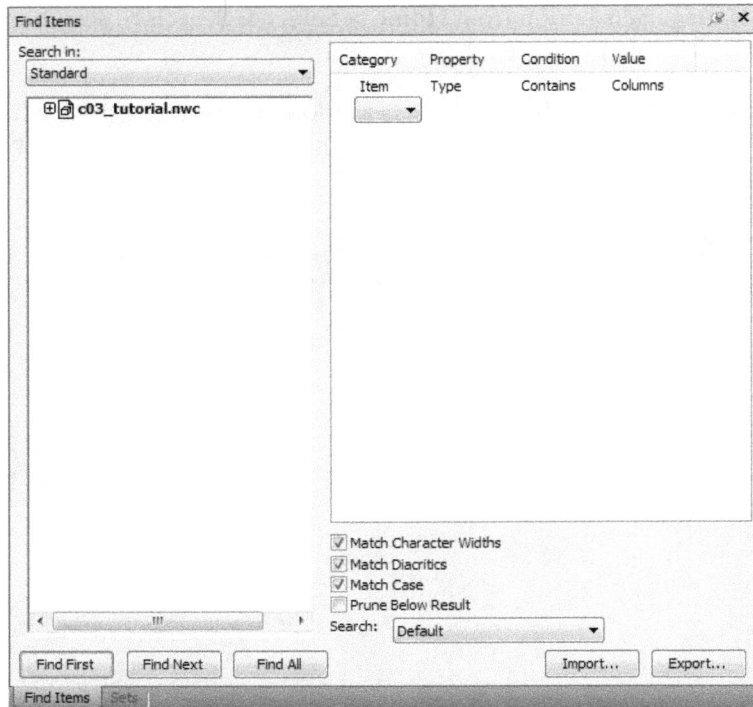

Figure 3-57 The **Find Items** *window displaying various settings for the search operation*

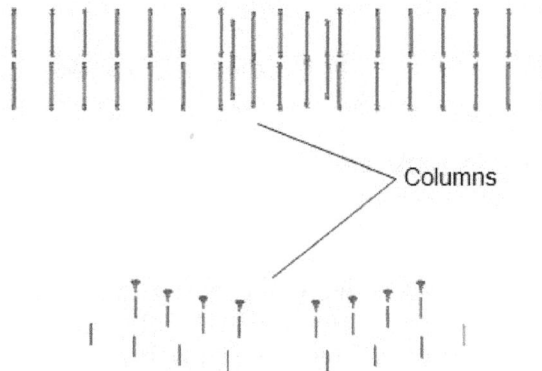

Figure 3-58 *Columns displayed in Scene View*

5. Choose the **Sets** tab in the grouped window; the **Sets** window is displayed. Choose the **Save Search** button; a search set is created. Right-click on the created search set; a shortcut menu is displayed. Choose the **Rename** option from the menu and name the search set as **Columns** and press ENTER.

6. Choose the **Find Items** tab to display the **Find Items** window. Now, specify a different search criteria, as shown in Figure 3-59.

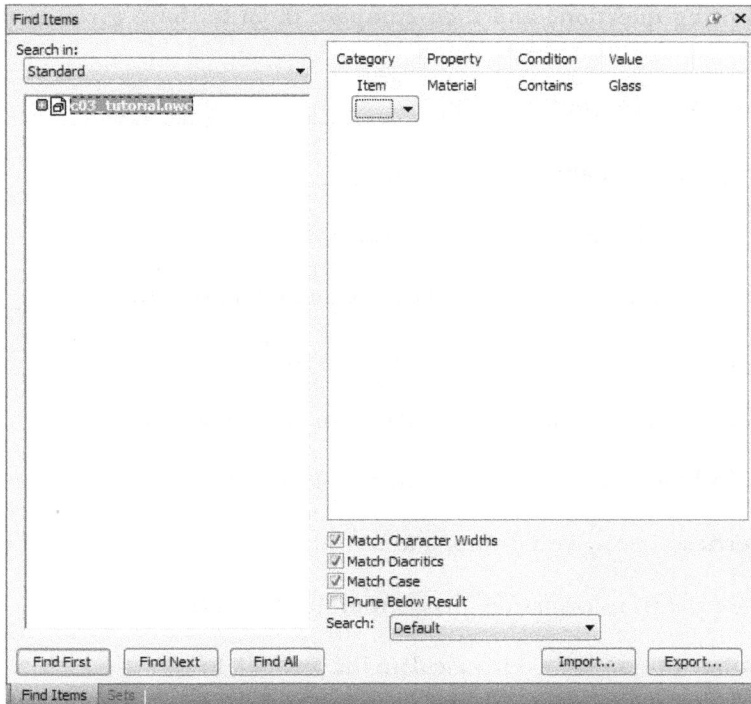

Figure 3-59 *The **Find Items** window displaying the search criteria*

7. Choose the **Find All** button; objects that have the glass material assigned are highlighted in the Scene View.

8. Repeat the procedure followed in step 5 and save the selected objects as search set with the name **Glass** and press ENTER.

Saving the Project
In this section, you will save the project.

1. To save the project with the current view, choose **Save As** from the Application Menu; the **Save As** dialog box is displayed.

2. Browse to *nws_2017/c03_nws_2017_tut* folder and enter ***c03_navisworks_2017_tut03*** in the **File name** edit box.

3. Next, select **Navisworks File Set (*.nwf)** file format from the **Save as type** drop-down list and then choose the **Save** button; the project is saved.

Self-Evaluation Test

Answer the following questions and then compare them to those given at the end of this chapter:

1. The **Selection Tree** is invoked from the _____ panel of the **Home** tab.

2. **Selection Inspector** contains a list of all the _____ objects.

3. The _____ option is used to reverse the selection.

4. The properties of an item in the model can be analyzed using the _____ window.

5. The _____ gizmo is used to move the selected item.

6. The **Select** tool is used to select a particular item in a model. (T/F)

7. The **Quick Find** tool is used to analyze the properties of an item in the model. (T/F)

8. In the **Properties** window, you cannot add a custom property to a selected item. (T/F)

9. You can invoke the transformation gizmo from the **Animation** tab. (T/F)

10. The **Accumulate** option is used to calculate the area. (T/F)

Review Questions

Answer the following questions:

1. Which of the following tabs is part of the **Selection Tree** window?

    a) **Standard**   b) **Compact**
    c) **Properties**   d) All of these

2. Which of the following tools is not a selection tool?

    a) **Same Name**   b) **Same Type**
    c) **Same Timeliner**   d) **Require**

3. Which of the following tools is used to control the visibility of objects in the model?

    a) **Append**   b) **Merge**
    c) **Find Items**   d) None of these

4. Which of the following tools is not a transformation tool?

    a) **Rotate**   b) **Scale**
    c) **Move**   d) **Text**

5. Which of the following tools is not used as the **Measure** tool?

    a) **Point to Point**   b) **Area**
    c) **Add Tag**   d) **Angle**

6. The **Measure** tools are invoked from the **Item Tools** tab. (T/F)

7. The **Point to Multiple Points** tool is used to calculate the total distance between multiple points. (T/F)

8. The **Add Link** tool is used to customize the appearance of a model. (T/F)

9. The **Sets** window contains only a list of saved viewpoints. (T/F)

10. The **Convert to Redlines** tool is used to convert measurements to redlines. (T/F)

EXERCISE

Exercise 1   Measuring Height

Download the *c03_navisworks_2017_ex1* file from *http://www.cadcim.com*. The path of the file is as follows: *Textbooks > Civil/GIS > Navisworks > Exploring Autodesk Navisworks 2017*. Open the file and then calculate the height of the stairs above the ground. Create tags across the four corners of the wall and calculate area of the wall. Figure 3-60 shows the Residence Building model to be used in this exercise.   **(Expected time : 30 min)**

The following steps are required to complete the exercise:

1. Open the file *c03_navisworks_2017_ex1*.
2. Display the saved views.
3. Calculate distance between point 1 and 2 using the **Perpendicular Lock** tool.
4. Create tags at the corners of the wall.
5. Calculate area of the wall.
6. Save the project.
7. Name of the file to be saved is *c03_navisworks_2017_ex01*.

Figure 3-60 The Residence Building

Answers to Self-Evaluation Test

1. **Select & Search**, **2**. selected, **3**. **Invert Selection**, **4**. **Properties**, **5**. **Move**, **6**. T, **7**. F, **8**. F, **9**. F, **10**. F.

Chapter 4

Viewpoints, Sections, and Animations

Learning Objectives

After completing this chapter, you will be able to:
- *Work with viewpoints*
- *Create cross-sections for reviewing models*
- *Create and share animations*

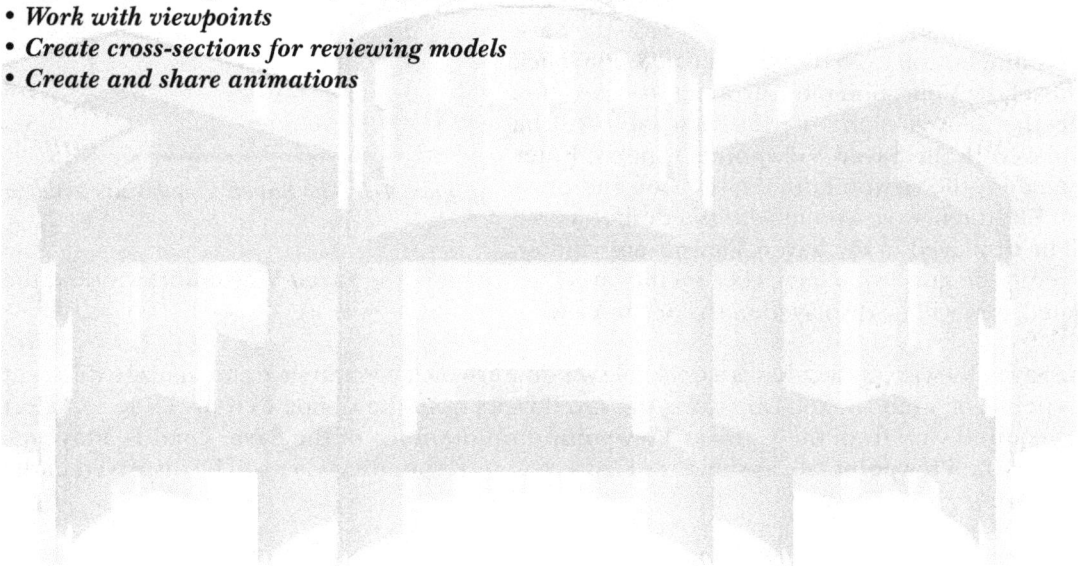

INTRODUCTION

In the previous chapter, you learned to select objects, control object attributes, and use review tools. In this chapter, you will learn various methods of saving and using viewpoints, creating cross sections, and the basic concepts of animation. All these features will be discussed sequentially in this chapter.

WORKING WITH VIEWPOINTS

In Navisworks, you can save views as a viewpoint in a project. The viewpoints are the snapshot of the views created in a scene. The saved viewpoints are used to manage and review the changes in a project. Each viewpoint retains the navigation mode that was used while creating it along with comments, measurements, markups, camera projection, and so on. You can also use viewpoints as links. Viewpoints are also a great way to quickly jump between different project views. The methods of saving, organizing, editing, and sharing viewpoints are discussed next.

Saving Viewpoints

In a project, sometimes you need to recall previous views. You can save model views by saving viewpoints. Before saving a particular view, you need to make right orientations and add mark up, tags, and comments. After making the right orientation, choose the **Saved Viewpoints Dialog Launcher** button; the **Saved Viewpoints** window will be displayed, as shown in Figure 4-1. To save the desired orientation as a viewpoint choose the **Save Viewpoint** button from the **Save, Load & Playback** panel of the **Viewpoint** tab. On doing so, a text box with the default name of the viewpoint will be displayed in the **Saved Viewpoints** window. Enter a name for the viewpoint in the text box and press ENTER; the new viewpoint with the desired name will be displayed in the **Saved Viewpoints** window.

Figure 4-1 The Saved Viewpoints window

To recall the saved view later, click on the saved viewpoint in the **Saved Viewpoints** window; the related view will be displayed in the Scene View.

The Saved Viewpoints window is a dockable window in which you can save and manage different viewpoints of a model. You can access the saved views from the ribbon as well. To do so, select the required view from the **Current Viewpoint** drop-down list of the **Save, Load & Playback** panel of the **Viewpoint** tab, as shown in Figure 4-2; corresponding view will be displayed in the Scene View.

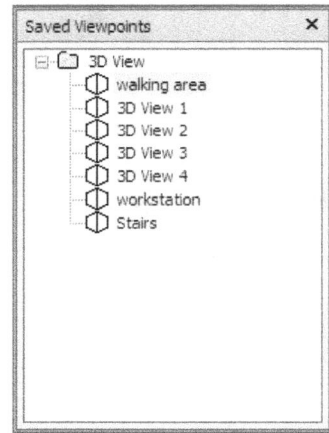

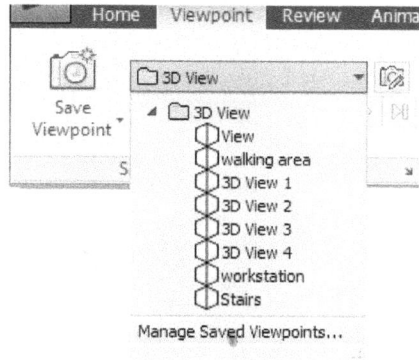

Figure 4-2 *Various options in the* **Current Viewpoint** *drop-down list*

Organizing Viewpoints

In the **Saved Viewpoints** window, you can arrange the saved views in a systematic manner by creating folders and sub folders. This helps you to easily locate them later. To create a folder, make sure that no other folder is selected in the **Saved Viewpoints** window and then right-click; a shortcut menu will be displayed. Select the **New Folder** option from the menu, as shown in Figure 4-3; a folder will be created in the **Saved Viewpoints** window. Specify a name for the folder and press ENTER. To organize various views in this folder, drag and drop the views in it.

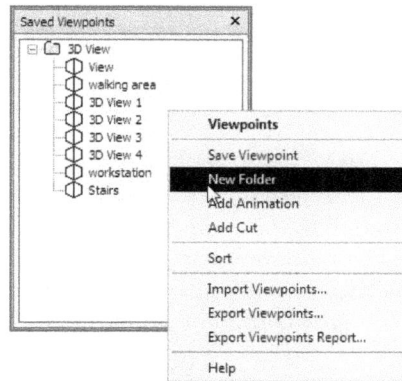

Figure 4-3 *Choosing the* **New Folder** *option from the shortcut menu*

You can change the default name of a viewpoint. To rename a particular viewpoint, select the required viewpoint in the **Saved Viewpoints** window and right-click on it; a shortcut menu will be displayed, as shown in Figure 4-4. Choose the **Rename** option from the menu and specify a new name of the viewpoint in the text box, and press ENTER. To delete a viewpoint from the **Saved Viewpoints** window select the desired viewpoint, right-click and then choose the **Delete** option from the shortcut menu displayed.

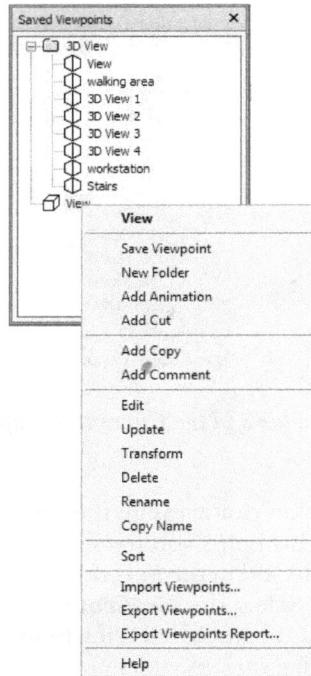

Figure 4-4 *The shortcut menu*

You can also create an exact copy of a viewpoint in the **Saved Viewpoints** window. To do so, select the viewpoint that you want to copy, right-click and then choose the **Add Copy** option from the shortcut menu displayed, refer to Figure 4-4; the same view will be added to the **Saved Viewpoints** window with a different name.

You can also add comments to the saved viewpoints. To add a comment to a saved viewpoint, choose the **Add Comment** option from the shortcut menu, refer to Figure 4-4; the **Add Comment** dialog box will be displayed, as shown in Figure 4-5. Specify the comment in the **Add Comment** text box. Select the required status for the comment from the **Status** drop-down list and choose the **OK** button to close the dialog box; the comment will be added to the saved view. Similarly, you can edit the previously added comments.

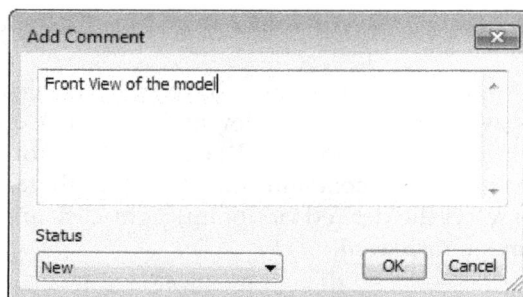

Figure 4-5 *The **Add Comment** dialog box*

Note

*To view the added comments, you can use the **View Comments** tool from the **Comments** panel of the **Review** tab.*

Editing Viewpoints

You can configure several attributes of the saved viewpoints such as camera position, field of view, linear speed, and angular speed. To modify these attributes, select the desired viewpoint from the **Saved Viewpoints** window and then right-click on it; a shortcut menu will be displayed. Choose the **Edit** option from this menu, refer to Figure 4-4; the **Edit Viewpoint - Current View** dialog box will be displayed, as shown in Figure 4-6. Alternatively, choose the **Edit Current Viewpoint** tool from the **Save, Load & Playback** panel in the **Viewpoint** tab. The options in this dialog box are discussed next.

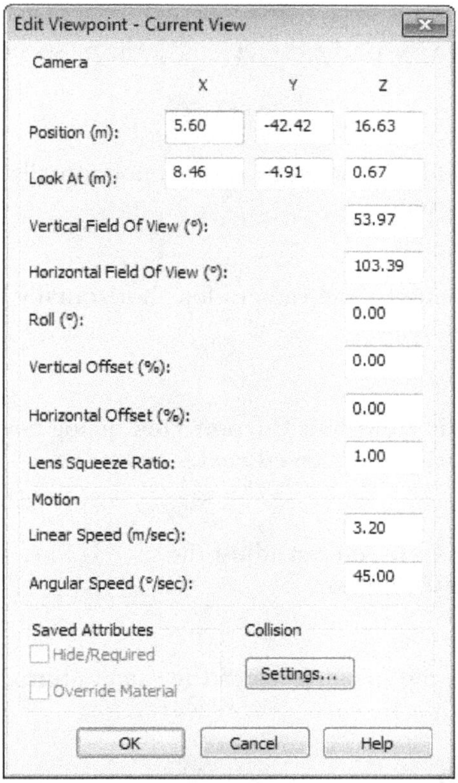

*Figure 4-6 The **Edit Viewpoint - Current View** dialog box*

Camera Area

The options in the **Camera** area are used to modify various camera settings. These options are discussed next.

Position

You can adjust the camera position in a project as per your requirement by specifying the X, Y, and Z coordinates in the edit boxes corresponding to the **Position** option.

Look At

In a project, you can adjust the focal point of the camera by specifying the values in X, Y, and Z edit boxes corresponding to the **Look At** option.

Vertical and Horizontal Field Of View

In a project, you can define the area of a scene to be viewed through the camera. To do so, specify the values for horizontal and vertical angle of view in the **Vertical Field Of View** and **Horizontal Field Of View** edit boxes, respectively. These options will be enabled only when you are in the camera viewpoint.

Roll

In a project, the camera can be rotated around its front-to-back axis by specifying a value for rotating the camera in the **Roll** edit box.

Vertical Offset

You can shift the field of view of the camera vertically by specifying a value in the **Vertical Offset** edit box.

Horizontal Offset

You can also shift the field of view of the camera horizontally by specifying a value in the **Horizontal Offset** edit box.

Lens Squeeze Ratio

In a project, you can compress the camera lens horizontally by specifying a value in the **Lens Squeeze Ratio** edit box.

Motion Area

In the **Motion** area of the **Edit Viewpoint-Current View** dialog box, you can adjust the walking speed. The options in this area are discussed next.

Linear Speed

While navigating in a project, you can adjust the speed of motion by specifying the desired value in the **Linear Speed** edit box.

Angular Speed

Similarly, you can adjust the speed at which the camera turns by specifying a value in the **Angular Speed** edit box.

Saved Attributes Area

You can use the options in the **Saved Attributes** area of the **Edit Viewpoint-Current View** dialog box for modifying saved viewpoints. The options in this area are discussed next.

Hide/Required

In a project, you can save any hidden markup information about the model by selecting the **Hide/Required** check box.

Override Appearance

You can select the **Override Appearance** check box to save the material override information. On selecting this check box, any appearance information in the viewpoint is also saved.

Note

*In the **Collision** area, you can specify the parameters related to the third person views which were discussed in Chapter 2.*

The changes made in this dialog box will be saved in the nwf and nwd files and will be loaded whenever you open that particular file. When you open a new file in Navisworks then the default viewpoint settings will be loaded. You can change the default settings using the **Options Editor** dialog box. To invoke this dialog box, choose the **Options** button from the Application Menu; the **Options Editor** dialog box will be displayed. In this dialog box, expand the **Interface** node in the left pane and select the **Viewpoint Defaults** option; various options will be displayed in the right pane of the dialog box. These options are discussed next.

The **Save Hide/Required Attributes** check box is used to save any hidden information about the objects in a viewpoint. For example, if you hide some objects in a model and save it as a viewpoint with the **Save Hide/Required Attributes** check box selected then the objects will remain hidden in that particular viewpoint. Select the **Override Appearance** check box to save the information of the appearance. Select the **Override Linear Speed** check box to set the default linear speed. On selecting this check box, the **Default Linear Speed (m/sec)** edit box will be activated. Specify the required value in this edit box. Similarly, specify a value in the **Default Angular Speed** edit box. In the **Collision** area, choose the **Settings** button to invoke the **Default Collision** dialog box and make the required changes for the third person view in this dialog box. The changes made in this dialog box will be saved as default settings.

Sharing Viewpoints

While working in a project, you can share the created viewpoints with other users. To use the viewpoints from another model, you need to import it via XML file. Similarly, you can export the viewpoints from the current project file to be used in other models and applications. These viewpoints are exported from Navisworks to the XML file. The process of exporting and importing of viewpoints is discussed next.

Exporting Viewpoints

In a project, you can export the viewpoints data to the XML file so that you can use these viewpoints in other applications. To export viewpoints to the XML file, right-click in the **Saved Viewpoints** window; a shortcut menu will be displayed. Choose the **Export Viewpoints** option from the shortcut menu; the **Export** dialog box will be displayed. In the **Export** dialog box, browse to the location where you need to save the file. Specify the file name and choose the **Save** button; the viewpoints file will be saved at the specified location. You can also invoke the **Export** dialog box from the ribbon. To do so, choose the **Viewpoints** tool from the **Export Data** panel in the **Output** tab; the **Export** dialog box will be displayed.

Importing Viewpoints

You can import viewpoints to Navisworks via XML file. To import the viewpoints, right-click in the **Saved Viewpoints** window; a shortcut menu will be displayed. Choose **Import Viewpoints** option from the menu; the **Import** dialog box will be displayed. In the dialog box, select the required viewpoint XML file and choose the **Open** button; the file will be loaded on the screen.

SECTIONING IN Navisworks

In Navisworks, sectioning is used to create cross-section views of the model so that you can get a clear picture of the internal view of the model. Using this feature, you can create different views in a model and can link them together. To use this feature, first you need to enable the sectioning mode. To do so, choose the **Enable Sectioning** button from the **Sectioning** panel in the **Viewpoint** tab; the **Sectioning Tools** tab will be displayed in the Ribbon.

You can use this tab to section the model in any of the two sectioning modes: Planes and Box. These two modes are discussed next.

Creating Section Views Using Planes Mode

In the Planes mode, you can enable a maximum of six sectional planes for creating the cross-section views of the model. To create a cross-section view of the model, first you need to enable the Planes mode by choosing the **Planes** tool from **Sectioning Tools > Mode > Planes** drop-down. Next, select the desired plane from the **Current Plane** drop-down list in the **Plane Settings** panel of the **Sectioning Tools** tab, refer to Figure 4-7. When you select an option from this drop-down list, the bulb icon next to it becomes active, indicating that it is currently enabled and is displayed in the Scene View. Now, select the required alignment option from the **Alignment** drop-down list in the **Planes Settings** panel; the selected plane will be aligned as per the option selected.

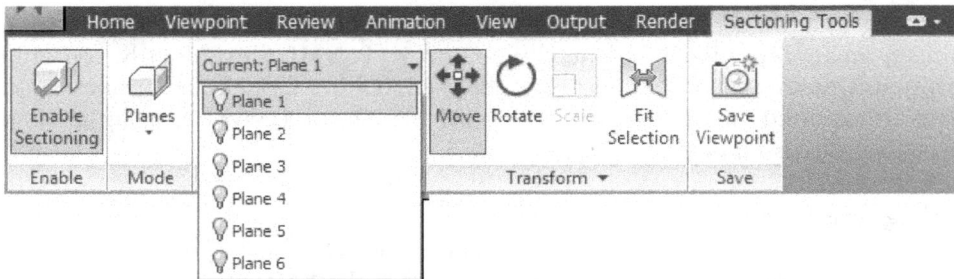

*Figure 4-7 The **Current Plane** drop-down list in the **Sectioning Tools** tab*

Note

*If you select the **Top** alignment option from the **Alignment** drop-down list, the currently selected plane will be aligned to the top plane of the model.*

You can also select the required section planes and alignment from the **Section Plane Settings** window. To display this window, choose the **Section Plane Dialog Launcher** button in the **Planes Settings** panel of the **Sectioning Tools** tab; the **Section Plane Settings** window will be displayed, as shown in Figure 4-8.

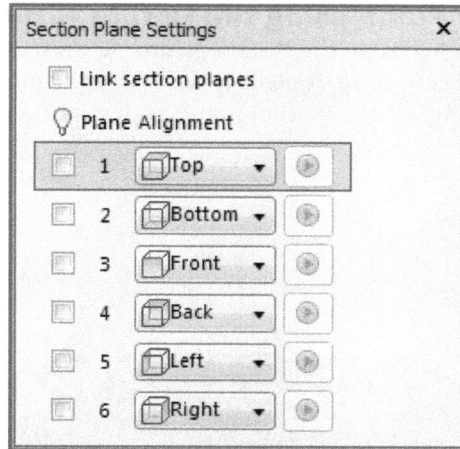

Figure 4-8 The Section Plane Settings window

Transforming Section Planes

The section planes can be transformed using transformation gizmos such as **Move** and **Rotate**. They can also be transformed numerically. Next, you will learn to transform these section planes using different methods.

Transforming Section Planes Using the Move Gizmo

When you select a section plane from the **Current Plane** drop-down list, refer to Figure 4-7, the **Move** gizmo is displayed by default. If not so, then choose the **Move** tool from the **Transform** panel in the **Sectioning Tools** tab; the **Move** gizmo will be displayed, as shown in Figure 4-9. Now, you can drag the gizmo to move the plane as required.

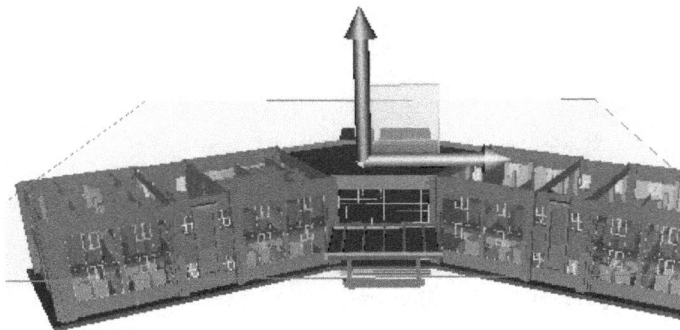

*Figure 4-9 Transforming section plane using the **Move** gizmo*

You can also move the active section plane to the edge of the selected objects in the Scene View. To do so, first select a section plane from the **Current Plane** drop-down list. Next, select the desired object in the Scene View and then choose the **Fit Selection** tool from the **Transform** panel in the **Sectioning Tools** tab; the section plane will move to the edge of the selected object in the Scene View.

Transforming Section Plane using the Rotate Gizmo

You can rotate a section plane by using the **Rotate** gizmo. To do so, choose the **Rotate** tool from the **Transform** panel in the **Sectioning Tools** tab; the **Rotate** gizmo will be displayed, as shown in Figure 4-10. Now, you can rotate the section plane as per your requirement by using the arcs in the gizmo.

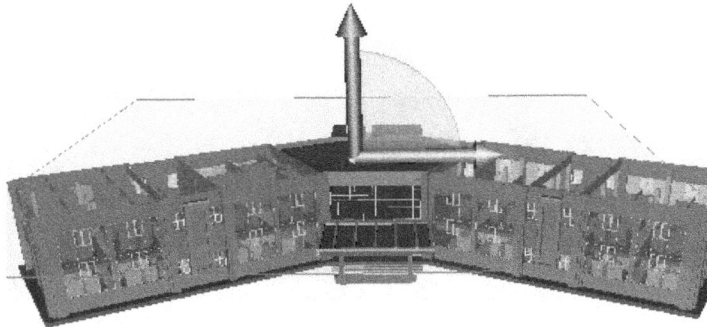

*Figure 4-10 Transforming section planes using the **Rotate** gizmo*

Transforming Section Plane Numerically

You can also transform a section plane numerically by specifying its position and rotation parameters. To do so, expand the **Transform** panel in the **Sectioning Tools** tab. Specify the X, Y, and Z co-ordinates in the edit boxes corresponding to the **Position** and **Rotation** options in this panel, as shown in Figure 4-11.

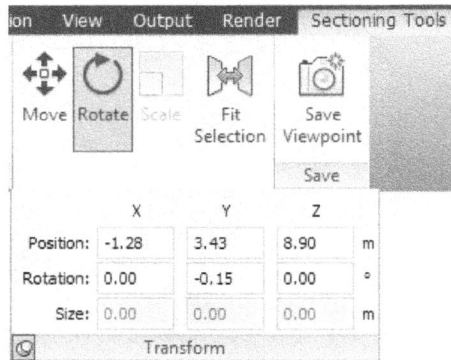

Figure 4-11 Transforming section planes numerically

Note
*In the Planes mode, the **Scale** tool is inactive.*

Linking Section Planes

In a project, you can link all the section planes together using the **Link Section Planes** option to move them as one section. To do so, first select the desired planes from the **Current Plane** drop-down list in the **Planes Settings** panel of the **Sectioning Tools** tab; the selected section planes are enabled and they will cut through the model in the Scene View. Next, choose the **Link Section Planes** tool from the **Plane Settings** panel in the **Sectioning Tools** tab; all the selected section planes will be linked together and the **Move** gizmo will be displayed in the Scene View. Drag the gizmo to move all the section planes together, refer to Figure 4-12.

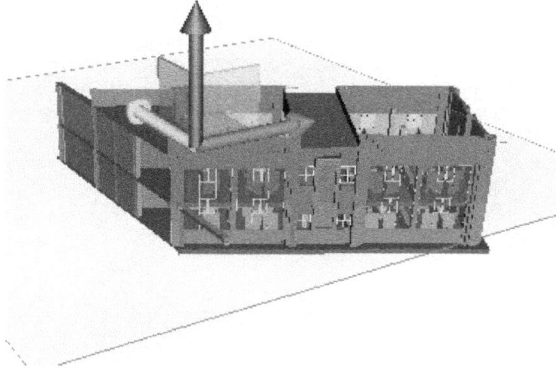

Figure 4-12 Moving all the linked section planes together

To rotate all the section planes together, choose the **Rotate** tool from the **Transform** panel in the **Sectioning Tools** tab; the **Rotate** gizmo will be displayed in the Scene View. Now, use the arcs in the gizmo to rotate the planes together, as shown in Figure 4-13.

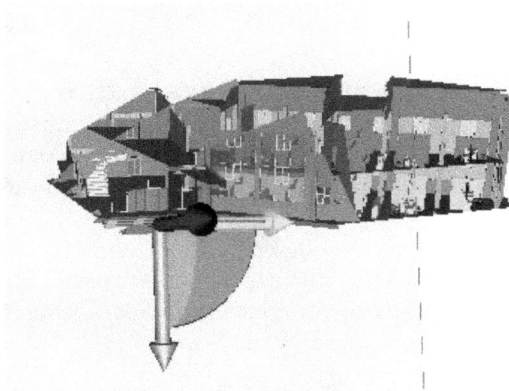

Figure 4-13 Rotating all the section planes together

Creating Section Views Using the Box Mode

In the Box mode, a section box is created around the model. Using this section box, you can focus on a specific area of the model by limiting its geometry. You can clip the unwanted portion of the model in the Scene View by using the Section Box mode. To enable the Section Box mode, choose the **Box** tool from **Sectioning Tools > Mode > Planes** drop-down; a Section Box will be displayed around the model along with the **Move** gizmo, as shown in Figure 4-14. This Section Box can be transformed using the **Move**, **Rotate**, and **Scale** gizmos. Note that on enabling the **Box** mode, the options in the **Plane Settings** panel will be disabled.

Figure 4-14 *Transforming section box using the **Move** gizmo*

Transforming Section Box

The section box can be transformed using transformation gizmos such as **Move**, **Rotate**, and **Scale**, and it can also be transformed numerically. The different methods to transform section box are discussed next.

Transforming Section Box Using the Move Gizmo

When you enable the Box mode, the **Move** gizmo will be displayed inside the section box by default. If not so, then choose the **Move** tool from the **Transform** panel in the **Sectioning Tools** tab. Next, drag the gizmo to move the section box along the required axis.

You can also move the section box to the edge of the selected objects in the Scene View. To do so, after displaying the section box, select the desired object in the Scene View and then choose the **Fit Selection** tool from the **Transform** panel in the **Sectioning Tools** tab; the section box will move to the edge of the selected object in the Scene View.

Transforming Section Box Using the Rotate Gizmo

You can also rotate the section box by using the **Rotate** gizmo. To rotate the section box, choose the **Rotate** tool from the **Transform** panel in the **Sectioning Tools** tab; the **Rotate** gizmo will be displayed. Use the gizmo arms or arcs to rotate the section box, as shown in Figure 4-15.

Figure 4-15 *Rotating section box using the **Rotate** gizmo*

Transforming the Section Box Using the Scale Gizmo

You can also scale the section box by using the **Scale** gizmo. To modify the scale of the section box, choose the **Scale** tool from the **Transform** panel in the **Sectioning Tools** tab; the **Scale** gizmo will be displayed inside the box. To resize the section box along a single axis, drag the gizmo arm along the X, Y, or Z direction. To resize the section box across two axes, use the colored triangles, refer to Figure 4-16.

Figure 4-16 *Transforming section box using the **Scale** gizmo*

Transforming Section Box Numerically

You can also transform the section box numerically by specifying its position, rotation, and size parameters. To do so, expand the **Transform** panel in the **Sectioning Tools** tab. In this panel, specify the **X**, **Y**, and **Z** co-ordinates in the edit boxes corresponding to the **Position**, **Rotation**, and **Size** options respectively, refer to Figure 4-11.

ANIMATION IN Navisworks

In Navisworks, you can create animations by using the following two methods: Viewpoint animation and Object animation. The Viewpoint animation is an efficient way of creating animation by recording views while navigating through a model or by saving specific views of the model. In the Object animation, after animating objects, you can interact with them by adding scripts such as opening doors or closing doors. In this animation, when you move an object to the desired location, many keyframes will be generated between the start point and end point, thus making the transitions smooth and speeding up the process of animation. The **Animator** and **Scripter** features are used for creating object animation, which will be discussed in the later chapters. In the next section, you will learn about creating viewpoint animations, and playing, editing, and sharing animations.

Creating Viewpoint Animations By Recording

In Navisworks, you can create viewpoint animation while navigating through a model by using the **Record** tool. It will record each movement while navigating, and create frames for each movement. All these frames are saved in the animation folder in the **Saved Viewpoints** window. When you play the animation, all these frames will be played in a sequence. To record the viewpoints, choose the **Record** tool from **Viewpoint > Save, Load & Playback > Save Viewpoint** drop-down; the **Recording** panel will be displayed in the Ribbon, as shown in Figure 4-17. Alternatively, choose the **Record** tool from the **Create** panel in the **Animation** tab. Now, start navigating using the navigation tools. You can also create sections while navigating; in such a case everything will be recorded in the viewpoint animation. To change the position while navigating, you can pause the recording by choosing the **Pause** option from the **Recording** panel. Alternatively, you can choose the **Pause** button from the **Playback** panel in the **Animation** tab. To continue recording, choose the **Pause** button again. When you pause the animation, cuts are inserted automatically during the recording. These cuts will be displayed in the **Saved Viewpoints** window. To stop the recording, choose the **Stop** button from the **Recording** panel. Alternatively, choose the **Stop** button from the **Playback** panel in the **Animation** tab; the animation will be saved under the **Animation** folder in the **Saved Viewpoints** window.

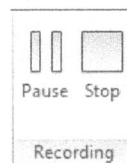

Figure 4-17 The Recording panel

You can also change the name of the added **Animation** folder. To do so, right-click on the folder; a shortcut menu will be displayed. Choose **Rename** from the menu and specify a new name for the folder and then press ENTER.

Creating Viewpoint Animations Frame By Frame

You can also create viewpoint animation frame by frame. To do so, first invoke the **Saved Viewpoints** window. Next, right-click in the **Saved Viewpoints** window; a shortcut menu will be displayed. Choose the **Add Animation** option from the shortcut menu; a new viewpoint animation folder with the name **Animation** will be created in the **Saved Viewpoints** window. You can assign a name other than the default one to the folder, if required. To create an animation, navigate to the model as desired. Select the previously created animation folder and right-click; a shortcut menu will be displayed. Choose the **Save Viewpoint** option from the menu; the corresponding view will be saved in the animation folder. Repeat these steps to create more frames. You can also pause the animation by inserting cuts directly in the **Saved Viewpoints** window. To do so, right-click below the animation frame where you want to add a pause; a shortcut menu will be displayed. Choose the **Add Cut** option from the menu; a cut will be added and displayed in an edit box. Specify a name for the cut and press ENTER; the cut will be added after the selected animation frame.

> **Note**
> *In an animation, you can change the sequences of viewpoints by dragging the required viewpoints in the **Saved Viewpoints** window. To remove a viewpoint from an animation, drag the required viewpoint and drop it outside the animation folder in the **Saved Viewpoints** window. Similarly, to add a viewpoint in an animation, drag and place the viewpoint under the animation folder.*

Playing Viewpoint Animations

After creating an animation, you can play it by selecting it from **Animation > Playback > Animations** drop-down, refer to Figure 4-18; the selected animation will be highlighted in the **Saved Viewpoints** window. Choose the **Play** button from the **Playback** panel; the animation will start playing. While playing animation, you will notice that progress of the animation is shown in the **Playback Time** and **Percentage** edit boxes in the panel. You can adjust the playback time by adjusting the playback slider or by specifying a value using the spinner. To control the animation, use the Multimedia buttons in the **Playback** panel, refer to Figure 4-18.

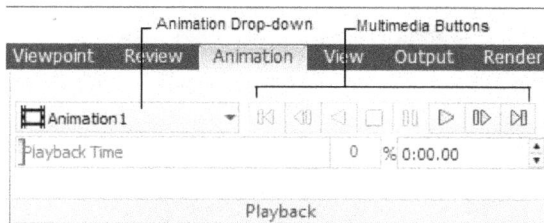

*Figure 4-18 The **Playback** panel in the **Animation** tab*

Combining Multiple Animations

In Navisworks, multiple animations can be combined and played together as a single animation. These animations are created either by using the **Record** tool or they can be created frame by frame. To combine multiple animations, first create an animation folder in the **Saved Viewpoints** window. Next, drag and drop the required viewpoint animations in this folder. These animations can be played in the sequence in which they were inserted in the animation folder.

Editing Viewpoint Animations

After creating an animation, you can edit the playback duration. You can also edit the smoothness of the animation playback by reducing the jerk. To edit an animation, right-click on the required animation in the **Saved Viewpoints** window; a shortcut menu will be displayed. Choose the **Edit** option from the shortcut menu; the **Edit Animation** dialog box will be displayed, as shown in Figure 4-19.

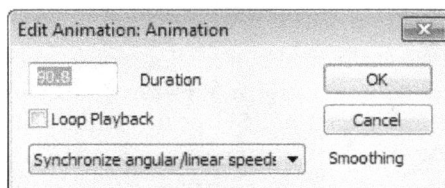

*Figure 4-19 The **Edit Animation** dialog box*

In this dialog box, specify the desired duration value in the **Duration** edit box; the duration of the animation will be modified. To play the animation continuously, select the **Loop Playback** check box. If you do not want to play the animation continuously, then clear the check box. To reduce the jerk in animation, select the **Synchronize angular/linear speeds** option from the **Smoothing** drop-down list. This will smoothen the difference between the speeds of each frame in the animation. To ignore smoothing, select **None** from the **Smoothing** drop-down list. After making all the changes, choose the **OK** button to close the dialog box.

Sharing Animations

In Navisworks, you can share the animation by using the export feature. To use it, choose the **Export Animation** tool from the **Export** panel in the **Animation** tab; the **Animation Export** dialog box will be displayed, as shown in Figure 4-20. The options in this dialog box are discussed next.

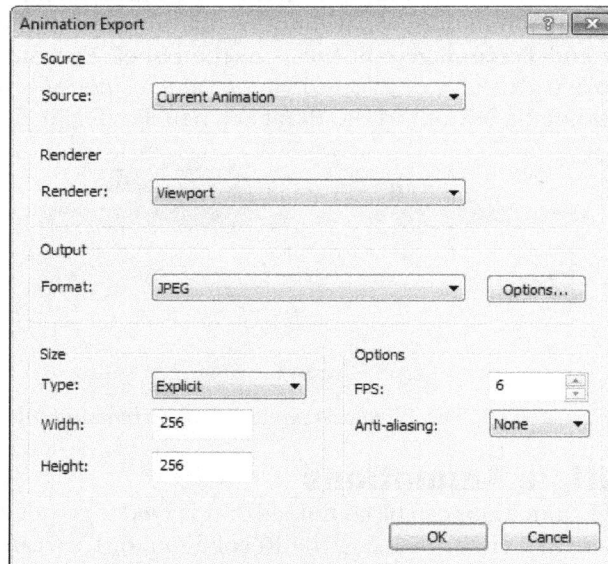

*Figure 4-20 The **Animation Export** dialog box*

Source Area

The **Source** area contains the **Source** drop-down list, refer to Figure 4-20. You can specify the source to export the animation from this drop-down list. The options available in this drop-down list are: **Current Animation**, **Current Animator Scene**, and **TimeLiner Simulation**. Select the **Current Animation** option to export the currently selected viewpoint animation. Select the **Current Animator Scene** option to export the currently selected object animation. Select the **TimeLiner Simulation** option to export the currently selected TimeLiner sequence.

Renderer Area

In the **Renderer** area, you can select the required renderer options from the **Renderer** drop-down list. This drop-down list contains two options: **Viewport** and **Autodesk**. Select the **Viewport** option for quickly rendering an animation. Select the **Autodesk** option when you need the highest rendering quality of Autodesk materials.

Output Area

The **Output** area contains the **Format** drop-down list. You can select the options from this drop-down list to specify an output format in which you want to export the animation. The options available in the drop-down list are: **JPEG**, **PNG**, **Windows AVI**, and **Windows Bitmap**.

You can also configure the desired output format such as **Compression**, **Smoothing**, **Interlacing**, and so on. To configure the selected output format, choose the **Options** button located next to the **Format** drop-down list; a dialog box related to the option selected in the **Format** drop-down list will be displayed in which you can modify the format parameters. The options in the dialog box depend upon the option that you select from the **Format** drop-down list in the **Animation Export** dialog box. Note that the **Options** button will be disabled on selecting the **Windows Bitmap** option from the **Format** drop-down list.

Size Area

In the **Size** area of the **Animation Export** dialog box, you can specify the size of the animation to be exported. To specify the size of the exported animation, select the required options from the **Type** drop-down list. Next, enter the width and height in pixels in the **Width** and **Height** edit boxes, respectively.

Options Area

In the **Options** area of the **Animation Export** dialog box, you can specify the value for the number of frames per second in the **FPS** edit box. Higher the FPS value, smoother will be the animation, but the rendering time will increase. To smoothen the edges of the exported images, select appropriate value from the **Anti-Aliasing** drop-down list. Choose the **OK** button to close the dialog box.

TUTORIALS

General instructions for downloading tutorial files:

1. Download the *c04_nws_2017_tut* zip file for the tutorial from *http://www.cadcim.com*. The path of the file is as follows: *Textbooks > Civil/GIS > Navisworks > Exploring Autodesk Navisworks 2017.*

2. Now, save and extract the downloaded folder at the following location:
    C:\ nws_2017\c04_nws_2017_tut

Tutorial 1   Creating Viewpoints

In this tutorial, you will open the *c04_navisworks_2017_tut1* file, create viewpoints and save them in the **Saved Viewpoints** window.   **(Expected time: 20 min)**

The following steps are required to complete this tutorial:

a. Open the *c04_navisworks_2017_tut1.nwf* file.
b. Display the **Saved Viewpoints** window.
c. Display the 3D view of the model.
d. Create viewpoints.
e. Save the project.

Opening the Model

In this section, you will open the model created in Revit software.

1. To open the existing model, choose the **Open** button from the Quick Access Toolbar; the **Open** dialog box is displayed.

2. In this dialog box, browse to the following location:
   C:\nws_2017\c04_nws_2017_tut

3. Select the **Navisworks File Set (*.nwf)** file format from the **Files of type** drop-down list.

4. Next, select *c04_navisworks_2017_tut1* from the displayed list of files in the dialog box, and choose the **Open** button; the model is displayed in the Scene View, as shown in Figure 4-21.

Figure 4-21 Model opened in Navisworks

Displaying the Saved Viewpoints Window

In this section, you will display the **Saved Viewpoints** window.

1. Choose the **Saved Viewpoints Dialog Launcher** button from the **Save, Load & Playback** panel in the **Viewpoint** tab; the **Saved Viewpoints** window is displayed.

Displaying the 3D View of the Model

In this section, you will display the 3D view of the model.

1. Expand the **3D View** node in the **Saved Viewpoints** window.

2. Select the **North Elevation** viewpoint in the **3D View** node; the current view changes and the north elevation view of the model is displayed, as shown in Figure 4-22.

Figure 4-22 North elevation of the model

Note

*If an avatar is displayed with the **North Elevation** view then clear the **Third Person** check box from the **Realism** drop-down in the **Viewpoint** tab.*

Creating Viewpoints

In this section, you will create viewpoints.

1. Invoke the **Orbit** tool from the Navigation Bar; the **Orbit** cursor appears on the screen.

2. Press and drag the left mouse button and rotate the current view to align it, as shown in Figure 4-23.

Figure 4-23 The model after rotating

3. Invoke the **Zoom** tool from the Navigation Bar; the Zoom cursor appears on the screen.

4. Place the **Zoom** cursor on the model, as shown in Figure 4-24.

*Figure 4-24 Placing the **Zoom** cursor on the model*

5. Enlarge the viewing scale of the model by clicking and dragging the mouse forward, as shown in Figure 4-25.

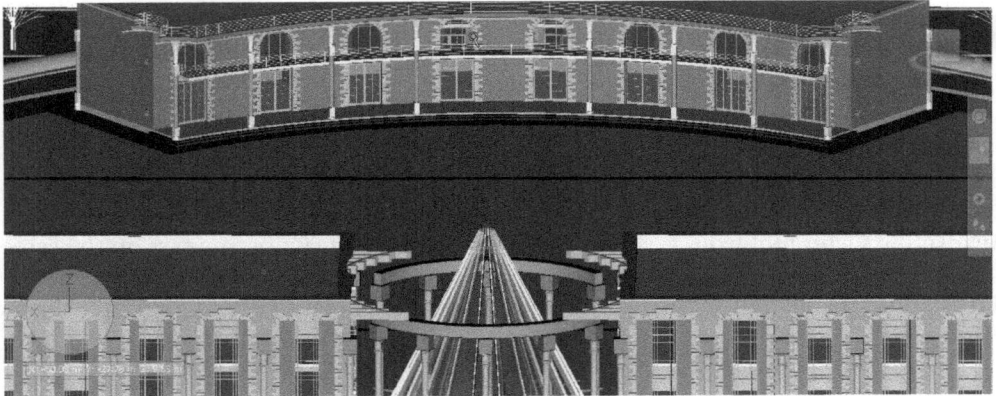

Figure 4-25 The model after zooming

6.   Again, place the **Zoom** cursor on the door, as shown in Figure 4-26.

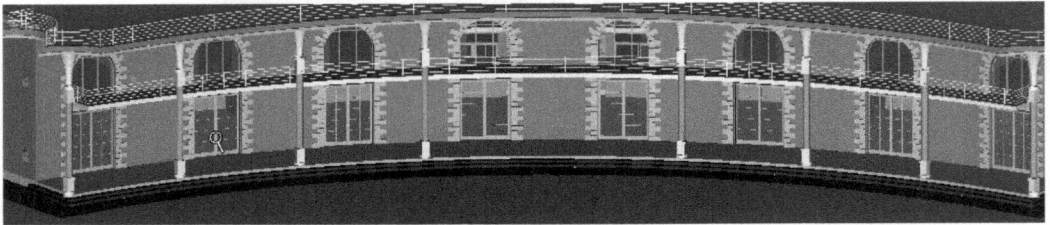

Figure 4-26 Cursor placed on the door

7.   Enlarge the viewing scale of the door by dragging the mouse forward, as shown in Figure 4-27.

Figure 4-27 Partial view of the door after zooming

8.   Right-click in the **Saved Viewpoints** window; a shortcut menu is displayed.

9. Choose the **New Folder** option from the menu; a folder is created and an edit box is displayed in the **Saved Viewpoints** window.

10. Rename the folder created as **Saved Views** and press ENTER.

11. Next, right-click in the **Saved Views** folder; a shortcut menu is displayed.

12. Choose the **Save Viewpoint** option from the displayed menu; a viewpoint is created. Rename the saved viewpoint as **Measurements** and press ENTER. Figure 4-28 shows the **Measurement** viewpoint saved in the **Saved Viewpoints** window.

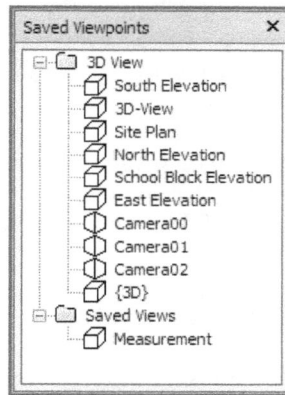

*Figure 4-28 The **Measurement** viewpoint saved in the **Saved Viewpoints** window*

13. Next, using the **Zoom** tool, zoom out the model by clicking and dragging the mouse backward, as shown in Figure 4-29.

Figure 4-29 Model after zooming out

14. Next, move and place the cursor on the **Top** face of the ViewCube; the top face of the ViewCube is highlighted.

15. Click on the **Top** face of the ViewCube; the view of the model changes to the top view, as shown in Figure 4-30.

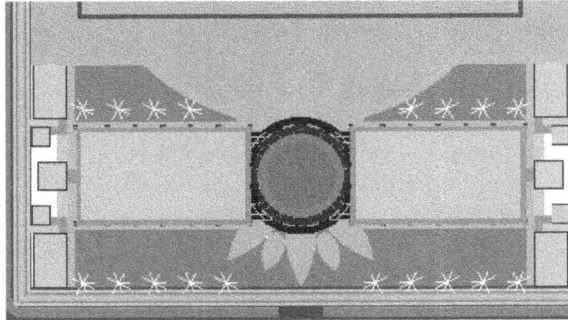

Figure 4-30 Top view of the model

Note

You can zoom out the model to get a view similar to the one shown in Figure 4-30.

16. Now, right-click in the **Saved Views** folder in the **Saved Viewpoints** window; a shortcut menu is displayed.

17. Choose the **Save Viewpoint** option from the menu.

18. Repeat the procedure followed in steps 11 and 12 and save the viewpoint with the name **Area**. Figure 4-31 shows the **Saved Viewpoints** window with the **Area** viewpoint.

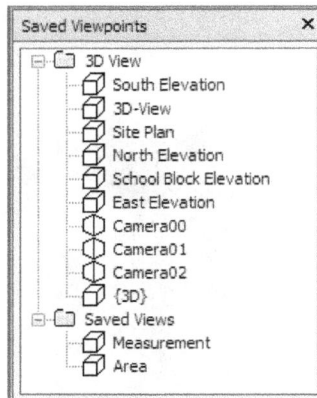

*Figure 4-31 The **Saved Viewpoints** window*

19. Select the **3D View** in the **Saved Viewpoints** window; the view is displayed in the Scene View. Now, close the **Saved Viewpoints** window.

Saving the Project

In this section, you will save the project.

1. To save the project with the current view, choose **Save As** from the Application Menu; the **Save As** dialog box is displayed.

2. Browse to *nws_2017\ c04_nws_2017_tut* folder and enter **c04_navisworks_2017_tut01** in the **File name** edit box. Choose the **Navisworks File Set (*.nwf)** file format from the **Save as type** drop-down list, and choose the **Save** button; the project is saved.

Tutorial 2   Using Planes Mode

In this tutorial, you will open the *c04_navisworks_2017_tut2* file and create cross-section of the model in the Planes mode to get the inside view.   **(Expected time: 30 min)**

The following steps are required to complete this tutorial:

a. Open the *c04_navisworks_2017_tut2.nwf* file.
b. Enable sectioning.
c. Create sections using planes mode and various transforming tools.
d. Save the project.

Opening the Model

In this section, you will open the model created in Revit software.

1. To open the model, choose the **Open** button from the Quick Access Toolbar; the **Open** file dialog box is displayed on the screen.

2. In this dialog box, browse to the following location:
   C:\nws_2017\c04_nws_2017_tut.

3. Ensure that the **Navisworks File Set (*.nwf)** file format is selected in the **Files of type** drop-down list. Next, select the *c04_navisworks_2017_tut2* from the displayed list of files and choose the **Open** button; the model is displayed in the Scene View, as shown in Figure 4-32.

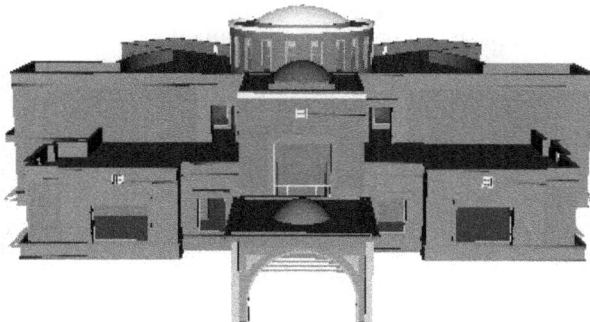

Figure 4-32 Model opened in Navisworks

Enabling Cross-section Mode

In this section, you will enable the cross-section mode of the current view of the model.

1. Choose the **Enable Sectioning** option from the **Sectioning** panel in the **Viewpoint** tab; the **Sectioning Tools** contextual tab is displayed in the ribbon.

Creating Sections using Planes Mode and Transforming Tools

In this section, you will use the **Planes** tool to create cross-sections of the model.

1. Ensure that the **Planes** tool is chosen from the **Planes** drop-down in the **Mode** panel of the **Sectioning Tools** tab.

2. Ensure that the **Plane 1** option is selected in the **Current: Plane** drop-down list in the **Planes Settings** panel of the **Sectioning Tools** tab.

3. Choose the **Move** tool from the **Transform** panel in the **Sectioning Tools** tab; the section plane is displayed, as shown in Figure 4-33.

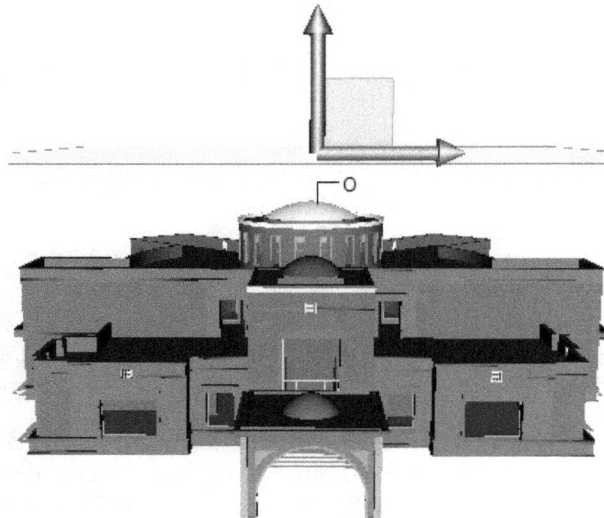

Figure 4-33 Model with the section plane

4. Invoke the **Select** tool from the Quick Access Toolbar.

5. Select the building element marked O, refer to Figure 4-33; the selected object is highlighted in blue.

6. Choose the **Fit Selection** tool from the **Transform** panel in the **Sectioning Tools** tab; the section plane moves to the highlighted building element, as shown in Figure 4-34.

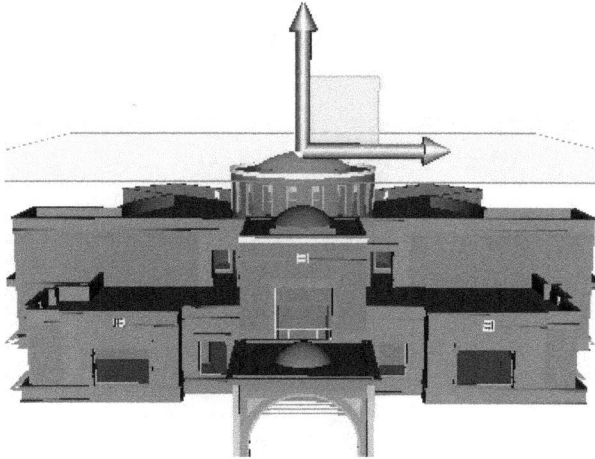

Figure 4-34 Section plane moved to the selected building element

7. Choose the **Left** alignment option from **Sectioning Tools > Plane Settings > Alignment** drop-down list; the section plane changes, as shown in Figure 4-35.

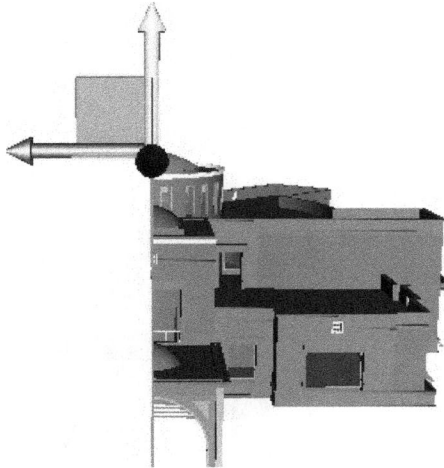

Figure 4-35 Model with the left aligned section plane

8. Next, choose the **Rotate** tool from the **Transform** panel in the **Sectioning Tools** tab; the **Rotate** gizmo is displayed, as shown in Figure 4-36.

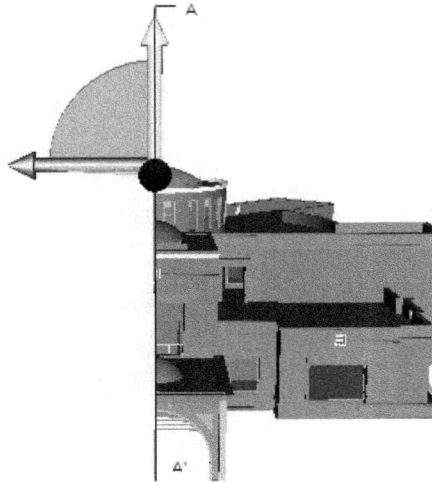

Figure 4-36 The **Rotate** *gizmo*

9. Place the cursor on the red curve between the Y and Z axis; the curve gets highlighted, as shown in Figure 4-37.

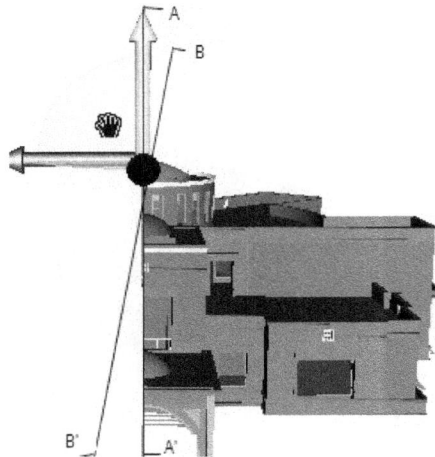

Figure 4-37 Placing the cursor on the curve

10. Press and hold the left mouse button, drag the **Rotate** gizmo in the right direction, and release it when the section plane meets the BB' axis, refer to Figure 4-37; the section plane rotates, as shown in Figure 4-38.

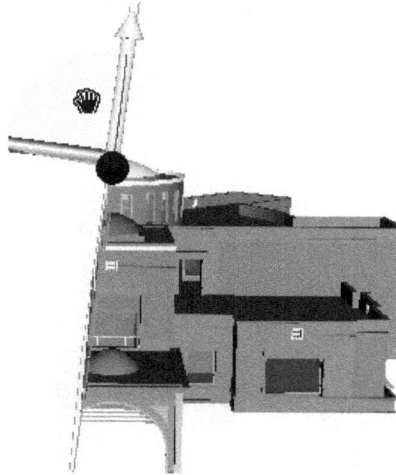

Figure 4-38 *The model after rotating*

11. Deselect the **Rotate** tool from the **Transform** panel in the **Sectioning Tools** tab; the **Rotate** gizmo and section plane disappears, as shown in Figure 4-39.

12. Next, invoke the **Orbit** tool from **Viewpoint > Navigate > Orbit** drop-down; the Orbit cursor appears in the Scene View.

13. Press and hold the left mouse button, drag the cursor in the right direction and release the cursor when the section view of the model looks similar to the one shown in Figure 4-40.

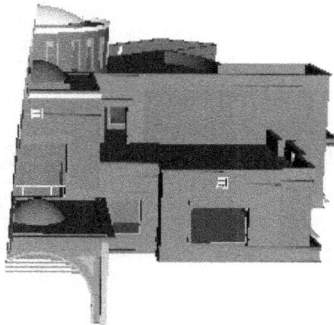

Figure 4-39 *Rotated and sectioned view of the model*

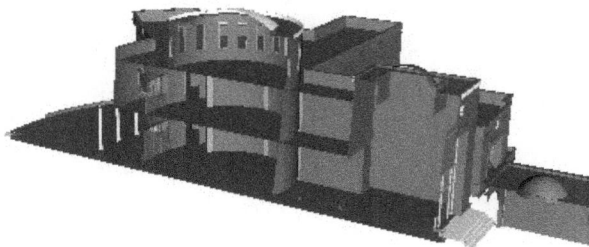

Figure 4-40 *Section view of the model*

14. Now, choose the **Save Viewpoint** option from the **Save** panel of the **Sectioning Tools** tab; the **Saved Viewpoints** window is displayed in the Scene View.

15. Enter **Section View** in the edit box displayed in the **Saved Viewpoints** window, and press ENTER; the view is saved as Section View.

16. Close the **Saved Viewpoints** window and choose the **Enable Sectioning** button in the **Sectioning** panel of the **Viewpoint** tab; the sectioned model is displayed, as shown in Figure 4-41.

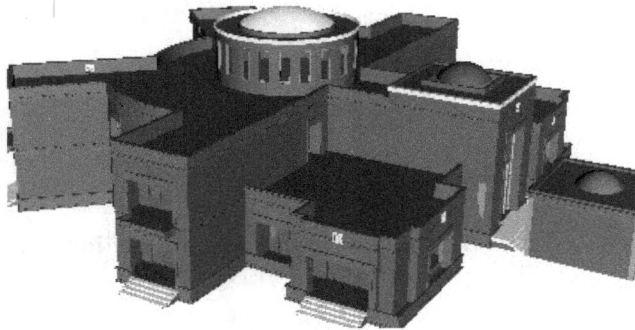

Figure 4-41 The sectioned model displayed in the Scene View

Saving the Project
In this section, you will save the project.

1. To save the project with the current view, choose **Save As** from the Application Menu; the **Save As** dialog box is displayed.

2. Browse to *nws_2017\c04_nws_2017_tut* folder and enter **c04_navisworks_2017_tut02** in the **File name** edit box. Choose **Navisworks File Set (*.nwf)** file format from the **Save as type** drop-down list, and then choose the **Save** button; the project is saved.

Tutorial 3   Using Section Box Mode

In this tutorial, you will open the *c04_navisworks_2017_tut3* file and use the Section Box mode to focus on specific area of the model.   **(Expected time: 30 min)**

The following steps are required to complete this tutorial:

a. Open the file *c04_navisworks_2017_tut3* in Navisworks.
b. Enable Sectioning.
c. Use section box and transforming tools.
d. Save the viewpoint.
e. Save the project.

Opening the Model

In this section, you will open the model created in Revit software.

1. Choose the **Open** button from the Quick Access Toolbar; the **Open** dialog box is displayed.

2. In this dialog box, browse to the following location:
   C:\nws_2017\c04_nws_2017_tutorial.

3. Ensure that the **Navisworks File Set (*.nwf)** option is selected in the **Files of type** drop-down list.

4. Select the file **c04_navisworks_2017_tut3** from the displayed list of files; the file name appears in the **File name** edit box.

5. Choose the **Open** button in the dialog box; the model is displayed in the Scene View, as shown in Figure 4-42.

Figure 4-42 Model opened in Navisworks

Enabling Cross-Section Mode

In this section, you will enable the cross-sectioning mode of the current view of the model.

1. Choose the **Enable Sectioning** option from the **Sectioning** panel in the **Viewpoint** tab; the **Sectioning Tools** contextual tab is displayed in the ribbon.

Using the Section Box Mode and Transforming Tools

In this section, you will use the **Section Box** mode option for the cross-sectioning of the model.

1. Choose the **Box** tool from the **Sectioning Tools > Mode > Plane** drop-down; the Section Box is displayed along with the **Move** gizmo in the Scene View, as shown in Figure 4-43.

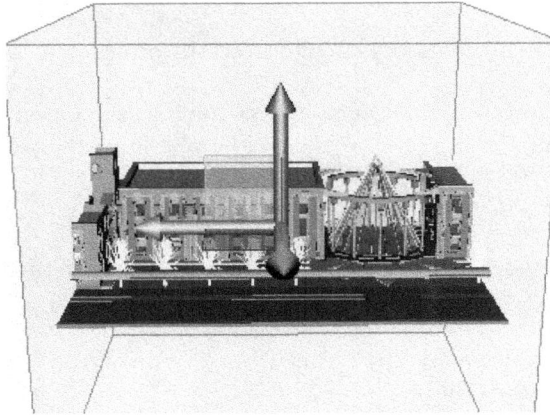

Figure 4-43 *The Section Box with the **Move** gizmo*

Note

*If the Section Box and the **Move** gizmo are not displayed, then choose the **Move** tool from the **Transform** panel in the **Sectioning Tools** tab. Also, note that the direction of the gizmo may vary.*

2. Place the cursor on the red arm (x-axis); the axis is highlighted and a hand symbol appears, as shown in Figure 4-44.

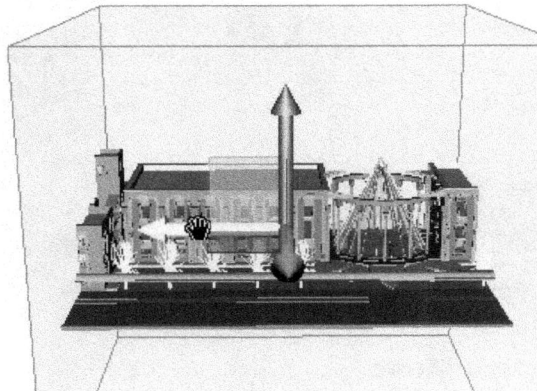

Figure 4-44 *Placing the cursor on the red arm of the gizmo*

3. Press and hold the left mouse button and then drag the gizmo in the left direction. Release the mouse button when the model looks similar to the one shown in Figure 4-45.

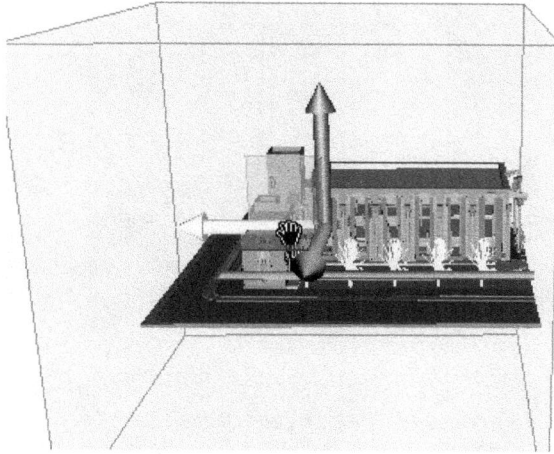

Figure 4-45 Model after dragging the gizmo in the left direction

4. Place the cursor on the green arm; the axis is highlighted and a hand symbol appears, as shown in Figure 4-46.

Note
*If the green arm arrow is not visible, invoke the **Orbit** tool and rotate the model in the right direction.*

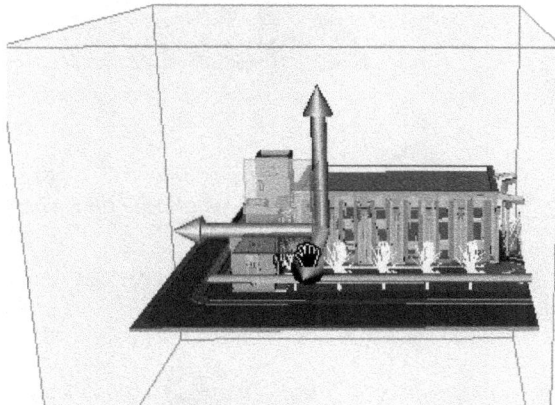

Figure 4-46 Placing the cursor on the green arm of the **Move** gizmo

5. Press and hold the left mouse button and then drag the gizmo in the backward direction. Release the mouse button when the model looks similar to the one shown in Figure 4-47.

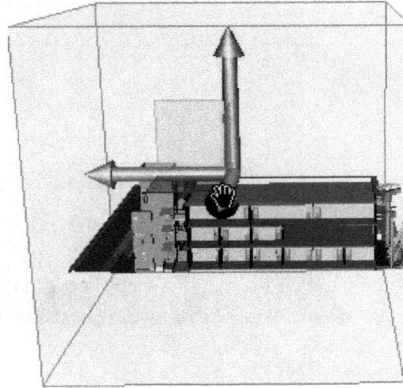

Figure 4-47 *Model after dragging the gizmo in the backward direction*

6. Choose the **Scale** tool from the **Transform** panel in the **Sectioning Tools** tab; the **Scale** gizmo appears inside the **Section Box**, as shown in Figure 4-48.

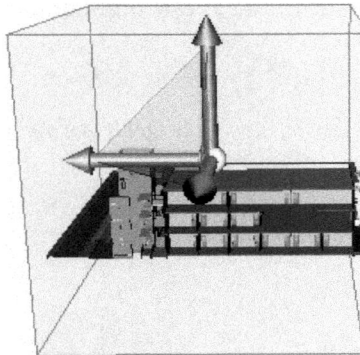

Figure 4-48 *The Scale gizmo inside the section box*

7. Place the cursor on the green triangle (triangle between X and Z axes); the triangle gets highlighted, as shown in Figure 4-49.

Note
Using the curve between the X and Z axes will enlarge the section box along the Y axis.

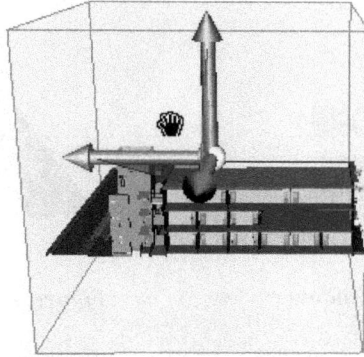

Figure 4-49 Placing the cursor on the green triangle of the **Scale** *gizmo*

8. Press and hold the left mouse button and then drag the cursor in the backward direction. Release the button when the model looks similar to the one shown in Figure 4-50.

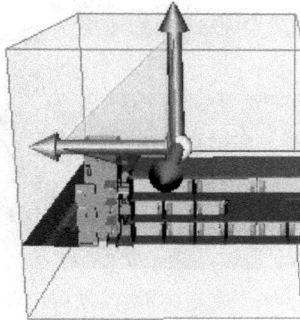

Figure 4-50 Model after using the **Scale** *gizmo*

9. Choose the **Scale** tool from the **Transform** panel of the **Sectioning Tools** tab; the section box and the **Scale** gizmo disappear and the model is displayed, as shown in Figure 4-51.

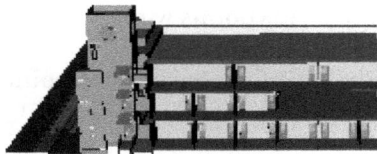

Figure 4-51 The model after the **Scale** *gizmo disappears*

10. Choose the **Zoom Window** tool from the **Zoom** drop-down in the **Navigation Bar**; the shape of the cursor changes.

11. Place the cursor at the corner in the sectioned model, as shown in Figure 4-52.

12. Press and hold the left mouse button and then drag the cursor to draw a window, as shown in Figure 4-53.

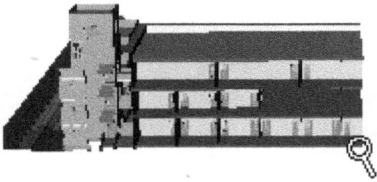

Figure 4-52 *Placing the **Zoom Window** cursor*

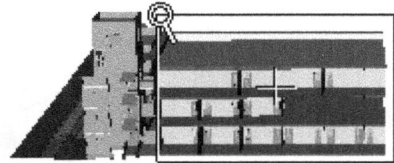

Figure 4-53 *Drawing the zoom window*

13. Release the mouse button; the model is enlarged, as shown in Figure 4-54.

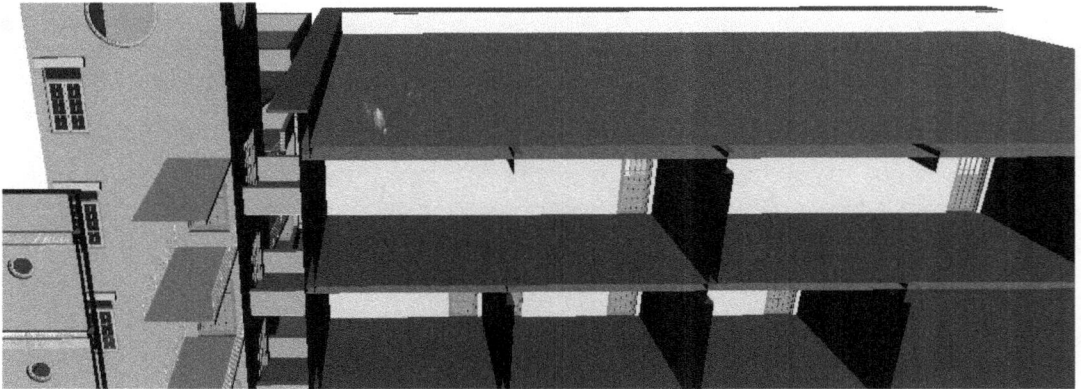

Figure 4-54 *The enlarged view of the sectioned model*

Saving the Viewpoint

In this section, you will save the current sectioned viewpoint.

1. Choose the **Save Viewpoint** button from the **Save** panel of the **Sectioning Tools** tab; the current view is saved in the **Saved Viewpoints** window.

2. Specify **Section View** in the edit box of the **Saved Viewpoints** window and press ENTER.

Saving the Project

In this section, you will save the project.

1. To save the project with the current view, choose **Save As** from the Application Menu; the **Save As** dialog box is displayed.

2. Browse to the *nws_2017\ c04_nws_2017_tut* folder and enter **c04_navisworks_2017_tut03** in the **File name** edit box. Then, choose **Navisworks File Set (*.nwf)** file format from the **Save as type** drop-down list. Now, choose the **Save** button to save the project.

Self-Evaluation Test

Answer the following questions and then compare them to those given at the end of this chapter:

1. The saved viewpoints are displayed in the_____ window.

2. The _____ dialog box is used to modify various viewpoint properties.

3. The two sectioning modes available in Navisworks are _____ and _____.

4. The **Link Section Plane** option is available in the _____ panel.

5. The _____ mode is used to focus on a specific area of the model.

6. The **Add Comment** dialog box is used to modify viewpoints. (T/F)

7. The viewpoints are exported from Navisworks to the XML file. (T/F)

8. In the Section Box mode, six sectional cuts are created in the model. (T/F)

9. The **Record** tool is used to record the animation while navigating through a model. (T/F)

10. Navisworks does not allow you to combine multiple animations into a single animation. (T/F)

11. You cannot save any hidden information in a viewpoint. (T/F)

Review Questions

Answer the following questions:

1. Which of the following tabs contains the **Record** tool?

    a) **Viewpoint** tab   b) **Animation** tab
    c) **Home** tab   d) None of these

2. Which of the following tools is used for displaying the **Section Plane Settings** window?

    a) **Record**   b) **Section Plane Dialog Launcher**
    c) **Box**   d) None of these

3. Which of the following options is used as an animation progress indicator?

    a) **Playback Time**   b) **Percentage**
    c) **Viewport**   d) None of these

4. Which of the following is a type of animation?

     a) Viewpoint Animation   b) Object Animation
     c) Play Animation   d) None of these

5. Which of the following options is used to create six sectional cuts in a model?

     a) **Box**   b) **Record**
     c) **Planes**   d) None of these

6. The **Loop Playback** check box is used to play the animation continuously. (T/F)

7. The **Fit Selection** option is used to move the section plane to the selected object. (T/F)

8. The **Rotate** gizmo is not used with section box. (T/F)

9. The **Edit Animation** dialog box is used to make changes in animation. (T/F)

10. The **Link Section Plane** option is used to link all the selected section planes together. (T/F)

EXERCISE

Exercise 1   Creating Sections

Download and open the *c04_navisworks_2017_ex1* file from *http://www.cadcim.com*. The path of the file is as follows: *Textbooks > Civil/GIS > Navisworks > Exploring Autodesk Navisworks 2017*. Figure 4-55 shows the model to be used for this exercise. Section the model using the **Planes** and **Box** tools. **(Expected time: 45 min)**

The steps required to complete this exercise are given below:

1. Open the file *c04_navisworks_2017_ex1*.
2. Adjust the model using the **Orbit** and **Zoom** tools.
3. Enable sectioning.
4. Create planes using the **Planes** mode.
5. Transform section planes using transformation tools.
6. Create sections using the **Box** mode.
7. Transform section box using transformation tools.
8. Save the viewpoints.
9. Save the project as *c04_navisworks_2017_ex01*.

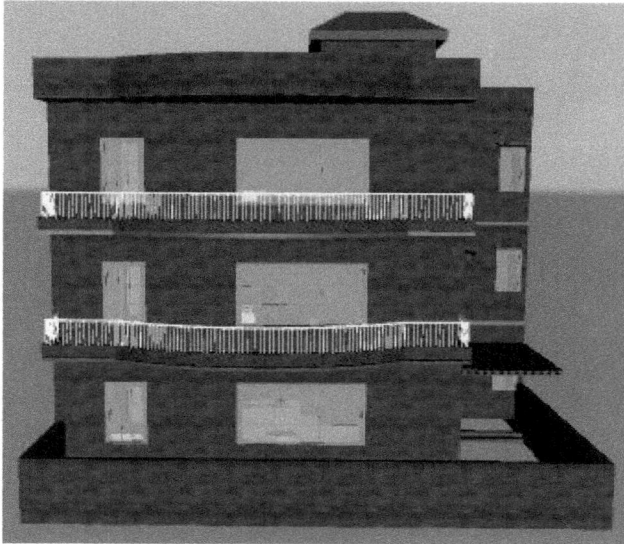

Figure 4-55 *The Residence building*

Exercise 2   Saving Viewpoints

Download and open the *c04_navisworks_2017_ex1* file from *http://www.cadcim.com*. The path of the file is as follows: *Textbooks > Civil/GIS > Navisworks > Exploring Autodesk Navisworks 2017*. Save the viewpoints using the **Save Hide/Required Attributes** option. You will observe both the viewpoints. Figures 4-56 and 4-57 show the model and the **Saved Viewpoints** window after saving the viewpoints.   **(Expected time: 45 min)**

The steps required to complete this exercise are given below:

1. Open the file *c04_navisworks_2017_ex1*.
2. Invoke the **Saved Viewpoints** window and create two folders with the name Saved Views1 and Saved Views2.
3. Select the **Save Hide/Required Attributes** check box in the **Options Editor** dialog box.
4. Select the wall, hide it, and save it as a viewpoint in the Saved Views1 folder.
5. Name the viewpoint as Wall Hidden.
6. Save a viewpoint with the wall shown and name it as Wall in the Saved Views folder.
7. Clear the **Save Hide/Required Attributes** check box in the **Options Editor** dialog box.
8. Save two viewpoints one with the wall hidden and one with the wall shown in Saved Views2 folder.
9. Compare the differences between the viewpoints.
10. Save the project as *c04_navisworks_2017_ex02*.

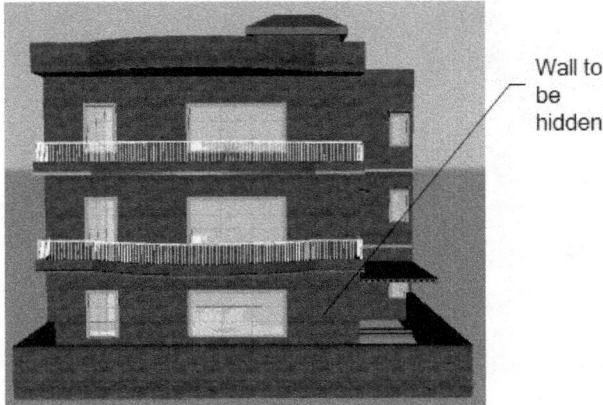

Figure 4-56 Model used in the exercise

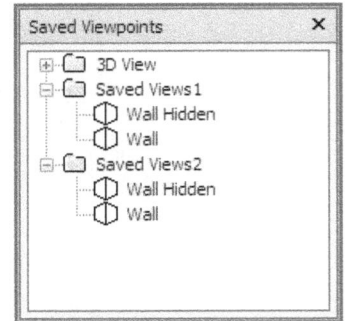

Figure 4-57 Saved viewpoints

Answers to Self-Evaluation Test

1. Saved Viewpoints, 2. Edit Viewpoint, 3. Planes and Box, 4. Plane Settings, 5. Box, 6. F, 7. T, 8. F, 9. F, 10. T 11. F

Chapter 5

TimeLiner

After completing this chapter, you will be able to:
- *Understand the concept of TimeLiner*
- *Create tasks in TimeLiner*
- *Attach items to tasks*
- *Arrange tasks in a hierarchy*
- *Configure tasks parameters*
- *Assign cost to tasks*
- *Play simulation*
- *Add animation to TimeLiner*
- *Export animation from TimeLiner*
- *Link TimeLiner to external scheduling software*

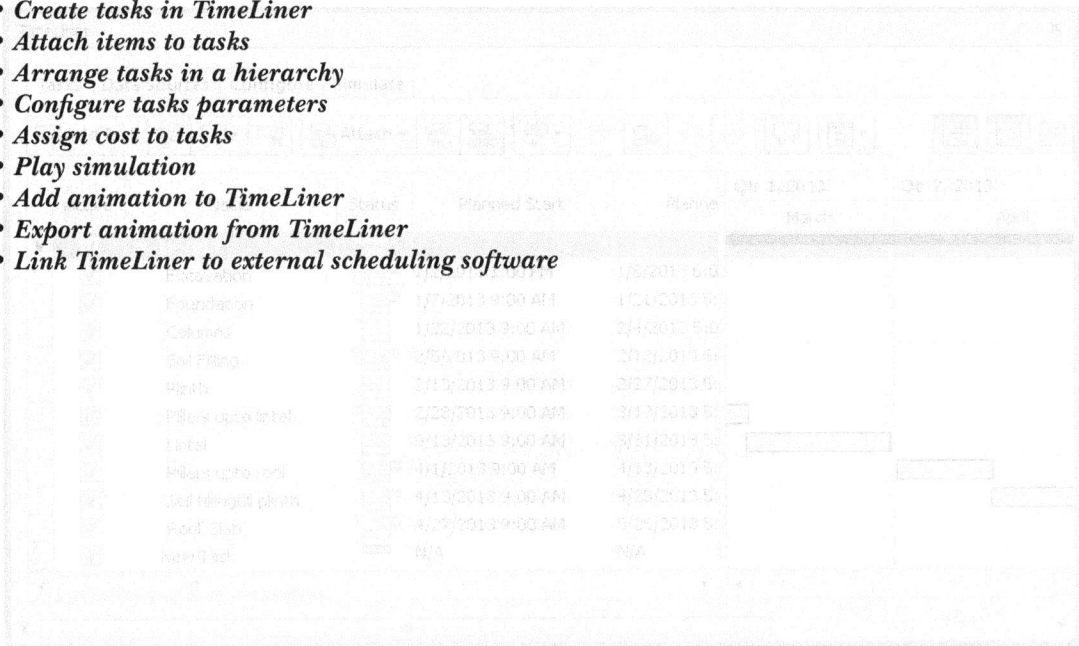

INTRODUCTION

In the previous chapter, you learned about creating viewpoints, cross-sections, and viewpoint animation. In this chapter, you will learn about the concept of Timeliner, creating tasks, preparing schedules, assigning cost, attaching objects to tasks, and creating 4D simulation.

INTRODUCTION TO TimeLiner

In a construction project, various activities are involved such as excavation, laying foundation, erecting columns, walls, roof slab, and so on. These activities can occur concurrently or sequentially. While carrying out these activities, delays may occur due to some co-ordination reasons. In order to minimize the delay in achieving the targets in a project, you need to monitor the activity at each stage. In Navisworks, you can monitor the activities at each stage by using the Timeliner feature.

The Timeliner feature in Navisworks is used to visualize the construction sequence of the model before it begins. To plan a construction sequence, you need the schedules that include duration of various tasks necessary to complete the work. Using this schedule, you can estimate the total cost and resources allocated to each task. Thus, for visualizing the construction sequence of a model, you need to create a 4D simulation. To create a 4D simulation, you need a 3D model and the construction schedule. In 4D simulation, the planning sequence of construction activities will be displayed with respect to time. While playing the 4D simulation, you will be able to see the effects of schedule on the model and can compare planned dates against actual dates. This comparison helps you assess the delay in project completion. You can also identify the areas that will be affected during multiple activities of construction. Thus, the total required time to complete a project can be calculated easily.

The construction schedule can be entered manually in the **TimeLiner** window or you can import it from a variety of sources such as Primavera, Microsoft Project, or any other such project management software. You can also add cost data to the Timeliner simulation for a better understanding of the accrued cost anytime during the construction process. Thus, the total project cost can be tracked throughout its schedule. You can also use the Gantt chart view which shows the project status. Timeliner creates simulation according to the added schedules. If you change the model or make changes in the schedule, Timeliner automatically updates the simulation.

THE TimeLiner WINDOW

The **TimeLiner** window is a dockable window that can be used to add tasks, attach objects in the model to the tasks, set durations, and simulate project schedules. To display the **TimeLiner** window, choose the **TimeLiner** tool in the **Tools** panel of the **Home** tab; the **TimeLiner** window will be displayed in the Scene View, as shown in Figure 5-1. Alternatively, select the **TimeLiner** check box from **View > Workspace > Windows** drop-down; the **TimeLiner** window will be displayed.

Tip
The options in the TimeLiner window will be activated only when you open an existing project.

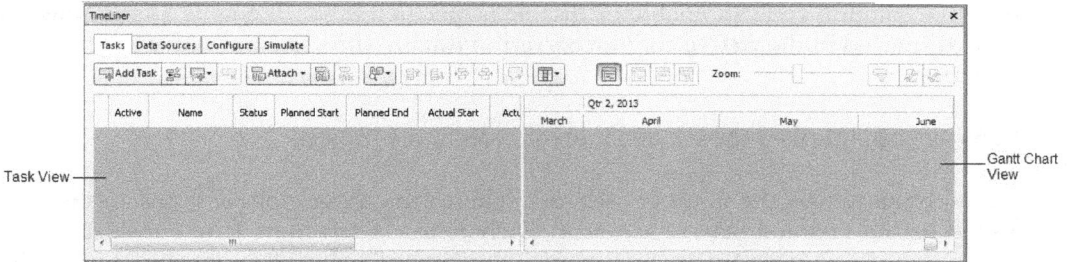

Figure 5-1 *The **TimeLiner** window*

The **TimeLiner** window consists of four tabs: **Tasks**, **Data Sources**, **Configure**, and **Simulate**. These tabs are discussed next.

Tasks Tab

You can use the **Tasks** tab to create and manage various tasks/activities involved in a project. The **Tasks** tab contains two panes: the Task View pane and the Gantt Chart View pane, refer to Figure 5-1. These two panes are discussed next.

Task View Pane

In the Task View pane of the **Tasks** tab, you can create and manage various tasks or activities involved in a project. These tasks are displayed in a multi-column table and can be arranged in a hierarchical structure. Figure 5-2 shows various tasks displayed in the **Tasks** tab. There are several columns in the Task View pane of the **Tasks** tab. These columns are discussed next.

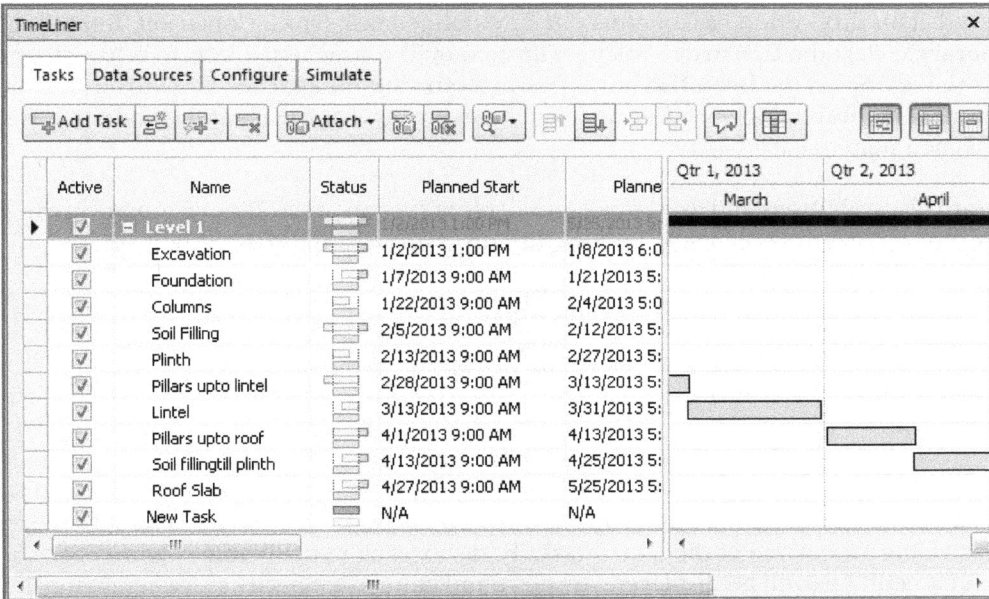

Figure 5-2 *The tasks displayed in the **Tasks** tab of the **TimeLiner** window*

In the **Active** column, you can enable or disable a specified task. To enable a task, select the check box displayed in the **Active** column corresponding to the task. To disable the task, clear the check box.

The **Name** column displays the list of names of tasks in the project.

The **Status** column shows the status of each task. There are various types of status for tasks such as Finished before Planned Start, Early Start Early Finish, Early Start Late Finish, Late Start Late Finish, Late Start Early Finish, and so on. The status of tasks will be displayed in two bars showing the planned dates against the actual dates relationship in the **Status** column.

The **Planned Start** column shows the planned start date of each task. To change the date of a particular task, double-click in its corresponding **Planned Start** date column; a drop-down arrow will be displayed. Click on the arrow; a calendar will be displayed. Select the date from this calendar; the selected date will be displayed in the **Planned Start** column. The **Planned End** column shows the planned end date of each task.

The **Actual Start** column shows the real start date of each task. The **Actual End** column shows the real end date of each task.

Note
*To change dates in any of the columns, follow the same process as given for the **Planned Start** column.*

The **Task Type** column shows the task types assigned to the tasks. To select the task types for a particular task, double-click in its corresponding **Task Type** column; a drop-down list will be displayed. This drop-down list contains three predefined task types: **Construct**, **Demolish**, and **Temporary**. Select the **Construct** task type if you want the attached objects to be constructed in the Scene View. Select the **Demolish** task type when you want the attached items to be demolished. Select the **Temporary** task type when attached items are temporary. The **Attached** column shows the attached item to each task.

There are several buttons and drop-downs available at the top of the Task View pane in the **Tasks** tab, as shown in Figure 5-3. These buttons and drop-downs are discussed next.

*Figure 5-3 The buttons and options in the **Tasks** tab*

Add Task

The **Add Task** button is used for adding a new task. This button can also be used to add a task at the bottom of an existing task. To add a task choose the **Add Task** button; the new task will be added in the **Tasks** tab, refer to Figure 5-4. You can also change the name of the added task. To do so, click in the **Name** column of the added task; an edit box will be displayed, as shown in Figure 5-4. Specify a new name for the task and press ENTER.

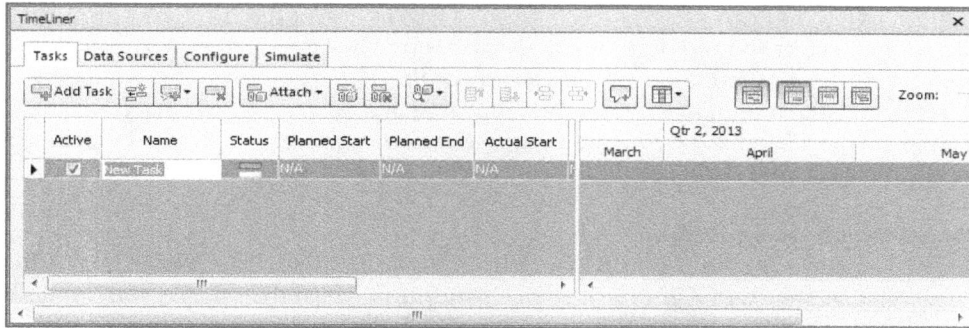

Figure 5-4 *New task added in the* **Tasks** *tab*

Insert Task

The **Insert Task** button is used to insert a new task above a selected task. To do so, select an existing task and then choose the **Insert Task** button; a new task will be added above the selected task in the Task View pane. Specify a name for the added task and press ENTER.

Auto-Add Tasks

In the **TimeLiner** window, you can automatically add a task for every topmost item, layer, selection, or search set. To automatically add a task, click on the **Auto-Add Tasks** button; a drop-down list will be displayed. This drop-down list contains three options: **For Every Topmost Layer**, **For Every Topmost Item**, and **For Every Set**. To add tasks for every topmost layer of the model, select the **For Every Topmost Layer** option from the drop-down list. To add tasks for every topmost item in the model, select the **For Every Topmost Item** option from the drop-down list. To add tasks for every search and selection set, select the **For Every Set** option from the drop-down list.

Delete Task

The **Delete Task** button is used for removing the selected task. To do so, select the task that you want to remove from the task list. Next, choose the **Delete Task** button; the selected task will be removed from the task list.

Attach

In Timeliner, before creating a 4D simulation, you need to attach the required objects to the tasks. To do so, first select the required object in the model and then select the required task. Next, choose the **Attach** button in the TimeLiner window; a drop-down list will be displayed. This drop-down list contains three options: **Attach Current Selection**, **Attach Current Search**, and **Append Current Selection**. Select the **Attach Current Selection** option from the drop-down list to attach the currently selected objects in the scene to the selected task. Select the **Attach Current Search** option from the drop-down list to attach all objects found after running a search. Select the **Append Current Selection** option to add the currently selected objects in the scene to the objects already attached to the selected tasks.

Note
*To enable the **Attach Current Search** option, a search set should be defined for the model.*

Alternatively, you can attach objects to the tasks by using the drag and drop method. To do so, first select the required object in the Scene View. Next, drag and place the selected object on the required task in the **Tasks** tab of the **TimeLiner** window. Note that if there is an existing attachment in the task, then that attached object will be replaced by the currently selected object. Similarly, you can attach selection or search sets from the **Sets** window by using the drag and drop method.

Auto-Attach Using Rules

The **Auto-Attach Using Rules** button is used to display the **TimeLiner Rules** dialog box, as shown in Figure 5-5. In the **TimeLiner Rules** dialog box, you can define, edit, and apply rules for automatically attaching the objects to the tasks. In this dialog box, there are three predefined rules. You can configure these predefined rules and can also define new rules.

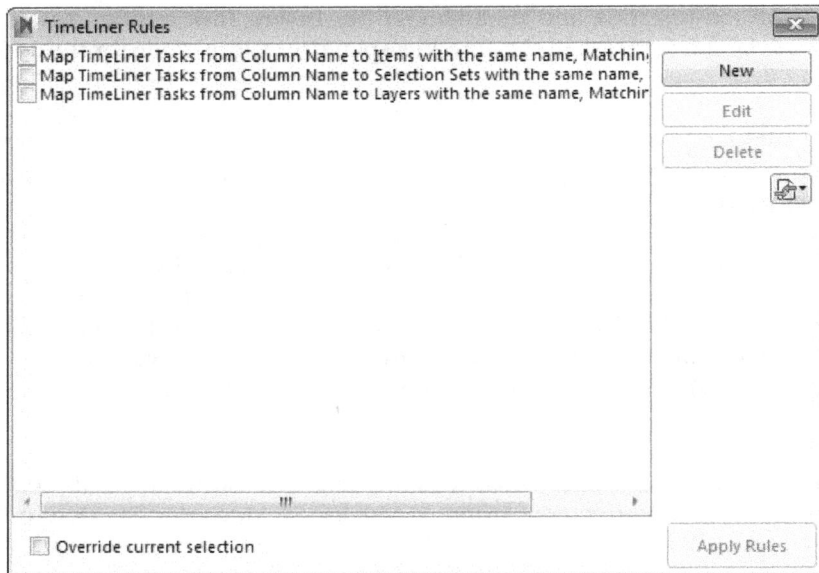

*Figure 5-5 The **TimeLiner Rules** dialog box*

The buttons used for modifying rules and adding new rules in the **TimeLiner Rules** dialog box are discussed next.

The **New** button available on the right of the **TimeLiner Rules** dialog box is used to create a new rule. On choosing the **New** button, the **Rules Editor** dialog box will be displayed, as shown in Figure 5-6. In this dialog box, specify a name for the required rule in the **Rule Name** edit box. Select the template which you want to use from the **Rule Templates** area. On selecting a template, the description related to the selected template will be displayed in the **Rule description** area, refer to Figure 5-6. To define the rule, click on the underlined value in the **Rule description** area; another **Rules Editor** dialog box will be displayed. In this dialog box, select the required option from the **Enter Value** drop-down list and then choose the **OK** button. After specifying the rules, choose the **OK** button; the **Rules Editor** dialog box will be closed, and a new rule will be added in the **TimeLiner Rules** dialog box.

The **Edit** button in the **Timeliner Rules** dialog box is used to edit the pre-defined or new rule, refer to Figure 5-5. Note that this button is inactive unless you select an existing rule. Select a rule from the list and choose the **Edit** button; the **Rules Editor** dialog box will be displayed, refer to Figure 5-6. In this dialog box, you can make the required changes for the selected rule and choose the **OK** button.

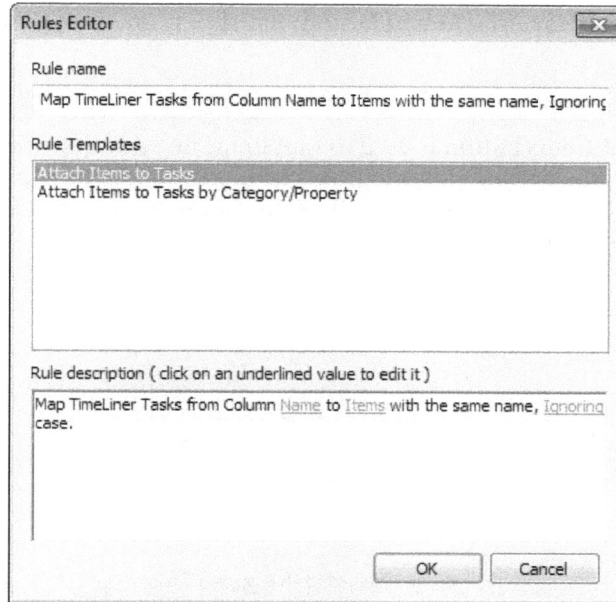

*Figure 5-6 The **Rules Editor** dialog box*

The **Delete** button in the **TimeLiner Rules** dialog box is used to delete the selected rule. To do so, select a rule from the list and then choose the **Delete** button; the selected rule will be deleted from the list in the **TimeLiner Rules** dialog box.

The **Import/Export Attachment Rules** button in the **TimeLiner Rules** dialog box is used for importing or exporting the rules. To do so, choose this button; a drop-down list will be displayed. This drop-down contains two options: **Import Attachment Rules** and **Export Attachment Rules**. To import rules from the XML file, select the **Import Attachment Rules** option; the **Import Attachment Rules** dialog box will be displayed. Browse and select the required file and ensure that the (.xml) extension is selected. Next, choose the **Open** button; the selected file will be imported. Similarly, to export rules to the XML file, select the **Export Attachment Rules** option from the list; the **Export Attachment Rules** dialog box will be displayed. In this dialog box, specify the file name and location and then choose the **Save** button; the file will be exported and saved at the specified location.

The **Override current selection** check box available at the bottom in the **TimeLiner Rules** dialog box is used to override the selection. On selecting this check box, it will replace the existing attached objects when the rules are applied to the tasks. If you clear this check box, rules will be applied to the tasks without replacing the attached objects. After specifying all the parameters, choose the **Apply Rules** button available at the bottom in the **TimeLiner Rules** dialog box; the rules will be applied to all the relevant tasks.

Clear Attachment

The **Clear Attachment** button in the Task View pane is used to detach the object(s) from the task. To do so, first select the task from which you want to detach the object(s). Next, choose the **Clear Attachment** button; the objects attached to the selected task will be removed.

Find Items

The **Find Items** button is used to find items in a schedule which are not included in tasks or attached in multiple tasks, or in overlapping tasks. When you choose the **Find Items** button, a drop-down list will be displayed, as shown in Figure 5-7. The options in this list are discussed next.

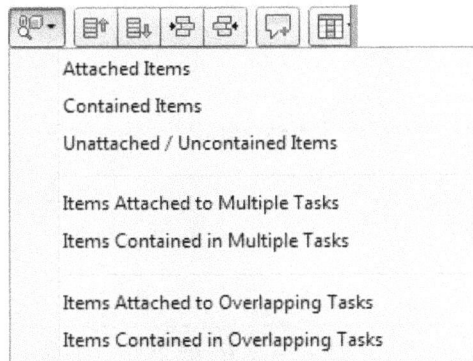

*Figure 5-7 The **Find Items** drop-down list*

Note
*If the **Find Items** button is not available in the **Tasks** tab of the **TimeLiner** window, invoke the **Options Editor** dialog box. In this dialog box, expand the **Tools** node and select the **TimeLiner**; options will be displayed in the right pane of the dialog box. Then, select the **Enable Find** check box; the **Find Items** button will be displayed in the **Tasks** tab.*

The **Attached Items** option is used to select the objects in the Scene View that are directly attached to a task. When you select this option, the objects that are directly attached to the tasks will be highlighted in the Scene View.

The **Contained Items** option is used to select objects that are directly attached to a task or contained within an attached object.

The **Unattached/Uncontained Items** option is used to select those objects that are not attached to tasks.

The **Items Attached to Multiple Tasks** option is used to select those objects that are directly attached to more than one task.

The **Items Contained in Multiple Tasks** option is used to select those objects that are contained within the object which is attached to more than one task.

The **Items Attached to Overlapping Tasks** option is used to select those objects which are attached to more than one task and the task duration overlaps.

The **Items Contained in Overlapping Tasks** option is used to select those objects that are contained within the objects attached to more than one task and where the task duration overlaps.

Move Up

The **Move Up** button in the Task View pane is used to move the selected task up in the list. To do so, select the task from the Task View pane and choose the **Move Up** button; the selected task will be shifted up in the list.

Move Down

The **Move Down** button is used to shift the selected task down in the list. To do so, select the task from the Task View pane and choose the **Move Down** button; the selected task will be shifted down in the list.

Indent

The **Indent** button is used to indent the selected task by one level in the hierarchy.

Outdent

The **Outdent** button is used to outdent the selected task by one level in the hierarchy.

Add Comment

The **Add Comment** button is used to add a comment to the task. When you choose this button, the **Add Comment** dialog box will be displayed. Specify a comment in the dialog box and choose the **OK** button; the comment will be added to the selected task.

Columns

The **Columns** button is used to control the column display in the task view. When you choose the **Columns** button, a drop-down list will be displayed. This drop-down list contains four options: **Basic, Standard, Extended**, and **Custom**. To create a customized column set, select the **Choose Columns** option from the drop-down list; the **Choose TimeLiner Columns** dialog box will be displayed, as shown in Figure 5-8.

The dialog box displays a list of all the available columns. To add a column in the multi-column table of the Task View pane, select the check box corresponding to that particular column from the **Choose TimeLiner Columns** dialog box. In this dialog box, the **Move Up** button is used to shift the selected column up in the column list. The **Move Down** button is used to shift the selected column down in the column list. The **Show All** button is used to select all the check boxes in the columns list. The **Hide All** button is used to clear all the check boxes in the columns list. Choose the **OK** button to close the dialog box.

Note

You can also access the discussed buttons by right-clicking in the Task view pane of the **TimeLiner** *window. On right-clicking, a shortcut menu will be displayed which contains a list of all these buttons.*

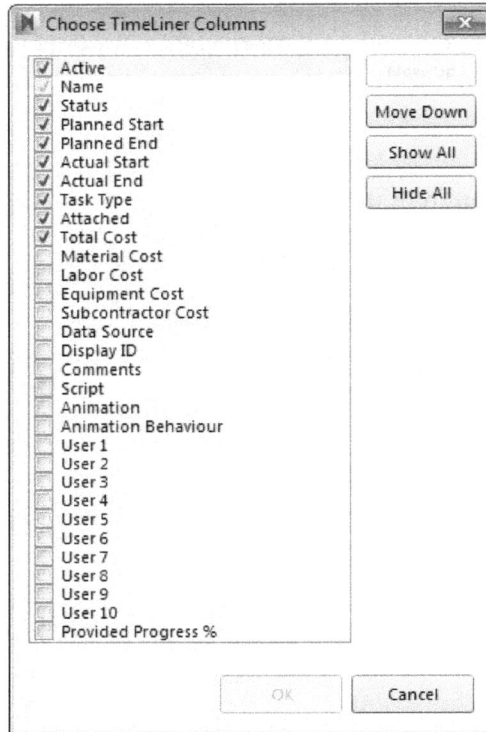

*Figure 5-8 The **Choose TimeLiner Columns** dialog box*

Filter by Status

The **Filter by Status** button is used to filter tasks based on their status. When you choose this button, a menu will be displayed. This menu contains options which are grouped in two areas: **Start** and **Finish**. To choose an option, select the check box corresponding to that option from the **Start** and **Finish** areas.

Export To Sets

The **Export To Sets** button is used to create selection sets of the objects attached to the tasks in the Timeliner. When you choose this button, a folder will be created in the **Sets** window which will contain the selection sets.

Export the schedule

The **Export the schedule** button is used to export a Timeliner schedule in CSV or Microsoft Project XML format. To do so, choose the **Export the schedule** button; a drop-down list will be displayed. This drop-down list contains two options: **Export CSV** and **Export MS Project XML**. Select the required option to export the Timeliner schedule in the **CSV** or **Microsoft Project XML** file; the **Export** dialog box will be displayed. In this dialog box, specify the file name and location and then choose the **Save** button; the file will be saved to the specified location. CSV files will be saved in *.csv format and Microsoft Project XML files will be saved in *.xml format.

Gantt chart View Pane

The Gantt chart view pane of the **TimeLiner** window displays Gantt chart. A Gantt chart is used to show the project status, refer to Figure 5-9. The time span of a project is displayed on the top in the Gantt chart view pane. The tasks in the Gantt chart view are represented as bars. You can modify task dates by moving and resizing bars. The changes made in the Gantt chart view will be updated in the Task view of the window. You can also view the task information and such as planned start and end dates, and the duration in the Gantt chart view. To do so, hover the cursor on the required bar in the Gantt chart view; the task information will be displayed in a rectangular box.

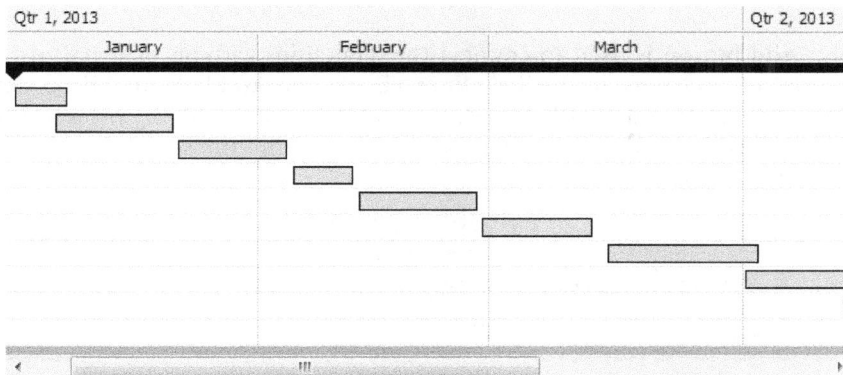

Figure 5-9 *The Gantt Chart view in the **TimeLiner** window*

You can also control the display of tasks in Gantt chart view by using the buttons that are discussed next.

Show or hide the Gantt chart

The **Show or hide the Gantt chart** button is used to display or hide the Gantt chart view. By default, the Gantt chart view pane is displayed in the **TimeLiner** window. To hide it, choose the **Show or hide the Gantt chart** button; the Gantt chart view pane will be hidden.

Show Planned Dates

The **Show Planned Dates** button is used to view the planned dates in the Gantt chart view. On choosing this button, the planned dates task bars will be displayed in grey color in the Gantt chart view.

Show Actual Dates

The **Show Actual Dates** button is used to view the actual dates in the Gantt chart view. On choosing this button, the actual date task bars will be displayed in blue color in the Gantt chart view.

Show Planned vs Actual Dates

The **Show Planned vs Actual Dates** button is used to view planned dates vs actual dates in the Gantt chart view. On choosing this button, the actual dates and planned dates task bars will be displayed in the Gantt chart view.

Data Sources Tab

In the **Data Sources** tab, you can import schedules from other project management softwares such as Microsoft Project, Primavera, and Asta Powerproject. When you choose the **Data Sources** tab, it will be empty. Once you import the schedules, the sources of the added schedules will be displayed in a tabular format in this tab. In this table, the **Name** column displays the name of the new data source. The **Source** column displays the source of the added data such as the name of the source software. The **Project** column displays the name and location of the added data. The buttons used for adding the schedules and managing them are discussed next.

Add

The **Add** button is used to connect the Timeliner with an external project file. To import a data source in the Timeliner, choose the **Add** button; a drop-down list will be displayed, as shown in Figure 5-10. This drop-down list contains all the project software from which you can import the schedules. The process of importing the schedules will be discussed in detail later in the chapter.

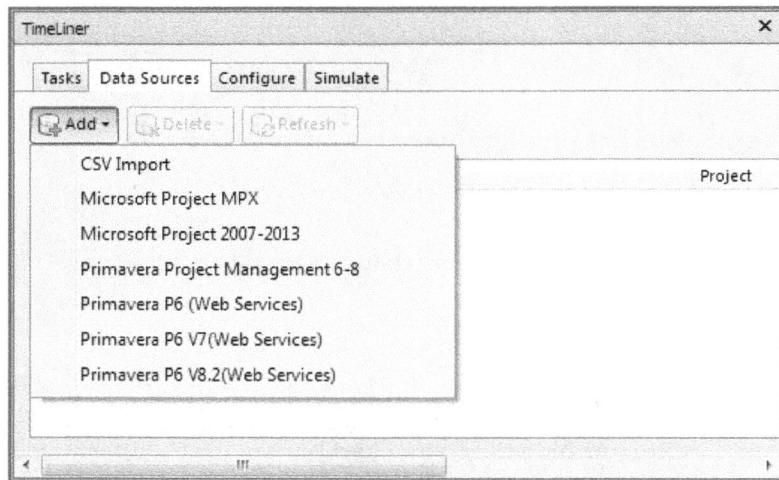

Figure 5-10 The Add drop-down list in the Data Sources tab

Delete

The **Delete** button is used to delete the selected data source. To do so, select the required data source and then choose the **Delete** button; the selected data source will be removed from the table.

Refresh Button

The **Refresh** button is used to refresh the data source(s). When you choose the **Refresh** button, a drop-down list will be displayed. This drop-down list contains two options: **Selected Data Source** and **All Data Sources**. Select the **Selected Data Source** option to refresh the selected data source added in the **Data Sources** tab. Select the **All Data Sources** option to refresh all the added data sources in the tab. When you select the **All Data Sources** option, the **Refresh from Data Source** dialog box will be displayed. In this dialog box, select the **Rebuild Task Hierarchy** radio button to rebuild the task hierarchy with latest data and tasks. Select the **Synchronize** radio button to update the existing tasks with latest data from the selected data source.

When you right-click in the table area of the **Data Sources** tab, a shortcut menu will be displayed. The options in this shortcut menu are discussed next.

Rebuild Task Hierarchy Option
The **Rebuild Task Hierarchy** option is used to rebuild the task hierarchy in the **Tasks** tab with the latest task and data.

Synchronize Option
The **Synchronize** option is used to update the existing tasks with latest data from the selected data source.

Delete Option
The **Delete** option is used to delete the current selected data source. To do so, first select the required data source and then choose the **Delete** option from the menu.

Edit Option
The **Edit** option is used to invoke the **Field Selector** dialog box. In this dialog box, you can create a new data source or edit the existing data source. This will be explained later in the chapter.

Rename Option
The **Rename** option is used to change the name of the selected data source. On choosing this option; the name field of the selected data source will be highlighted. Specify a new name and then press ENTER.

Configure Tab
In the **Configure** tab, you can set up the task parameters such as task types and appearance of tasks. This tab displays a list of all the task types in a tabular format, as shown in Figure 5-11. The various columns in the **Configure** tab are discussed next.

The **Name** column displays the names of all task types. There are three predefined tasks types: **Construct**, **Demolish**, and **Temporary**. The **Construct** task type is used for the tasks when the attached objects are to be constructed during simulation. The **Demolish** task type is used for the tasks when the attached objects are to be demolished. The **Temporary** task type is used for tasks when the attached objects are temporary.

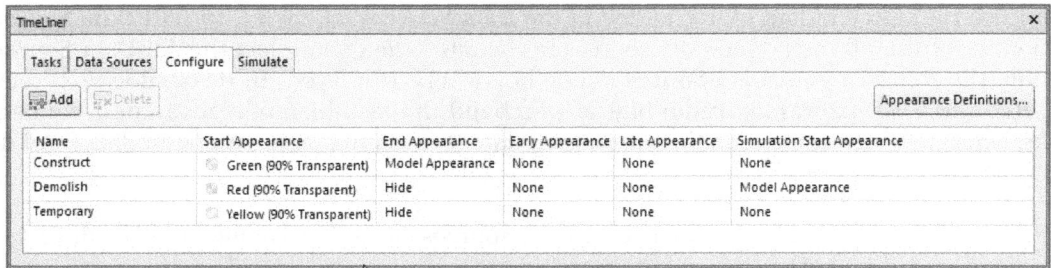

*Figure 5-11 The **Configure** tab in the **TimeLiner** window*

In the **Start Appearance** column, you can specify how the objects attached to the tasks will appear at the start of task during simulation. To do so, click in the **Start Appearance** column; a drop-down list will be displayed. The options in this list are discussed next.

The **Hide** option is used to hide the objects attached to the tasks during simulation. The **Model Appearance** is used to display the attached objects as they are defined in the model. You can also assign a color for the object appearance during simulation.

In the **End Appearance** column, you can specify how the objects attached to the tasks will appear at the end of the task during simulation. To do so, click in the **End Appearance** column; a drop-down list will be displayed. This drop-down list contains the same options: **Hide, Model Appearance**, and **colors**.

Similarly, in the **Early Appearance** and **Late Appearance** columns, you can specify the appearance of the objects attached to the tasks, occurring early or late in the simulation.

In the **Simulation Start Appearance** column, you can specify how the model will appear at the start of the simulation.

You can also create custom task types using various buttons in the **Configure** tab. These buttons are discussed next.

Add Button

The **Add** button is used to add a new task type. To do so, choose the **Add** button; a new task type will be added in the list in the **Configure** tab.

Delete Button

The **Delete** button is used to delete the selected task type from the list in the **Configure** tab. To do so, first select the task type from the list and then choose the **Delete** button; the selected task type will be removed.

Appearance Definitions Button

The **Appearance Definitions** button is used to invoke the **Appearance Definitions** dialog box, refer to Figure 5-12. In this dialog box, there are ten predefined appearance definitions which are displayed in a tabular format. You can use these definitions to modify the task types. There are three columns in the **Appearance Definitions** dialog box, which are discussed next.

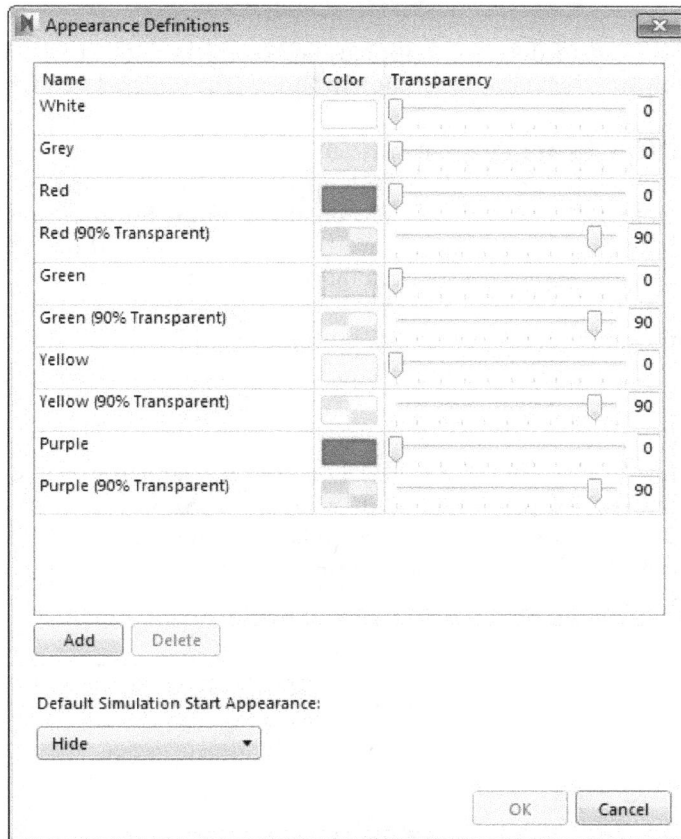

Figure 5-12   *The **Appearance Definitions** dialog box*

The **Name** column displays a list of names of all definitions. To change the name of definitions, double-click on the required name; the selected definition will turn into an edit box. Specify a new name and press ENTER; the selected definition will be renamed.

The **Color** column contains a list of colors for all definitions. To change the color for a particular definition, double-click on the required color; the **Color** dialog box will be displayed. In this dialog box, select the required color and then choose the **OK** button; the color of definitions will be changed.

The **Transparency** column contains different transparency settings. To adjust the transparency, drag the **Transparency** slider or specify the value in the edit box next to the slider.

You can also create new definitions for new task types in the **Appearance Definitions** dialog box by using the buttons and options which are discussed next.

Add

The **Add** button is used to create a new appearance definition. To do so, choose the **Add** button; a **New Appearance** will be added in the list of appearance definitions. Specify the name, color, and transparency for the new appearance definition, as discussed above.

Delete

The **Delete** button is used to delete the appearance definition. To do so, select the required appearance definition and then choose the **Delete** button; the selected definition will be deleted.

Default Simulation Start Appearance

This drop-down list is used to specify the default appearance applied to all objects in the model at the start of the simulation. To do so, click on the **Default Simulation Start Appearance** drop-down list; various options will be displayed. Select the required option from this list. By default, the **Hide** option is applied to all the objects in the model.

After making all the changes, choose the **OK** button; the **Appearance Definitions** dialog box will be closed.

Simulate Tab

In this tab, you will create 4D simulation which illustrates the sequence of construction activities throughout the duration of the project schedule. Choose the **Simulate** tab; the first task and the duration of that task will be displayed in the **Simulate** tab, refer to Figure 5-13. The **Simulate** tab contains two panes: the Task View pane and the Gantt chart view pane, refer to Figure 5-13. These panes are discussed next.

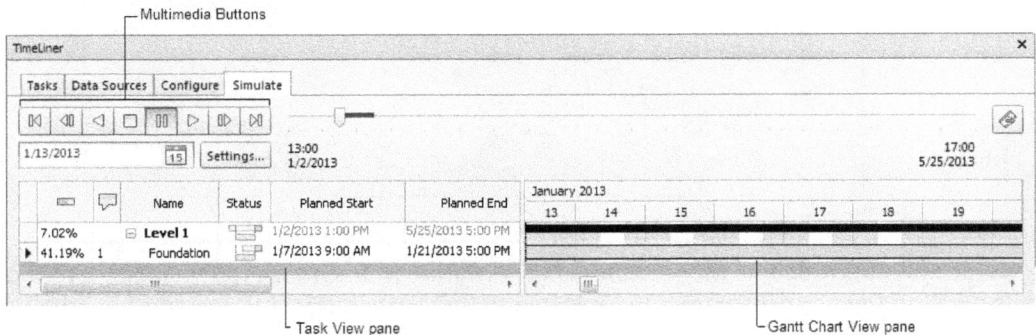

*Figure 5-13 The **Simulate** tab in the **TimeLiner** window*

Task View Pane

In the Task View pane, all the active tasks will be displayed in a multi column table. You can view the task progress in the **Simulation Progress** column. The progress of task completion will be displayed in percentage.

Multimedia Buttons

The multimedia buttons available at the top of Task View pane are used to control the animation, refer to Figure 5-13. The **Play** button is used to play the simulation from the current position. The **Forward** button is used to move forward in the simulation. The **Pause** button is used to pause the simulation. The **Play Backwards** button is used to playback the simulation. The **Rewind** button is used to rewind the simulation. The **Stop** button is used to stop the simulation. Below these multimedia buttons, there is an edit box that displays the planned start date of the project.

This shows the date throughout the project simulation. You can also view the project status for a particular date. To do so, click on the calendar icon next to the date edit box; a calendar will be displayed. Select the required date from the calendar; the simulation progress and the task will be displayed in the Task View pane.

Gantt Chart View Pane

The Gantt chart view pane shows the project status. The time span of a project is displayed at the top of the Gantt chart view pane. The active tasks in the Gantt chart view are represented as bars.

You can also configure the simulation by using the **Settings** button in the **Simulate** tab which is discussed next.

Settings Button

On choosing the **Settings** button, the **Simulation Settings** dialog box will be displayed, as shown in Figure 5-14. The options in this dialog box are discussed next.

In the **Start / End Dates** area of this dialog box, you can override the existing period for simulation run. To do so, first select the **Override Start / End Dates** check box in the **Start / End Dates** area. On selecting this check box, the **Start Date** and **End Date** edit boxes will become active. To change the dates, click on the down arrow button in the required edit box; a calendar will be displayed. Select the required dates from the calendar; the date in the edit box will be changed.

In the **Interval Size** area of the dialog box, you can configure the interval size of the simulation. To specify the interval size, enter the required value in the **Interval Size** edit box. Alternatively, you can use the spinner displayed in the **Interval Size** edit box to adjust the interval. Next, to specify the units for the interval, select the required option from the drop-down list displayed on the right of the **Interval Size** edit box. Note that the **Percent** option is selected as interval unit in the drop-down. You can also display

Figure 5-14 The Simulation Settings dialog box

the tasks being performed during the specified interval in the Task View pane of the **Simulate** tab. To do so, select the **Show all tasks in interval** check box. To define the time for complete simulation, specify a value in the **Playback Duration (Seconds)** edit box.

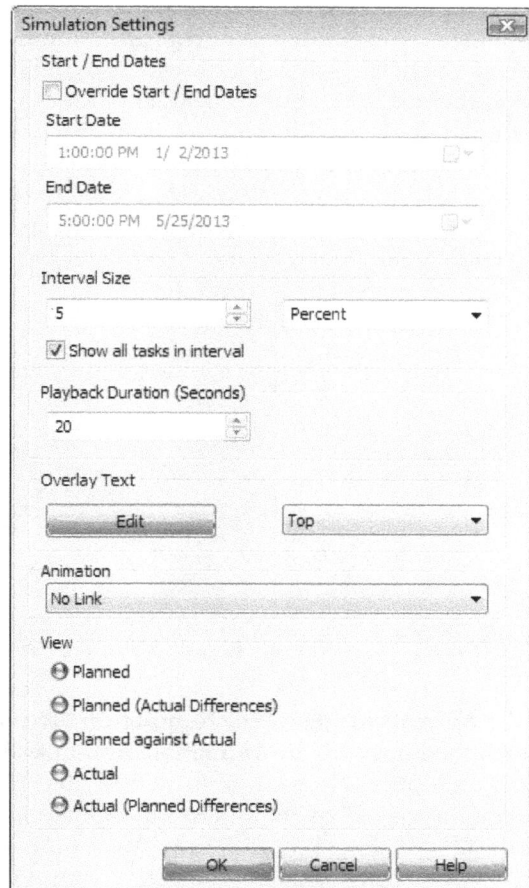

In the **Overlay Text** area, you can specify how the date, time, and cost information will be displayed in the Scene View. To do so, choose the **Edit** button; the **Overlay Text** dialog box will be displayed, as shown in Figure 5-15. In this dialog box, you can define the display format by choosing the **Date/Time**, **Cost**, and **Extra** buttons. On choosing these buttons, a shortcut menu will be displayed. Choose the required option from this menu. On choosing an option from this menu, the related text or keywords will be displayed in the **Overlay Text** edit box. Similarly, you can define the color of the text by choosing the **Colors** button. You can also modify the font size by choosing the **Font** button. After specifying all the options, choose the **OK** button; the dialog box will be closed.

You can specify the location where the overlay text will be displayed in the Scene View. To display the overlay text at a certain position in the Scene View, click on the arrow next to the **Edit** button in the **Simulation Settings** dialog box; a drop-down list will be displayed. This drop-down list contains three options: **None**, **Top**, and **Bottom**. The **None** option is used for not displaying the overlay text in the Scene View during simulation. The **Top** option is used to display the overlay text at the top in the Scene View. The **Bottom** option is used to display the overlay text at the bottom in the Scene View during simulation.

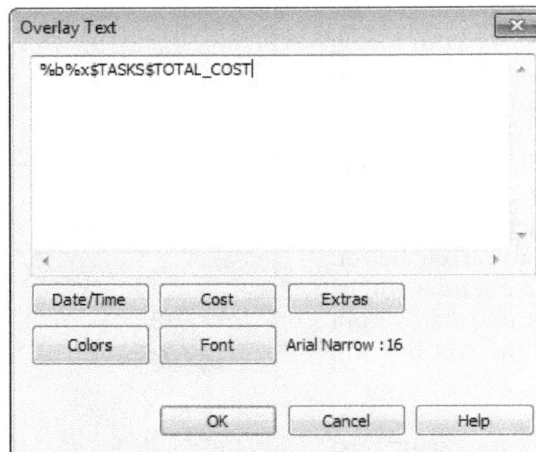

Figure 5-15 The **Overlay Text** *dialog box*

In the **Animation** area of the **Simulation Settings** dialog box, you can link an animation to the entire schedule using the **Animation** drop-down list, refer to Figure 5-16. To display no animation, select the option **No Link** from the list. To link the schedule to a viewpoint animation, select the option **Saved Viewpoints Animation**. To link schedule to any camera animation, select the **Scene X - > Camera** option from the drop-down list.

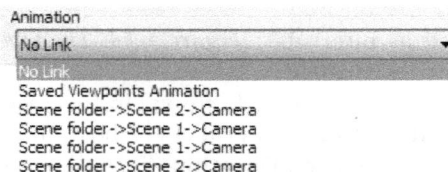

Figure 5-16 The **Animation** *drop-down list*

Note
*The camera animation is created using the **Animator** window and is discussed in detail in the next chapter. Note that the camera animations will not be available in this list if you have not created them.*

Tip
If you want to run the animator camera animation throughout the simulation, specify the start date of the task to which animation is assigned as the start date of simulation and the end date of task as the end date of simulation.

In the **View** area of the **Simulation Settings** dialog box, there are various options that show the actual and planned dates relationship. These options are discussed next.

The **Planned** option is used for simulating the planned schedules only. The **Planned (Actual Differences)** option is used for simulating the actual schedule against the planned schedules. The **Planned against Actual** option is used for simulating planned schedule against the actual schedules. The **Actual** option is used for simulating the actual schedules only. To assign a view, select the radio button corresponding to the required options.

After configuring all the settings, choose the **OK** button; the **Simulation Settings** dialog box will be closed and the changes will be applied.

In the **Simulate** tab, the position slider is used to move backward and forward through the simulation. While playing the simulation, this slider will move forward. If you want to move the simulation backward, drag the slider backward.

PLAYING SIMULATION
In Navisworks, simulation is a real-world process of visualizing the construction sequence of a project. After developing the model and linking the tasks with the construction elements, you can play the simulation. Before playing a simulation, ensure that all tasks are active in the **Tasks** tab of the **TimeLiner** window. To do so, in the **Tasks** tab, select the **Active** check box for the tasks that you want to include in simulation. Next, assign the task types to each task to be included in the simulation. To do so, click in the **Task Type** column corresponding to the required task; a drop-down list will be displayed. Select the required task types from this list. Then, attach items from the model in the Scene View to the active tasks. Next, choose the **Play** button in the **Simulate** tab; the simulation will start playing. The left pane of the **TimeLiner** window will display each task according to its occurrence, and the right pane will display the tasks in a Gantt chart view. The construction sequence will be displayed in the Scene View according to its task sequence.

EXPORTING ANIMATION FROM THE TIMELINER
You can export the simulation as an AVI file or a sequence of images. To do so, choose the **Export Animation** button on the right of the position slider; the **Animation Export** dialog box will be displayed, as shown in Figure 5-17. In this dialog box, select the **TimeLiner Simulation** option as the source from the **Source** drop-down list. Specify other parameters for exporting animation in this dialog box, as explained in Chapter 4 and then choose the **OK** button; the animation will be exported to the specified location.

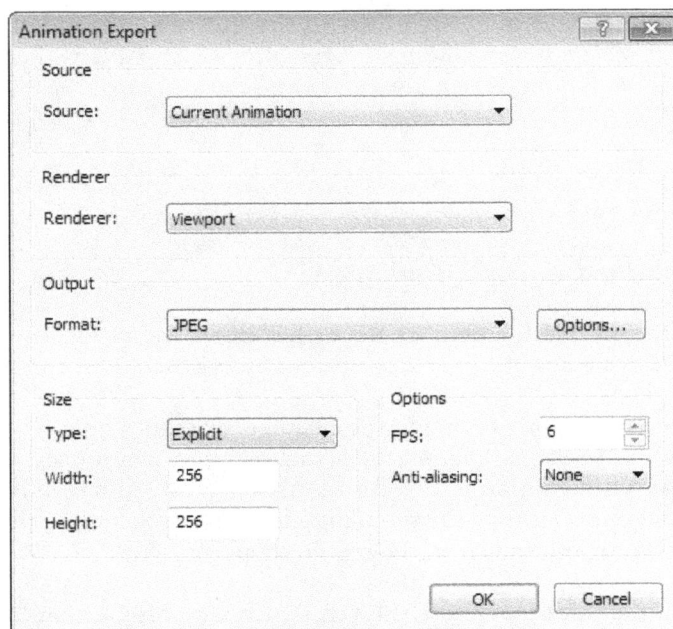

Figure 5-17 *The **Animation Export*** *dialog box*

ADDING ANIMATION TO THE TimeLiner

In the **TimeLiner** window, you can add the viewpoint animation and the camera animation to the individual tasks or to the entire schedule. The animation will be played during the simulation as it increases the quality of simulation. The method of adding animation to the entire schedule or to the individual tasks within that schedule is discussed next.

Adding Animation to the Entire Schedule

You can add a viewpoint animation to the entire schedule displayed in the **Tasks** tab of the **TimeLiner** window. To do so, invoke the **Saved Viewpoints** window by choosing the **Saved Viewpoints Dialog Launcher** button from the **Save, Load & Playback** panel of the **Viewpoint** tab. In the **Saved Viewpoints** window, select the desired viewpoint animation. Next, choose the **Settings** button in the **Simulate** tab of the **TimeLiner** window; the **Simulation Settings** dialog box will be displayed, refer to Figure 5-14. In this dialog box, select the **Saved Viewpoints Animation** option from the **Animation** drop-down list. Next, choose the **OK** button. Now, when you start the simulation, the selected viewpoint animation will be played in the Scene View. Similarly, the camera animation can be added to the entire schedule in the **TimeLiner** window. To add the camera animation, select the required camera animation option from the **Animation** drop-down list in the **Simulation Settings** dialog box and then choose the **OK** button.

Note
*The camera animations will be available in the **Animation** drop-down list only if you have created them. The method of creating camera animation is discussed in detail in Chapter 6.*

Adding Animation to Tasks

In Navisworks, you can also add animation to an individual task. To do so, first ensure that the **Animation** and **Animation Behavior** columns are displayed in the Task View pane of the **Tasks** tab. If not, then select the **Choose Columns** option from the **Column** drop-down list in the **Tasks** tab; the **Choose TimeLiner Columns** dialog box will be displayed. In this dialog box, select the **Animation** and **Animation Behavior** check boxes, refer to Figure 5-18. Choose the **OK** button; the dialog box will be closed and the columns will be added in the Task View pane of the **Tasks** tab. Next, click in the **Animation** column corresponding to the task in which you want to add the animation; a drop-down list containing all the recorded scenes will be displayed, as shown in Figure 5-18. Select the required animation scene or animation set from this drop-down list. The selected animation scene will be added to the selected task. Now when you play the simulation; the added animation will be played during the simulation of that particular task.

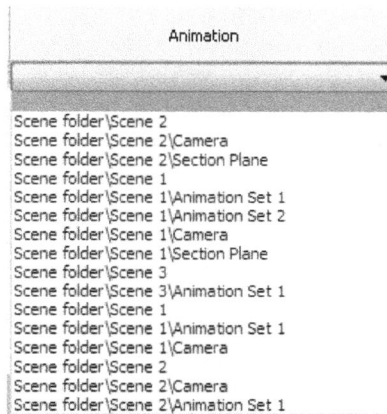

Figure 5-18 *The Animation drop-down list*

By default, the animation playback duration is in seconds and the duration of tasks can be in weeks, days or months. So, to create an effective simulation, you need to match the animation duration scale to the length of tasks. To do so, click in the **Animation Behavior** column corresponding to the task in which animation is added; a drop-down list will be displayed, as shown in Figure 5-19. This drop-down list contains three options: **Scale**, **Match Start**, and **Match End**. The **Scale** option is used to match the duration of animation to the duration of task. The **Match Start** option is used to match the start of animation with the start of task in the simulation. In this case, if the schedule is of shorter duration than the animation then the animation will continue to play until it completes. The **Match End** option is used to match the end of animation with the end of task in the simulation. In this case, the animation will stop at the same time when the simulation of the task completes.

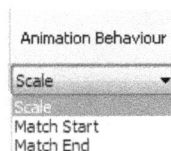

Figure 5-19 *The Animation Behavior drop-down list*

TIMELINER COSTS

There are five cost types available in **TimeLiner**: **Material Cost**, **Labor Cost**, **Equipment Cost** **Subcontractor Cost**, and **Total Cost**. These cost types are used to track the cost of a project throughout the schedule. These costs can also be imported from other softwares such as Primavera, MSP, and so on. The Total Cost is the sum of all the costs together and you cannot change it. These costs are not of any particular currency. To display these cost types in the Task View pane of the **Tasks** tab, choose the **Column** button; a drop-down list will be displayed. Select the **Choose Columns** option from this drop-down list; the **Choose TimeLiner Columns** dialog box will be displayed, as shown in Figure 5-20. In this dialog box, select the **Total Cost**, **Material Cost**, **Labor Cost**, **Equipment Cost**, and **Subcontractor Cost** check boxes to add these columns in the Task View pane. Choose the **OK** button; the dialog box will be closed, and all cost types columns will be displayed in the Task View pane of the **Tasks** tab.

You can also specify the cost value in their corresponding columns. For example, to specify the material cost, click in the required task cell in the **Material Cost** column; it will change into an edit box. Enter the amount in the edit box. Similarly,

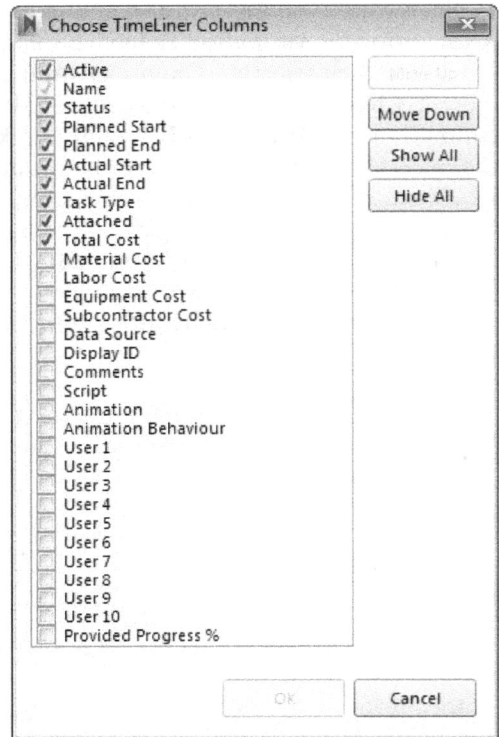

Figure 5-20 The Choose TimeLiner Columns dialog box

you can specify other costs in their corresponding columns. On specifying the cost of a task, you will notice that the total cost for that task will be displayed in the **Total Cost** column. If you make changes in a cost type, then the value in the **Total Cost** column will be automatically updated.

The costs can also be imported from external scheduling software to the Timeliner. To do so, you need to add a source in the **Data Sources** tab. To view the costs during simulation, you need to specify the cost keyword in the **Overlay Text** dialog box (as discussed earlier). In this dialog box, when you choose the **Cost** button, a menu will be displayed containing all the cost types. Select the cost type that you want to display during simulation and choose the **OK** button. Now, when you play the simulation, the corresponding cost type will be reflected in the Scene View.

IMPORTING SCHEDULES FROM EXTERNAL SCHEDULING SOFTWARE

In the **TimeLiner** window, you can import schedules from other software. For example, to import data from an external project schedule such as Microsoft Project, choose the **Add** button in the **Data Sources** tab in the **TimeLiner** window; a drop-down list will be displayed. Select **Microsoft Project 2007-2013** from the list; the **Open** dialog box will be displayed. Browse to the required folder and select the *.mpp file from the file list. Choose the **Open** button; the **Field Selector** dialog box will be displayed, as shown in Figure 5-21.

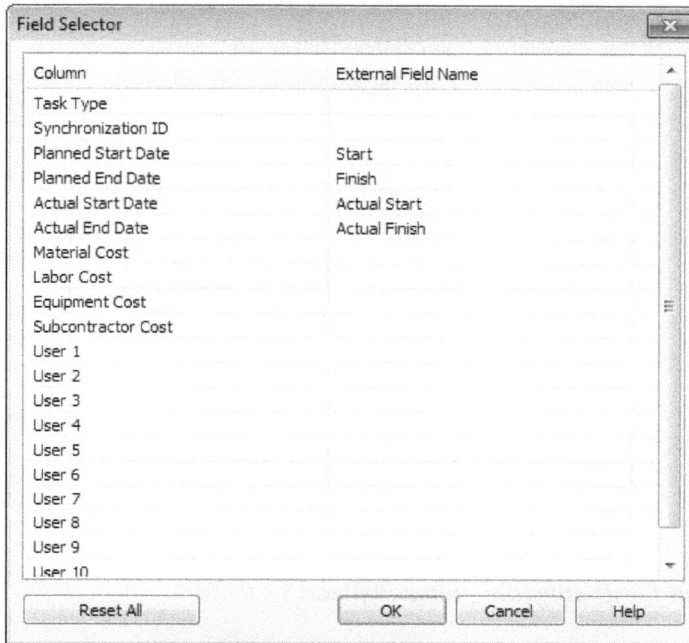

Figure 5-21 *The **Field Selector** dialog box*

In this dialog box, the left column is for the Timeliner schedule and the right column is for mapping incoming fields to the Timeliner columns. For example, to uniquely identify each imported task, click in the **External Field Name** column corresponding to the **Synchronization ID**; a drop-down list will be displayed. Select the **Unique ID** option from this list. Similarly, map other fields such as **Task Type, Planned Start Date, Planned End Date**, and so on, refer to Figure 5-21. After specifying all options, choose the **OK** button; dialog box will be closed. The data will be imported with the name of **New Data Source** in the **Data Sources** tab. To change name of the new data source, right-click on the new data source; a shortcut menu will be displayed. Select **Rename** from this menu, specify a new name, and press ENTER.

Similarly, you can import data from Primavera P6, Asta PowerProject, or CSV data files. These imported data sources can be configured. To do so, right-click on the imported data source in the **Data Sources** tab; a shortcut menu will be displayed. Choose **Edit** from this menu; the **Field Selector** dialog box will be displayed, refer to Figure 5-21. To remove the added data source from the **Data Sources** tab, select the source and choose the **Delete** button. To create tasks from a data source, choose the **Refresh** button in the **Data Sources** tab; a drop-down list will be displayed. In this drop-down list, select the **Selected Data Source** option to create tasks from a selected data source. Select the **All Data Sources** option to create tasks from all the data sources. On selecting any of the two options, the **Refresh from Data Source** dialog box will be displayed, as shown in Figure 5-22. There are two options in this dialog box: **Rebuild Task Hierarchy** and **Synchronize**. To import all the related data from the external schedule to the **Tasks** tab, select the **Rebuild Task Hierarchy** radio button. It will add all the external data to the **Task** tab. To update all the existing tasks with the new external data, select the **Synchronize** radio button. You can also access these options by right-clicking on the imported data source in the **Data Sources** tab. On right-clicking, a shortcut menu will be displayed. Select the **Rebuild Task Hierarchy** or **Synchronize** option from this menu as required.

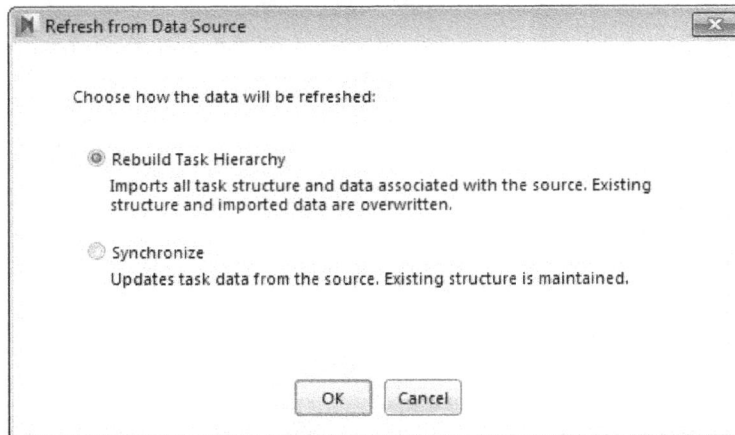

*Figure 5-22 The **Refresh from Data Source** dialog box*

TUTORIALS

General instructions for downloading tutorial files:

1. Download the *c05_nws_2017_tut* zip file for this tutorial from *http://www.cadcim.com*. The path of the file is as follows: *Textbooks > Civil/GIS > Navisworks > Exploring Autodesk Navisworks 2017.*

2. Now, save and extract the downloaded folder at the following location: *C:\ nws_2017\ c05_nws_2017_tut.*

Note
*The default unit system used in the tutorials is metric. To change the units to imperial, select the required units from **Options Editor > Interface > Display Units**.*

Tutorial 1	Creating 4D Simulation

In this tutorial, you will open the *c05_navisworks_2017_tut1.nwf* file and create 4D simulation by using various options available in the **TimeLiner** window. The schedule used in this tutorial has been created in Microsoft Project Professional 2010.   **(Expected time: 45min)**

The following steps are required to complete this tutorial:

a.   Open the file *c05_navisworks_2017_tut1.nwf.*
b.   Display the **TimeLiner** window.
c.   Import the Microsoft project file.
d.   Attach model objects to the tasks.
e.   Assign task types.
f.   Specify the simulation parameters.
g.   Play the simulation.
h.   Save the project.

Opening the File

In this section, you will open the file to be used in this tutorial.

1. Choose the **Open** button from the Quick Access Toolbar; the **Open** dialog box is displayed.

2. In this dialog box, browse to the following location:
    C:\nws_2017\c05_nws_2017_tut.

3. Select **Navisworks File Set (*.nwf)** from the **Files of type** drop-down list.

4. Select the file **c05_navisworks_2017_tut1** from the displayed list of files; the file name is displayed in the **File name** edit box.

5. Next, choose the **Open** button in the dialog box; the model is displayed in the Scene View, as shown in Figure 5-23.

Figure 5-23 *Model displayed in the Scene View*

Displaying the TimeLiner Window

In this section, you will display the **TimeLiner** window.

1. Choose the **TimeLiner** tool from the **Tools** panel of the **Home** tab; the **TimeLiner** window is displayed, as shown in Figure 5-24.

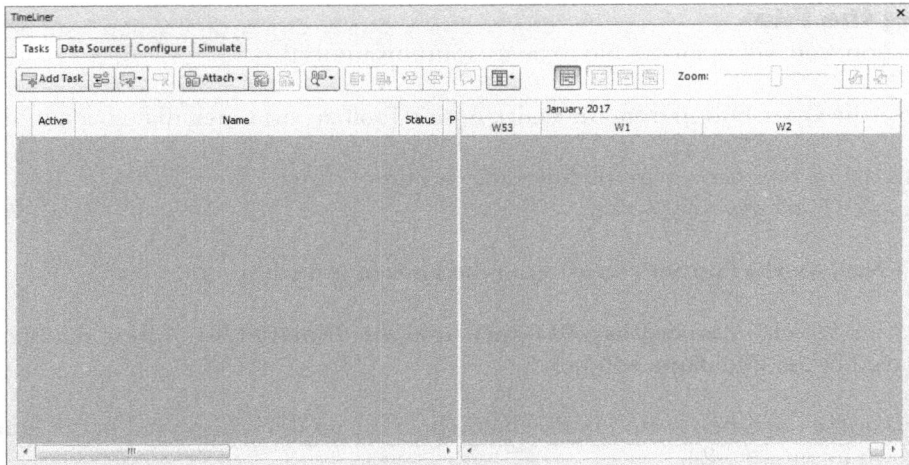

Figure 5-24 The TimeLiner window

Importing the Schedule Created in Microsoft Project Professional

In this section, you will import the construction schedule created in the Microsoft Project software.

1. In the **TimeLiner** window, choose the **Data Sources** tab.

2. In the **Data Sources** tab, choose the **Add** button; a drop-down list is displayed.

3. Select the **Microsoft Project 2007-2013** option from the drop-down list; the **Open** dialog box is displayed.

4. In this dialog box, browse to the following location: C:\nws_2017\c05_nws_2017_tut.

5. Select **Construction_Project** from the dialog box, as shown in Figure 5-25. Now, choose the **Open** button; the **Field Selector** dialog box is displayed.

> **Note**
> *You need to install the Microsoft Project 2010 or any other version on the system to perform this tutorial.*

6. In the **Field Selector** dialog box, click in the **External Field Name** column corresponding to **Synchronization ID**; a drop-down list is displayed, refer to Figure 5-26.

7. Next, select the **ID** option from the drop-down list, as shown in Figure 5-26.

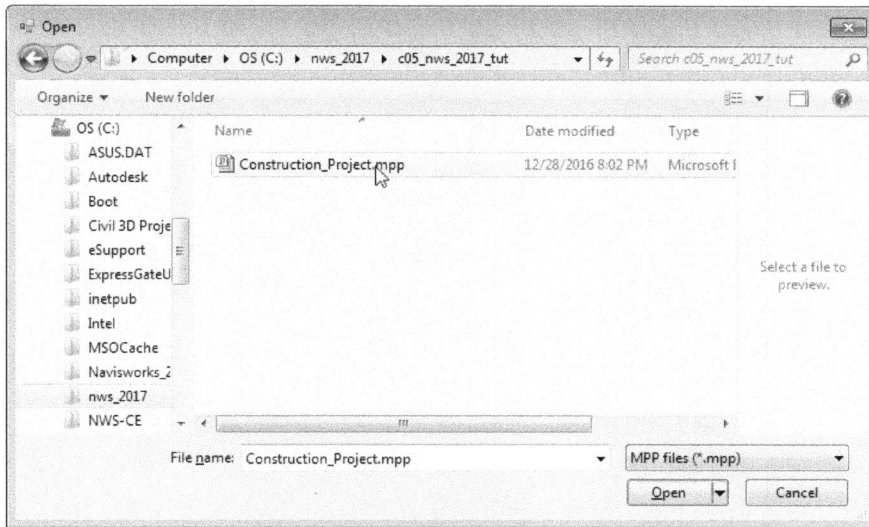

Figure 5-25 *Selecting the* ***Construction_Project*** *option from the* ***Open*** *dialog box*

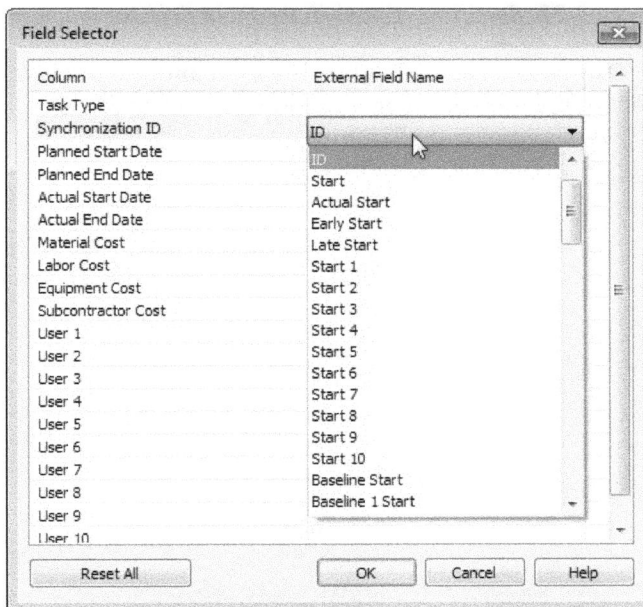

Figure 5-26 *Selecting the* ***ID*** *option for* ***Synchronization ID***

8. Similarly, specify the options for **Planned Start Date**, **Planned End Date**, **Actual Start Date**, **Actual End Date**, and **Material Cost**, refer to Figure 5-27.

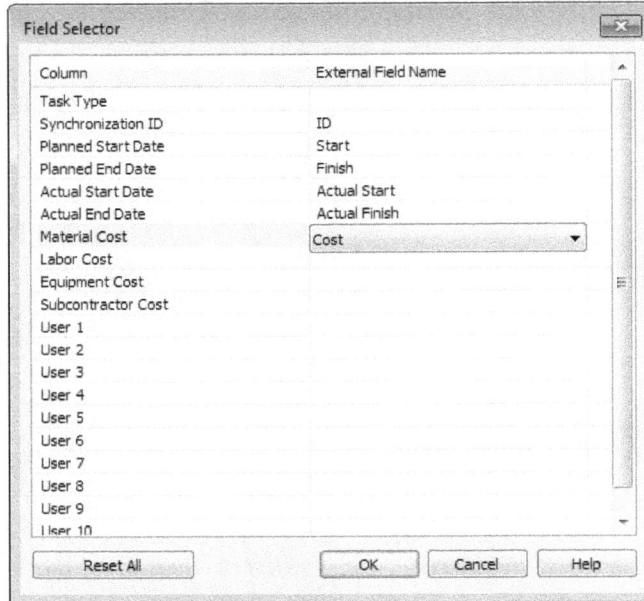

Figure 5-27 *Specifying options in the **Field Selector** dialog box*

9. Next, choose the **OK** button; the dialog box is closed and the new data source is added in the **Data Sources** tab of the **TimeLiner** window, refer to Figure 5-28.

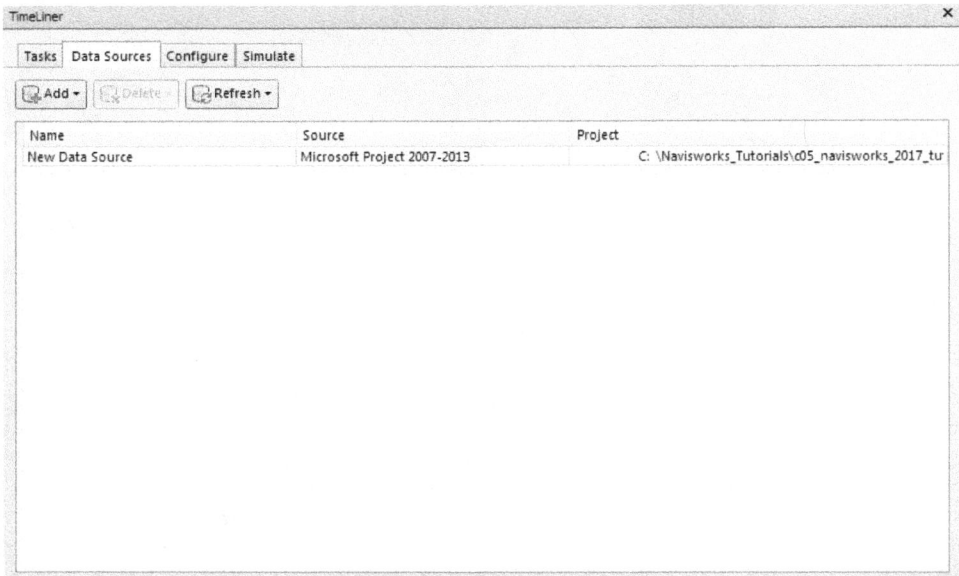

Figure 5-28 *New **Data Source** added in the **Data Sources** tab*

10. Next, select the **New Data Source** text under the **Name** column and right-click on it; a shortcut menu is displayed.

11. Choose the **Rebuild Task Hierarchy** option from the menu, as shown in Figure 5-29.

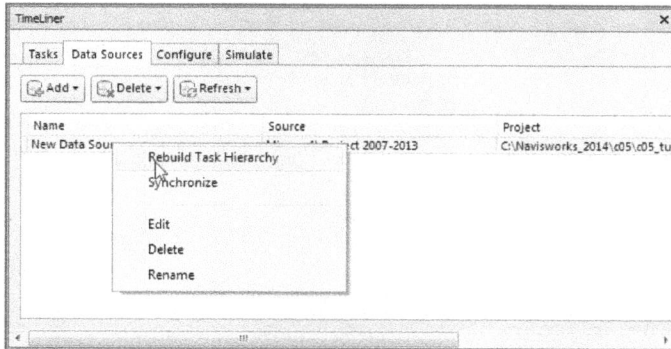

*Figure 5-29 Choosing the **Rebuild Task Hierarchy** option from the shortcut menu*

12. Next, choose the **Tasks** tab; the schedule is displayed in the **Tasks** tab, refer to Figure 5-30.

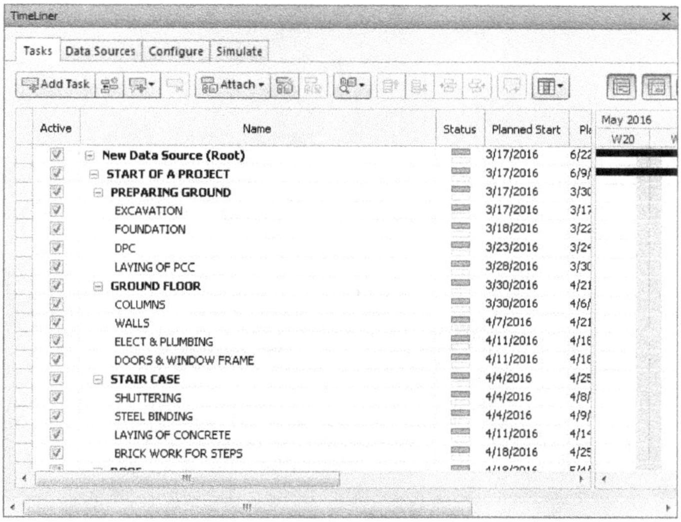

*Figure 5-30 Schedule displayed in the **Tasks** tab*

Attaching Model Objects to the Tasks
In this section, you will attach the model objects to the tasks.

1. Select the **Manage Sets** option from the **Sets** drop-down list in the **Select & Search** panel of the **Home** tab; the **Sets** window is displayed, as shown in Figure 5-31.

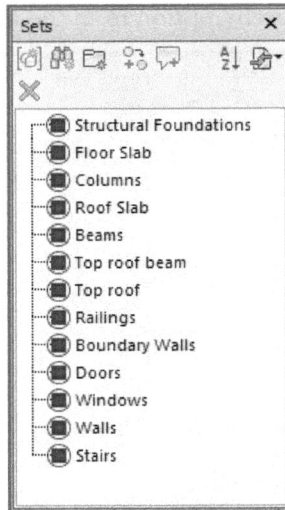

Figure 5-31 *The **Sets** window*

2. Select **Structural Foundations** in the **Sets** window; the selected objects are highlighted in the Scene View.

3. Choose the **Hide Unselected** button from the **Visibility** panel of the **Home** tab; the unselected objects are hidden in the Scene View and only the selected foundations are displayed, as shown in Figure 5-32.

Figure 5-32 *Selected foundations displayed in the Scene View*

4. In the **TimeLiner** window, select the **FOUNDATION** task under the **Name** column in the **Tasks** tab.

5. Now, select the **Attach Current Selection** option from the **Attach** drop-down list, refer to Figure 5-33; the selected structural foundations are attached to the **FOUNDATION** task.

6. Next, choose the **Hide Unselected** button from the **Visibility** panel of the **Home** tab; the unselected objects become visible in the Scene View.

7. Select **Floor Slab** in the **Sets** window.

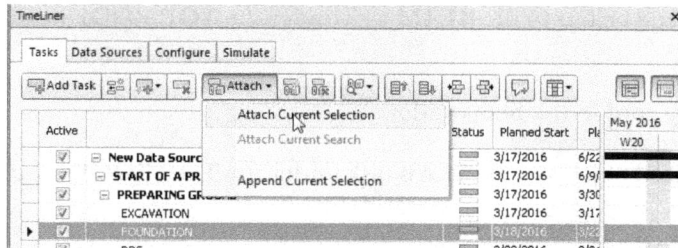

Figure 5-33 *Selecting the **Attach Current Selection** option from the list*

8. Select **DPC** and press the CTRL key, and then select the **LAYING OF PCC** task in the **TimeLiner** window.

9. Next, select the **Attach Current Selection** option from the **Attach** drop-down list; the selected objects are attached to the tasks.

10. Now, select **Columns** in the **Sets** window.

11. In the **TimeLiner** window, under the **Ground Floor** head, select the **Columns** sub task and then select the **Attach Current Selection** option from the drop-down list; the selected columns are attached to the task.

12. Next, select **Walls** in the **Sets** window.

13. Select the **Walls** sub task under the **Ground Floor** head in the **TimeLiner** window and then select the **Attach Current Selection** option from the drop-down list; the selected walls are attached to the task.

14. Select **Stairs** in the **Sets** window.

15. Select the **STAIR CASE** task in the **TimeLiner** window and then select the **Attach Current Selection** option from the drop-down list; the stairs are attached to the task.

16. Select **Beams** in the **Sets** window.

17. Attach the selected beams to the **BEAM** task under the **ROOF** head in the **TimeLiner** window.

18. Select **Roof Slab** in the **Sets** window; the roof slab is highlighted in the Scene View.

19. Next, attach the selected object to the **SLAB** task under the **ROOF** head in the **TimeLiner** window and select the **Attach Current Selection** option from the drop-down list; the roof slab is attached to the task.

20. Select **Top roof beam** in the **Sets** window.

21. Attach the selected beams to the **BEAM** task under the **TOP ROOF** head in the **TimeLiner** window.

22. Next, select **Top roof** in the **Sets** window.

23. Attach the selected objects to the **SLAB** task under the **TOP ROOF** head in the **TimeLiner** window.

24. Select **Railings** in the **Sets** window; the selected objects are highlighted in the Scene View.

25. Attach the selected railing to the **RAILING** task in the **TimeLiner** window.

26. Select **Boundary Walls** in the **Sets** window and assign it to **BOUNDARY WALL** task in the **TimeLiner** window.

27. Select **Doors** and **Windows** in the **Sets** window and assign them to **DOORS** and **WINDOWS GLASS** task, respectively under the **FINISHING** head in the **TimeLiner** window.

Assigning Task Types
In this section, you will assign the task type to the required tasks.

1. In the **TimeLiner** window, click in the **Task Type** column corresponding to the **FOUNDATION** task in the **Tasks** tab; a drop-down list will be displayed.

Note
*To make the **Task Type** column visible in the **TimeLiner** window, it is recommended to expand the table displayed in it.*

2. Select the **Construct** option from the list.

3. Repeat the procedure followed in steps 1 and 2 and assign the **Construct** task type to the following tasks: **DPC, LAYING OF PCC, COLUMNS, WALLS, STAIR CASE, BEAM** and **SLAB**; **BEAM** and **SLAB** under the **ROOF** task; **BEAM, SLAB, RAILING, BOUNDARY WALL** under the **TOP ROOF** task; **DOORS**, and **WINDOW GLASS** tasks under the **FINISHING** task.

Specifying the Simulation Parameters
In this section, you will specify the simulation parameters.

1. Choose the **Configure** tab in the **TimeLiner** window and ensure that **Green (90% Transparent)** is specified in the **Start Appearance** column corresponding to the **Construct** task type.

2. Next, choose the **Settings** button in the **Simulate** tab; the **Simulation Settings** dialog box is displayed.

3. Specify the playback duration as **20seconds** in the **Playback Duration** edit box, refer to Figure 5-34.

4. In the **Start/End Dates** area, select the **Override Start / End Dates** check box; the **Start Date** and **End Date** edit boxes are enabled.

5. In the **Start Date** edit box, click on the down arrow; a calendar is displayed.

6. Select the date 13th July, 2015 from the flyout; the date is displayed in the edit box.

7. Repeat the procedure followed in steps 5 and 6 and specify the date 17th October 2015 in the **End Date** edit box.

8. Specify the time **08:00:00 AM** and **5:00:00 PM** in the **Start Date** and **End Date** edit boxes, respectively, refer to Figure 5-34.

9. Next, select the **Actual** radio button in the **View** area.

Figure 5-34 The Simulation Settings dialog box

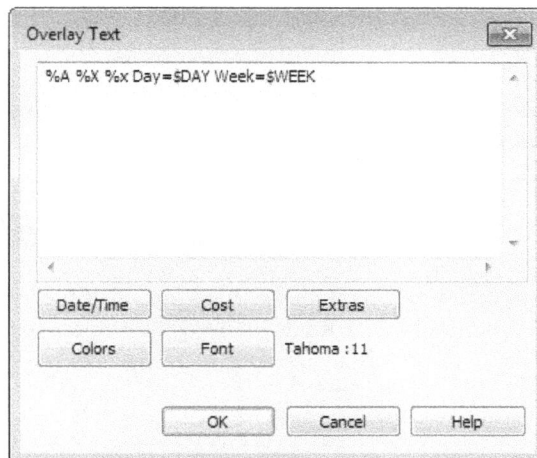

10. Next, choose the **Edit** button in the **Overlay Text** area; the **Overlay Text** dialog box is displayed, refer to Figure 5-35.

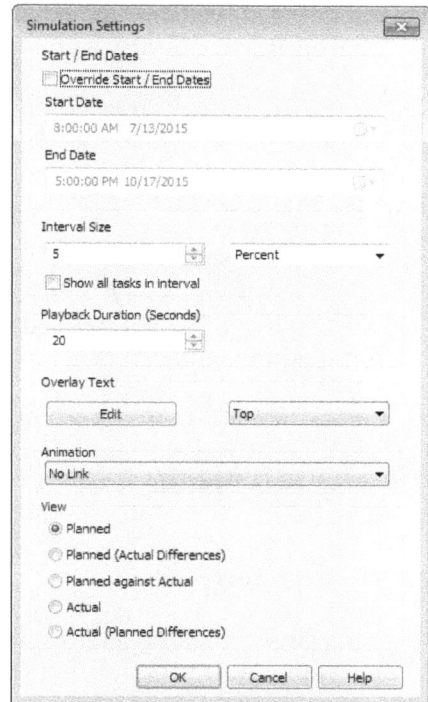

Figure 5-35 The Overlay Text dialog box

11. Delete the text displayed in the text box.

12. Choose the **Date/Time** button; a flyout is displayed. Select the **Date and time representation appropriate for locale** option from the flyout.

13. Next, choose the **Extras** button; a menu is displayed. Select the **Currently active tasks** and **Number of days since start** options from the menu.

14. Choose the **Cost** button; a menu is displayed. Select **Total Cost** from the menu.

15. Choose the **OK** button in the **Overlay Text** dialog box; the dialog box is closed.

16. Next, select the **Show all tasks in interval** check box in the **Interval Size** area of the **Simulation Settings** dialog box.

17. Choose the **OK** button; the **Simulation Settings** dialog box is closed.

Playing the Simulation
In this section, you will play the simulation of construction sequence.

1. Close the **Sets** window.

2. Now, choose the **Play** button available at the top in the **Simulate** tab; the simulation starts playing.

Saving the Project
In this section, you will save a project.

1. To save the project with the current view, choose **Save As** from the Application Menu, the **Save As** dialog box is displayed.

2. Browse to the *Navisworks_2017* folder and enter **c05_navisworks_2017_tut01** in the **File name** edit box.

3. Choose the **Navisworks File Set (*.nwf)** file format from the **Save as type** drop-down list, and then choose the **Save** button.

Tutorial 2   4D Sequence with Crane Animation

In this tutorial, you will open the *c05_navisworks_2017_tut2.nwf* file, and create 4D simulation of a building project, and link an object animation with the sequence.

(Expected time: 30min)

The following steps are required to complete this tutorial:

a. Open the *c05_navisworks_2017_tut2.nwf* file.
b. Create construction schedule in the TimeLiner window.
c. Link the crane animation with the task.

d. Specify the simulation parameters.
e. Play the simulation.
f. Save the project.

Opening the File

In this section, you will open the file to be used in this tutorial.

1. Choose the **Open** button from the Quick Access Toolbar; the **Open** dialog box is displayed.

2. In this dialog box, browse to the following location:
    C:\ nws_2017\c05_nws_2017_tut

3. Select **Navisworks File Set (*.nwf)** from the **Files of type** drop-down list; a list of files is displayed.

4. Select the file **c05_navisworks_2017_tut2** from the list; the file name appears in the **File name** edit box.

5. Next, choose the **Open** button in the dialog box; the model is displayed in the Scene View.

Creating Construction Schedule

In this section, you will create the construction schedule based on the objects saved as selection sets. Here, the tasks will be created using the Auto-Add Tasks method.

1. Choose the **TimeLiner** tool from the **Tools** panel of the **Home** tab; the **TimeLiner** window is displayed in the Scene View.

2. In the **Tasks** tab of the window, choose the **Auto-Add Tasks** button; a drop-down list is displayed.

3. Select the **For Every Set** option from this list; the tasks are created in the Task View area.

After creating the tasks, you need to arrange them in a construction order and assign the start and end dates of each task.

4. Select the task **Columns** in the Task View area, and click the **Move Up** button twice to move this task up in the list.

5. Repeat the procedure followed in step 4 and arrange the remaining tasks, refer to Figure 5-36.

6. Specify the start and end dates of each task, refer to Figure 5-36.

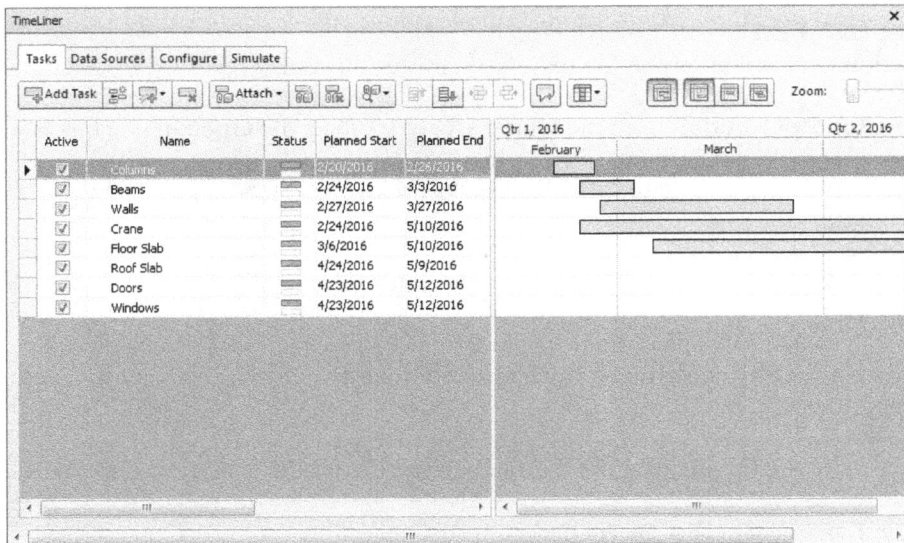

*Figure 5-36 Crane task added in the Task View area of the **TimeLiner** window*

Linking the Crane Animation with the Crane Task

In this section, you will link the crane animation which has been created using the options available in the **Animator** window.

1. Choose the **Columns** button and select the **Choose Columns** option from the list displayed; the **Choose TimeLiner Columns** dialog box is displayed.

2. In this dialog box, select the **Animation** and **Animation Behavior** check boxes and choose the **OK** button; the **Animation** and **Animation Behavior** columns are added in the Task View area.

3. Now, select the **Crane** task and click in the **Animation** column; a drop-down list is displayed.

4. Select the **Scene folder\Tower Crane** option from this list and ensure that the **Scale** option is selected in the **Animation Behavior** column.

5.   Next, click in the **Task Type** column for the **Crane** task; a drop-down list is displayed.

6. Select the **Temporary** option from the list.

Specifying the Simulation Parameters

In this section, you will specify the simulation parameters.

1. Choose the **Simulate** tab in the **TimeLiner** window and then choose the **Settings** button; the **Simulation Settings** dialog box is displayed.

2. In the **Overlay Text** area of this dialog box, choose the **Edit** button; the **Overlay Text** dialog box is displayed.

3. Choose the **Extras** button; a flyout is displayed. Select the **Number of days since start** option from the flyout.

4. Choose the **Colors** button and select the **Red** option from the flyout. Choose the **OK** button to apply the changes.

5. Retain the other settings in the **Simulation Settings** dialog box and choose the **OK** button to close it.

Playing the Simulation

In this section, you will play the simulation.

1. Choose the **Play** button in the **Simulate** tab of the **TimeLiner** window. Figure 5-37 shows various sequences while playing the simulation along with the animation.

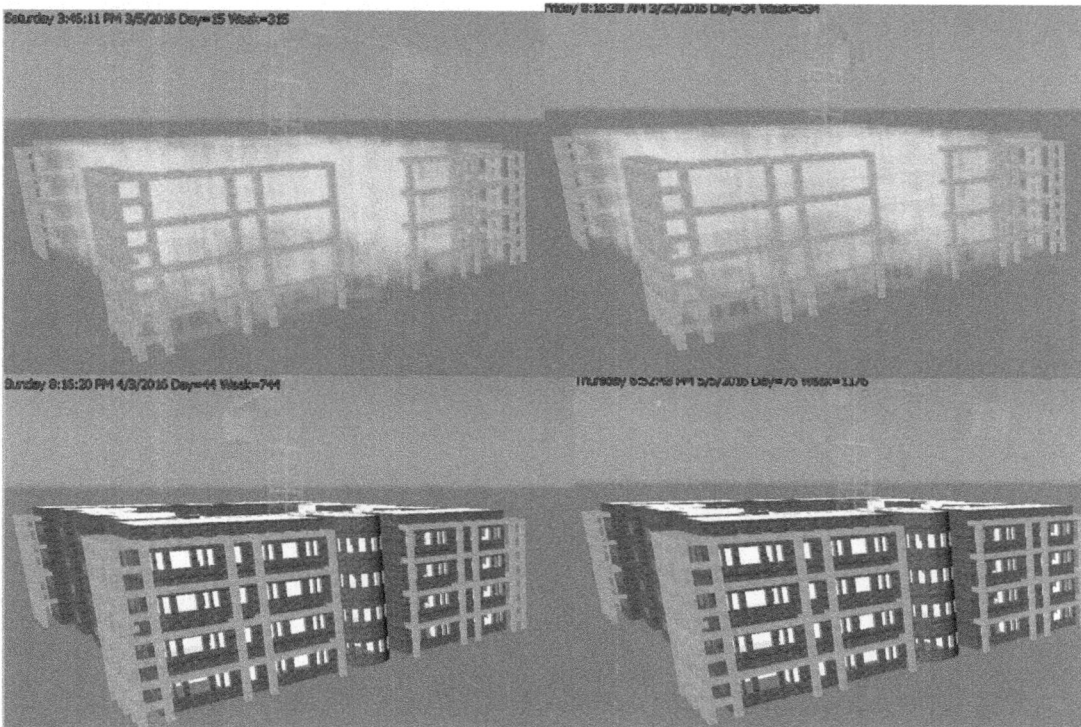

Figure 5-37 Sequence of the simulation and animation

Saving the Project

In this section, you will save the project.

1. To save the project with the current view, choose **Save As** from the Application Menu; the **Save As** dialog box is displayed.

2. Browse to the *nws_2017 \ c05_nws_2017_tut* folder and enter **c05_navisworks_2017_tut02** in the **File name** edit box.

3. Now, select the **Navisworks File Set (*.nwf)** file format from the **Save as type** drop-down list, and choose the **Save** button.

Self-Evaluation Test

Answer the following questions and then compare them to those given at the end of this chapter:

1. You can use the _____ option to export the Timeliner schedule in CSV format.

2. The _____ button is used to filter the tasks based on their status.

3. The _____ option is used to rebuild the task hierarchy with the latest data.

4. The _____ option is used to update the existing schedule with the latest data.

5. You can use the _____ task type for the tasks to be constructed during simulation.

6. In Navisworks, you can create selection sets of the objects attached in the Timeliner hierarchy. (T/F)

7. In the **TimeLiner** window, you can add comments to the tasks. (T/F)

8. You can attach same objects to multiple tasks. (T/F)

9. Timeliner rules cannot be imported or exported. (T/F)

10. Comments cannot be added to Timeliner tasks. (T/F)

Review Questions

Answer the following questions:

1. Which of the following options is used to add tasks in the **TimeLiner** window?

   a) **Refresh**   b) **Add Task**
   c) **Attach**   d) **Delete**

2. Which of the following tabs contains the options for importing the schedules?

   a) **Simulate**   b) **Data Sources**
   c) **Configure**   d) **Tasks**

3. Which of the following buttons is used to invoke the **Simulation Settings** dialog box?

   a) **Refresh**   b) **Clear Attachment**
   c) **Settings**   d) **Simulate**

4. Which of the following options is used to create selection sets of objects attached to the tasks?

   a) **Filter by Status**   b) **Export To Sets**
   c) **Add Comment**   d) **Update**

5. Which of the following options is used to define the task as temporary?

   a) **Construct**   b) **Demolish**
   c) **Temporary**   d) **Synchronize**

6. Model appearance during simulation cannot be configured. (T/F)

7. You cannot change the color of text displayed during the simulation. (T/F)

8. You cannot link camera animation to the tasks in Timeliner. (T/F)

9. In the Timeliner, you cannot import schedules from other software. (T/F)

10. The **Clear Attachment** button is used to detach the object from the task. (T/F)

EXERCISE

Exercise 1   Creating 4D Simulation

Download and open the *c05_navisworks_2017_ex1* file from *http://www.cadcim.com*. Create a 4D simulation by using the options available in the **TimeLiner** window. Figure 5-38 shows the model to be used in this exercise.   **(Expected time: 1hr)**

The following steps are required to complete this exercise:

a.   Open the file *c05_navisworks_2017_ex1*.
b.   Import the *Construction_exercise* schedule created in Microsoft Project Professional.
c.   Attach model objects to the tasks and specify the simulation parameters.
d.   Play the simulation.
e.   Save the project as *c05_navisworks_2017_ex01*.

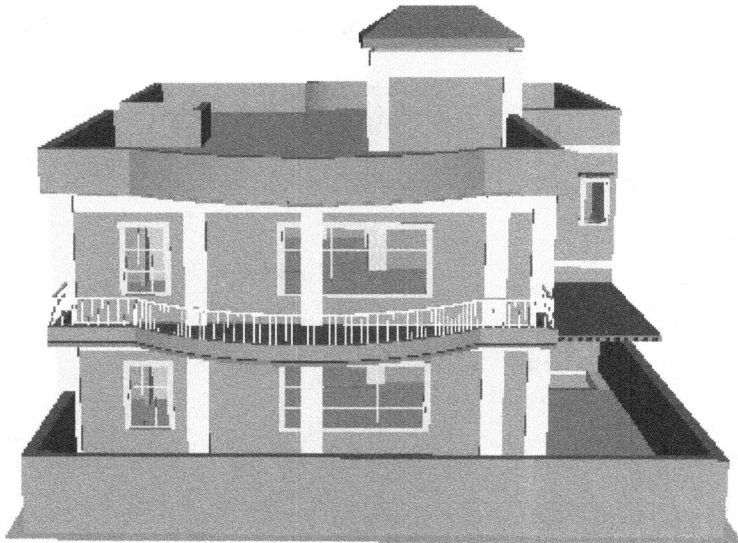

Figure 5-38 The Residence Building

Chapter 6

Working with Animator and Scripter

Learning Objectives

After completing this chapter, you will be able to:
- *Use the Animator window*
- *Use the Scripter window*
- *Enable scripts*

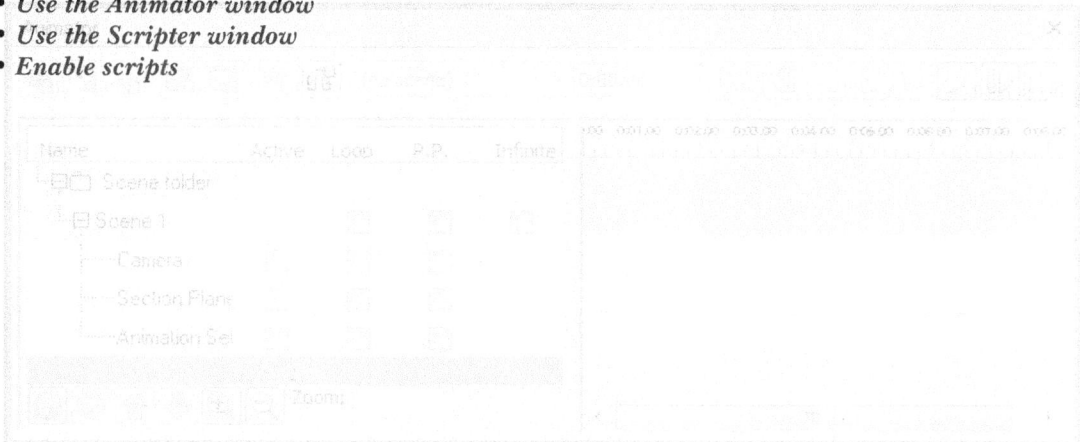

INTRODUCTION

In the previous chapter, you learned about saving viewpoints, creating viewpoint animation, and creating cross-sections of the model. In this chapter, you will learn to create object animation and adding scripts to the animation by using the options available in the **Animator** and **Scripter** windows. These two windows comprise several options that are used for creating object animation. The options in these windows are discussed in detail in this chapter.

THE ANIMATOR WINDOW

In Navisworks, the **Animator** window is provided to create object animation by capturing every movement of the given object. The options available in the **Animator** window allow you to create several types of animations. To invoke the **Animator** window, choose the **Animator** tool from the **Tools** panel of the **Home** tab; the **Animator** window will be displayed in the Scene View, as shown in Figure 6-1. There are three areas in this window: Animator toolbar, Animator tree view, and Animator timeline View. These areas are discussed next.

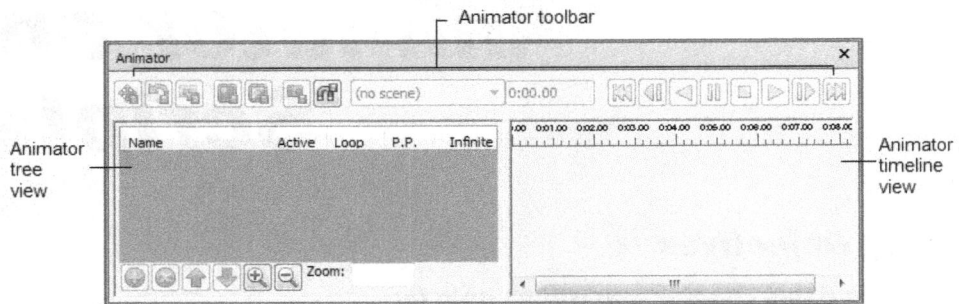

*Figure 6-1 The **Animator** window*

Animator Toolbar

The Animator toolbar is available at the top in the **Animator** window. It contains buttons and options used while creating object animation, as shown in Figure 6-2. These buttons are discussed next.

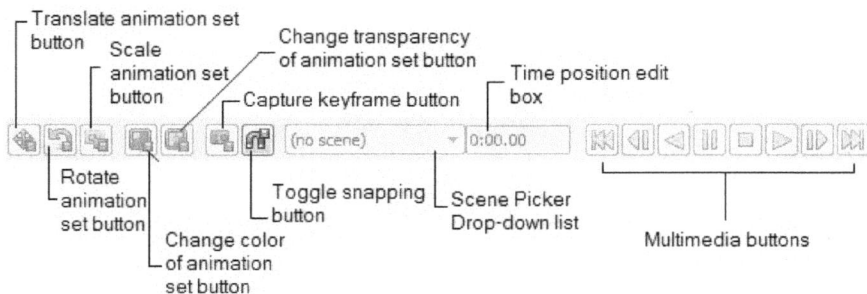

Figure 6-2 Buttons and options in the Animator toolbar

Translate animation set

The **Translate animation set** button is used to put the animator in translation mode. On choosing this button, a Move gizmo will be displayed in the Scene View and the **Translate** palette will be displayed at the bottom in the **Animator** window. Using the Move gizmo, you can make the required translational changes in the selected object. You can also specify the values of the X, Y, and Z coordinates in the respective edit boxes in the **Translate** palette to change the position of the selected objects.

Rotate animation set

The **Rotate animation set** button is used to make the rotational changes in the selected objects. On choosing this button, a Rotate gizmo will be displayed in the Scene View and the **Rotate** palette will be displayed at the bottom in the **Animator** window. Using the Rotate gizmo, you can rotate the selected object. Alternatively, specify values for the degree of rotation about X, Y, and Z axes in the **X**, **Y**, and **Z** edit boxes. To change the origin of rotation, enter the values in the **cX**, **cY**, and **cZ** edit boxes. To modify the orientation of the rotation, specify the values in the **oX**, **oY**, and **oZ** edit boxes.

Scale animation set

The **Scale animation set** button is used to modify the size of the selected object. On choosing this button, the Scale gizmo will be displayed in the Scene View and the **Scale** palette will be displayed at the bottom in the **Animator** window. Using the Scale gizmo, you can modify the size of the selected object. You can also specify the scale factor along X, Y, and Z axes in the respective edit boxes. To change the origin of scaling, specify the values in the **X**, **Y**, and **Z** edit boxes.

Change color of animation set

The **Change color of animation set** button is used to modify the color of the selected objects. On choosing this button, a **Color** palette will be displayed at the bottom in the **Animator** window. In this palette, select the **Color** check box to record color changes on capturing the keyframe or clear the check box to reset the color to its original state. To modify the color, specify the red, green, and blue values in their respective edit boxes or click on the color drop-down next to the edit boxes and select the desired color from it.

Change transparency of animation set

The **Change transparency of animation set** button is used to adjust the transparency of the selected object. On choosing this button, a **Transparency** palette will be displayed at the bottom in the **Animator** window. In this palette, select the **Transparency** check box to record transparency changes on capturing the keyframe. You can clear this check box to reset transparency. You can also specify the value for changing transparency in the edit box next to the **Transparency** check box or drag the slider to adjust the transparency level.

Note
The buttons discussed above are activated only after creating and selecting the animation set in the Animator tree view which is discussed in detail later in this chapter.

Capture keyframe

The **Capture keyframe** button is used to capture the current change to the model as a keyframe. On choosing this button, a keyframe will be created in the Animator timeline view.

Toggle snapping

The **Toggle snapping** button is used to enable or disable snapping while moving the object using gizmo.

Scene Picker

The **Scene Picker** drop-down list is available next to the **Toggle snapping** button, refer to Figure 6-2. It contains a list of all the scenes added in the **Animator** window. You can select an option from this drop-down list to make the selected scene active.

Time position

The **Time position** edit box is available next to the **Scene Picker** drop-down list, refer to Figure 6-2. This edit box shows the current position of the time slider in the Timeline view. You can modify the current position of the time slider by specifying a value in the **Time position** edit box.

Multimedia

You can use various multimedia buttons available in the Animator toolbar for controlling the animation, refer to Figure 6-2. These buttons are **Rewind**, **Pause**, **Stop**, **Step Forward**, **Step Backward**, and **To End**.

Note
The above discussed buttons are activated only after scenes are added to the Animator tree view.

Animator Tree View

The Animator tree view is available in the left pane of the **Animator** window, refer to Figure 6-1. The Animator tree view contains a list of all the added scenes and scene components such as animation set, camera, and section planes. These scenes and scene components can be arranged in a hierarchy and grouped in a scene folder. The scenes and scene folders are created by using the buttons available at the bottom of the Animator tree view. These buttons are discussed next.

Add scene

The **Add scene** button is used to add various scenes, scene folders, and scene components to the Animator tree view. To do so, choose the **Add scene** button; a flyout will be displayed with various options that are discussed next.

The **Add Scene Folder** option is used to create a scene folder. When you choose the **Add Scene Folder** option from the flyout, a folder with the name **Scene Folder** will be added to the tree view. The name of the folder will be in editable mode. Specify a new name in the edit box and press ENTER; the name of the folder will get updated.

The **Add Scene** option in the flyout can be used to add a scene to the scene folder. You can change the name of the scene. To do so, click on the scene name; an edit box will be displayed. Type the desired name in the edit box and press ENTER; the scene name will be updated.

Next, you can add various scene components to the added scene. To do so, first select the added scene and then choose the **Add scene** button; a flyout will be displayed. This flyout contains various options for adding the scene components such as **Add Camera**, **Add Animation Set**, and **Add Section Plane**. The **Add Camera** option is used to add a new camera to the added scene which will contain a list of viewpoints and keyframes. The **Add Animation Set** option is used to add an animation set to the added scene which will contain the objects to be animated and a list of keyframes. The **Add Section Plane** option is used to add a section plane set to the added scene which will contain a list of cross-sectional cuts and a list of keyframes. These options are discussed in detail further in the chapter.

After adding scenes and their components you will notice that some check boxes are displayed in the Animator tree view, refer to Figure 6-3. These check boxes are used to control the animation.

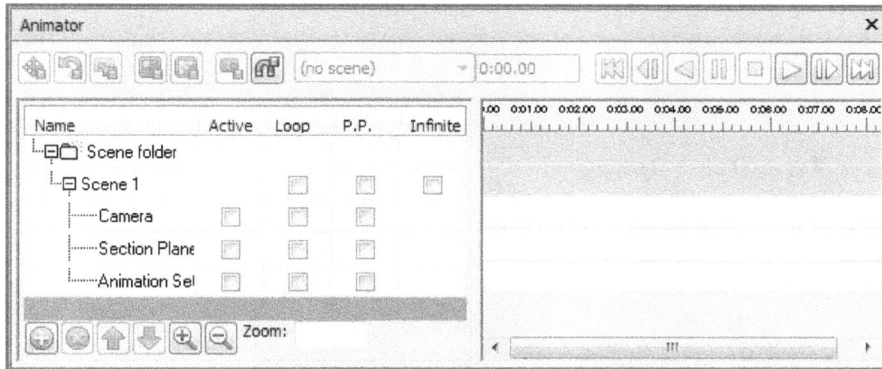

Figure 6-3 *Animator tree view in the* ***Animator*** *window*

Active
The check box in the **Active** column is used to make the animation active in the scene and only the activated animation will play in the Scene View. To make the animation active, select the check box corresponding to the required scene component in the **Active** column, refer to Figure 6-3.

Loop
The check box in the **Loop** column is used to activate the loop mode. In this mode, the animation will play continuously from start to end. To activate it, select the check box corresponding to the required scene component in the **Loop** column, refer to Figure 6-3.

P.P.
The check box in the **P.P.** column is used to activate ping-pong mode. In this mode, first the animation will play from start to end and then from end to start. To activate it, select the check box corresponding to the required scene component in the **P.P.** column, refer to Figure 6-3.

Infinite

The check box in the **Infinite** column is used to play the animation indefinitely. To play the animation indefinitely, select the check box corresponding to the required scene in the **Infinite** column, refer to Figure 6-3. Note that if this check box is selected, then the **Loop** and **P.P.** options will not work.

Delete

The **Delete** button is used to delete a scene folder, scene or any of its components. To do so, first select the required scene or scene components in the Animator tree view and then choose the **Delete** button; the selected scene will be removed.

Move up

The **Move up** button is used to move the selected scene up in the Animator tree view. To do so, select the required scene in the Animator tree view and then choose the **Move up** button; the selected scene will be shifted up.

Move down

The **Move down** button is used to move the selected scene down in the Animator tree view. To do so, select the required scene in the Animator tree view and then choose the **Move down** button; the selected scene will be shifted down.

Animator Timeline View

The right pane of the **Animator** window displays the Animator timeline view which refers to timeline with keyframes. It displays the length of overall animation in a bar chart view, as shown in Figure 6-4.

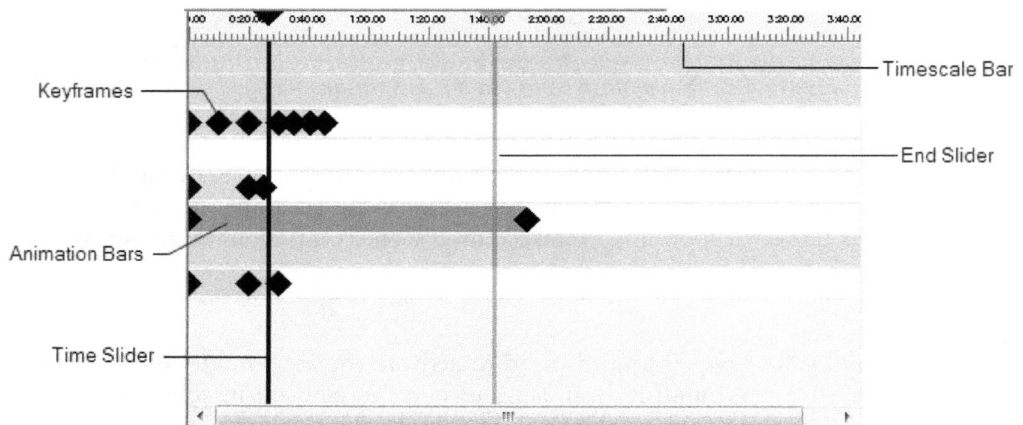

Figure 6-4 *The timeline view in the* ***Animator*** *window*

By default, a timescale bar is displayed at the top in the Animator timeline view, refer to Figure 6-4. After adding scenes and scene components you can create animation. When you create animation, keyframes, Time slider, End slider, and Animation Bars will be displayed in the Animator timeline view. These are discussed next.

Timescale Bar

In the timeline view, a timescale bar is displayed at the top of the timeline view, refer to Figure 6-4. This timescale bar displays time in seconds. You can enlarge the viewing scale of the time scale bar by choosing the **Zoom in** button available at the bottom in the Animator tree view. To reduce the viewing scale, choose the **Zoom out** button. Alternatively, you can specify a value in the **Zoom** edit box to reduce or enlarge the timescale bar.

Keyframes

While creating object animation, you need to capture the current changes in the model by choosing the **Capture keyframe** button from the Animator toolbar. The capturing generates keyframes in the Animator timeline view in the shape of black diamonds, refer to Figure 6-4. These keyframes are generated between the start time and the end time of the animation to provide a smooth transition when an animation is played. It defines the position of the model in the animation. The time duration of a keyframe can be changed by dragging it in the left or right direction.

> **Note**
> *You can also insert intermediate keyframes between two keyframes. To do so, drag and place the Time Slider between the two keyframes, adjust the viewpoint, and choose the* **Capture keyframe** *button; the keyframe will be generated.*

Time Slider

In the Animator timeline view, the time slider is a black vertical bar which shows the current position of an animation in the playback, refer to Figure 6-4. The time slider can be adjusted by dragging in the left or right direction. Alternatively, you can also use the multimedia buttons to change the position of the time slider.

End Slider

In the Animator timeline view, the end slider is a red vertical bar. The end slider shows the end position of an animation, refer to Figure 6-4. By default, end slider is always placed at the last keyframe. You cannot change the position of the end slider by dragging. To change its position, right-click on the end slider; a shortcut menu will be displayed. Choose the **Manually position endbar** option from the shortcut menu. Now, drag the slider in the left or right direction as required. If the **Infinite** check box is selected in the tree view, the end slider will be hidden in the Animator timeline view.

Animation Bars

The Animation bars are the colored bars located between consecutive keyframes, refer to Figure 6-4. Each animation type has a different color bar and you cannot edit these bars.

CREATING OBJECT ANIMATIONS

The options available in the **Animator** window are used for creating the object animations. Before creating an animation, you need to create a scene folder which will contain all the scene components. The method of creating a scene folder has been discussed earlier in this chapter. Next, you will create a scene in the scene folder. To create a scene, choose the **Add Scene** button; a flyout will be displayed. Choose the **Add Scene** option from the flyout; a scene will be created.

In this scene, you can create object animations using animation set, section plane, and camera. The procedure for creating object animation is discussed next.

Creating Object Animation Using the Animation Set

The animation set is created on the basis of currently selected objects in the Scene View. You can also create animation set by using the selection/search set previously created in the model. To create an animation set, first ensure that the added scene is selected in the Animator tree view. Now, select the required object(s) in the Scene View; the selected object(s) will be highlighted in blue color. Next, choose the **Add Scene** button; a flyout will be displayed. Choose **Add Animation Set > From current selection** from the flyout, as shown in Figure 6-5; the Animation Set will be added and displayed in an edit box in the Animator tree view, refer to Figure 6-6. If you want to change name of the animation set, specify the name in the edit box and press ENTER.

Alternatively, to create the animation set, right-click on the scene in the Animator tree view for which you want to create the animation; a shortcut menu will be displayed. Choose **Add Animation Set > From current selection** from the menu; the Animation Set will be created for the required scene.

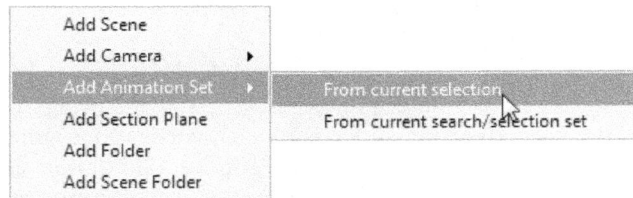

*Figure 6-5 Choosing **From current selection** from the menu*

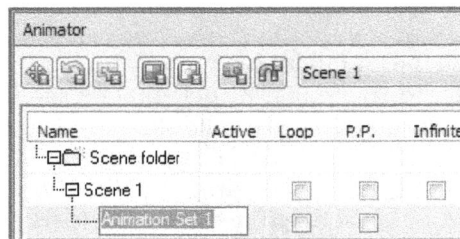

Figure 6-6 The Animation Set added in the selected scene in the Animator tree view

Similarly, you can also create the animation set by using the selection/search set. To do so, first ensure that a selection/search set is created in the model. If not, then create a selection/search set. Next, select the created search set or selection set from the **Sets** window. Now, select the scene in which the animation set is to be added in the Animator tree view. Next, choose the **Add Scene** button; a flyout will be displayed. Choose **Add Animation Set > From current search/selection set** from the flyout, refer to Figure 6-5; the Animation Set will be added to the selected scene in the Animator tree view. Note that if changes are made in the content of selection/search set, then content of the added animation set will be automatically updated. You can also manually update an animation set, which is discussed later in the chapter. Now, in the created animation set, you can create animation by setting the keyframes which is discussed next.

Setting Keyframes

In the Animation Set, the object animation can be created by using the buttons and options available in the Animator toolbar to transform the objects and its appearance. After transforming the object and its appearance, you can capture these changes as keyframes, which allow smooth animation of the object. The procedure of setting keyframes is discussed next.

First, select the Animation Set in the Animator tree view; the corresponding object will be highlighted in the Scene View and the buttons and options available in the Animator toolbar will be enabled. Now, choose the **Capture keyframe** button from the Animator toolbar to capture the start point of the animation; a keyframe will be generated at the start point in the Animator timeline view. Next, specify the required time for the next keyframe in the **Time position** edit box. You can also specify the time by dragging the Time slider in the timeline view. After specifying the time, choose any of the translation buttons from the Animator toolbar. For example, choose the **Translation animation set** button from the toolbar; a **Move** gizmo will be displayed in the Scene View. Drag the gizmo to change the position of the object, and choose the **Capture keyframe** button to capture the current position of the object; a keyframe will be generated at the specified time and the Time slider will move to the currently generated keyframe. Similarly, you can capture the rotation and scaling changes of an object by using the **Rotate animation set** and **Scale animation set** buttons, respectively. Repeat the steps to create more keyframes, and choose the **Stop** button when animation is complete in the Animator toolbar. Now, choose the **Play** button to play the animation.

Similarly, you can capture color and transparency changes in an object animation. To do so, first select the animation set in the Animator tree view; the corresponding object will be highlighted in the Scene View. Choose the **Capture keyframe** button to capture the start point of the animation. Next, specify the desired time in the **Time position** edit box. Now, choose the **Change color of animation set** button; a **Color** palette will be displayed at bottom of the **Animator** window. Select the **Color** check box to record color changes in the keyframe. Click on the color drop-down; a color list will be displayed. Select the required color and choose the **Capture keyframe** button to capture the final view of the object; a keyframe will be generated in the timeline view. Choose the **Stop** button when the animation is complete, and play the animation by choosing the **Play** button. Similarly, you can capture the transparency changes by using the **Change transparency of animation set** button.

Updating Animation Set

In the Animation Set, you can replace the currently selected object with a new object from the Scene View or from the selection/search set. To do so, select the required object from the Scene View. Next, right-click on the animation set; a shortcut menu will be displayed. Choose **Update Animation Set > From current selection** from the menu, as shown in Figure 6-7; the selected animation set will be updated after being replaced with the new object.

Similarly, you can modify the Animation Set based on the current selection/search set. To do so, select the desired selection/search set from the **Sets** window. Next, right-click on the Animation Set in the Animator tree view; a shortcut menu will be displayed. Choose **Update Animation Set > From current search/selection set** from the menu, refer to Figure 6-7; the selected Animation Set will be updated after being replaced with the new selection/search object.

Note
Updation will affect only the content of an animation set, not the keyframes.

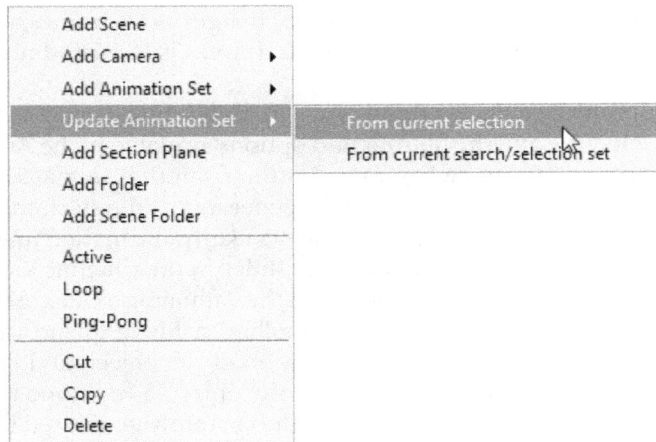

Figure 6-7 *Choosing the **Update Animation Set** option from the shortcut menu*

Creating Animation Using Camera

The object animation using Camera is created by using the existing viewpoints animation or by creating viewpoints using navigation tools. In a scene, either you can add a blank camera and then create the viewpoints or can copy an existing viewpoints animation to the camera. You can add only one camera in a scene in the Animator tree view. To add a blank camera, select the desired scene in the Animator tree view. Next, choose the **Add Scene** button; a flyout will be displayed. In this flyout, choose the option **Add Camera > Blank camera**, as shown in Figure 6-8; a **Camera** will be added to the selected scene and will be displayed in an edit box in the Animator tree view, refer to Figure 6-9. To change the name of the camera, specify the desired name in the edit box and press ENTER.

In the **Blank camera**, you can capture the camera viewpoints as keyframes. Before creating viewpoints, capture the start point of animation using the **Capture keyframe** button; a keyframe will be generated in the timeline view. Next, specify the required time for the next keyframe in the **Time position** edit box and then create the viewpoints such as walk inside the model using the **Walk** tool. Next, choose the **Capture keyframe** button to again capture the current view of the model; a keyframe will be generated at the specified time in the timeline view. Repeat these steps to create more keyframes and choose the **Stop** button to finish capturing the keyframes. Choose the **Play** button to play the animation.

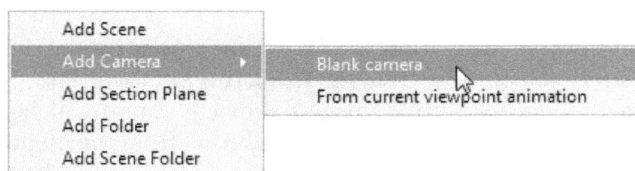

Figure 6-8 *Selecting the **Blank Camera** option from the menu*

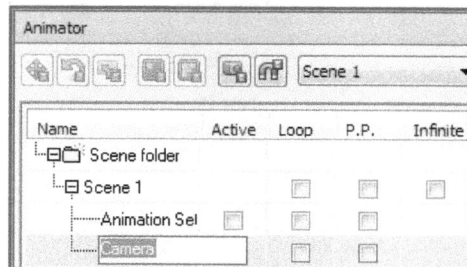

Figure 6-9 Blank camera added to the scene

Similarly, you can add a camera with an existing viewpoint animation. To do so, first select the required viewpoint animation from the **Saved Viewpoints** window. Select the required scene in the Animator tree view. Next, choose the **Add Scene** button; a flyout will be displayed. Choose **Add Camera > From current viewpoint animation** from the flyout, refer to Figure 6-8; all keyframes from the selected animation will be added to the timeline. Choose the **Play** button to play the animation.

You can also access the **Blank camera** and **From current viewpoint animation** options directly from the shortcut menu. To do so, right-click on the required scene in the Animator tree view; a shortcut menu will be displayed with both the options.

Note
The From current viewpoint animation option will be available in the menu only after you select the viewpoint animation from the Saved Viewpoints window.

Creating Animation Using the Section Plane

You can also animate the cross-section views of an object by using the **Add Section Plane** option. In this case, a section plane set is created which contains a list of all the cross-sectional cuts of the model and the keyframes. To add a section plane, first select the required scene in the Animator tree view. Next, choose the **Add Scene** button; a flyout will be displayed. In this flyout, select **Add Section Plane**; a section plane will be added to the selected scene and displayed in the edit box, refer to Figure 6-10. To change its name, specify a new name in the edit box and press ENTER. Now you can capture the cross-sectional cuts of the model. This process is discussed next.

To capture the cross sectional cuts by moving a section plane, select the added section plane in the tree view. Choose the **Enable Sectioning** button from the **Sectioning** panel in the **Viewpoint** tab; the **Sectioning Tools** contextual tab will appear in the ribbon. Select the **Planes** mode from **Sectioning Tools > Mode > Planes** drop-down. Select the required plane from **Sectioning Tools > Plane Settings > Current: Plane** drop-down list. Choose the **Move** or **Rotate** gizmo from the **Transform** panel in the **Sectioning Tools** tab; the selected gizmo will be displayed in the Scene View. Adjust the position of gizmo and choose the **Capture keyframe** button in the **Animator** window to capture the initial position of the section plane. Next, specify the required time for the next keyframe in the **Time position** edit box. Drag the gizmo for the cross-sectional cut and choose the **Capture keyframe** button to capture the current cross-sectional view. Repeat these steps to create more keyframes and choose the **Stop** button when animation is complete. Choose the **Play** button to play the animation.

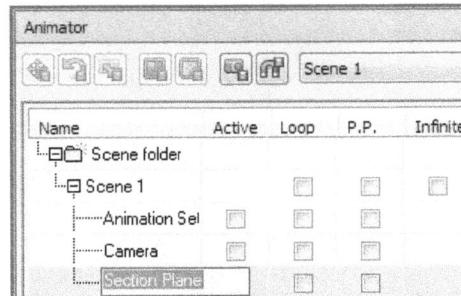

Figure 6-10 Section Plane added to the scene

Next, to capture cross-sectional cuts by moving a section box, select the added section plane set in the tree view. Choose the **Enable Sectioning** button from the tab; a **Sectioning Tools** tab will appear in the ribbon. Select **Box** from **Sectioning Tools > Mode > Planes** drop-down; a box will be displayed enclosing the model. Choose the **Move, Rotate** or **Scale** gizmo from the **Transform** panel in the **Sectioning Tools** tab; a gizmo will be displayed in the Scene View. Drag the gizmo to adjust the initial position of the section box. Choose the **Capture keyframe** button to capture the initial position of the section box. Next, specify time for the next keyframe in the **Time position** edit box. Drag the gizmo to adjust the depth for the cross-sectional cut, and choose the **Capture keyframe** button to capture the current view of the model. Repeat these steps to create more keyframes, and choose the **Stop** button when animation is completed. Now, choose the **Play** button to play the animation.

Note
In the Animation tree view, only one section plane set can be added in a scene

EDITING KEYFRAMES

After creating animation, you can modify the keyframes to change the position, time, field of view, and so on. The keyframes can be modified based on the method of creating animation: animation sets, camera, and section plane sets. The methods of editing keyframes are discussed next.

Editing Animation Set Keyframes

The keyframes are modified by changing their parameters in the **Edit Key Frame** dialog box. To invoke the **Edit Key Frame** dialog box, right-click on the desired animation set keyframe in the timeline view; a shortcut menu will be displayed, as shown in Figure 6-11. In this shortcut menu, choose the **Edit** option from the menu; the **Edit Key Frame** dialog box will be displayed, as shown in Figure 6-12. The various options in this dialog box are discussed next.

Time

The current time of the selected keyframe will be displayed in the **Time** edit box. You need to specify the value of time in seconds to change the time of the selected keyframe.

Translate

The current position of the selected object in the selected keyframe will be displayed in the **Translate** edit boxes. To move the selected object to a new position, specify the coordinates in the X, Y, and Z edit boxes corresponding to the **Translate** option.

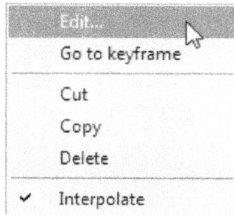

Figure 6-11 The shortcut menu

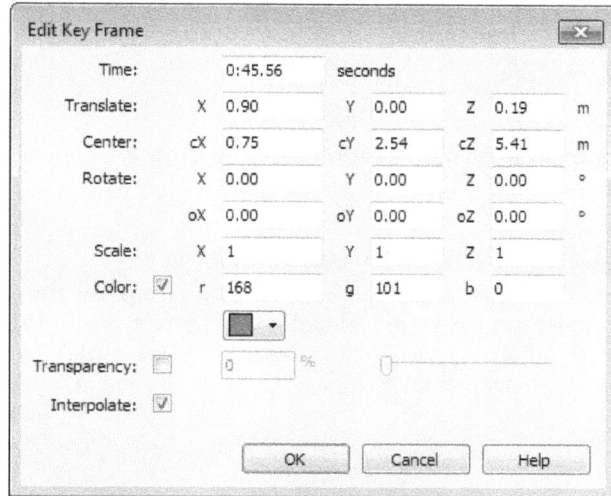

Figure 6-12 The **Edit Key Frame** dialog box

Center

The current center point of rotation or the scale of the selected object in a keyframe will be displayed in the **Center** edit boxes. To change the center point of rotation or scale, specify the coordinates in the edit boxes corresponding to the **Center** option.

Rotate

The degree of rotation along the X, Y, and Z axes of the selected object in a keyframe will be displayed in the **Rotate** edit boxes. To change the degree of rotation, specify the coordinates in the edit boxes corresponding to the **Rotate** option.

Scale

The scaling factor along the X, Y, and Z axis of the selected object in the selected keyframe will be displayed in the **Scale** edit boxes. To change the scaling factor along the X, Y, and Z axes, specify values in the edit boxes corresponding to the **Scale** option.

Color

If the color changes are recorded in the keyframe, then the **Color** check box will be selected in the **Edit Key Frame** dialog box. To change the color, click on the color drop-down; a color palette will be displayed. Select the required color from the palette. You can also specify the color value manually in the **r**, **g**, and **b** edit boxes.

Transparency

If transparency changes are recorded in the keyframe, then the **Transparency** check box will be selected in the **Edit Key Frame** dialog box. To adjust the transparency level, specify a value in the **Transparency** edit box or drag the slider.

Interpolate

If the **Interpolate** check box is selected, Navisworks will automatically interpolate between the current and the next keyframe. If the check box is cleared, there will no movement between the two keyframes.

After making all modifications, choose the **OK** button; the changes will be applied to the selected keyframe.

Editing Camera Keyframes

To edit a camera keyframe, you need to invoke the **Edit Key Frame** dialog box. To do so, right-click on the required camera keyframe in the timeline view; a shortcut menu will be displayed, refer to Figure 6-11. Choose the **Edit** option from the menu; the **Edit Key Frame** dialog box will be displayed, as shown in Figure 6-13. The options in this dialog box are discussed next.

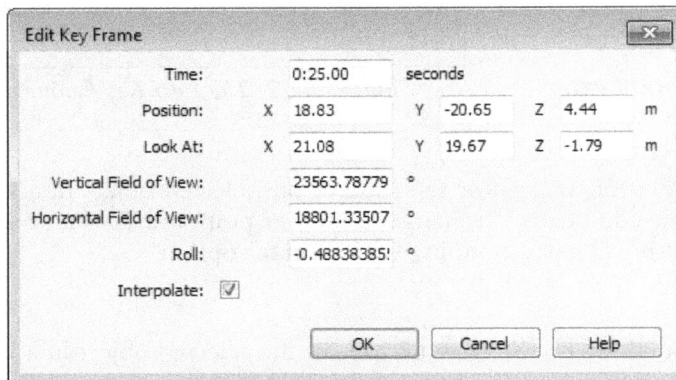

*Figure 6-13 The **Edit Key Frame** dialog box for editing camera keyframes*

Time

The current time of the selected keyframe will be displayed in the **Time** edit box. You can specify the value in seconds in this edit box to change the time of the selected keyframe.

Position

The edit boxes corresponding to the **Position** option will display the current position of the camera. To move the camera to a new position, specify the required values in the edit boxes.

Look At

The edit boxes corresponding to the **Look At** option will display the current focal point of the camera. To change the focal point of camera, specify the required values in the edit boxes.

Vertical Field of View

The current vertical field of view of the camera will be displayed in the **Vertical Field of View** edit box. To modify the vertical angle of view, enter a value in this edit box.

Horizontal Field of View

When you change the vertical angle, the horizontal angle will be adjusted automatically in the **Horizontal Field of View** edit box to match the aspect ratio.

Roll

The current rotation angle of the camera will be displayed in the **Roll** edit box. To modify the rotation value, specify a value in the **Roll** edit box. A positive value will rotate the camera counterclockwise and a negative value will rotate the camera clockwise.

Editing Section Plane Keyframes

The section plane set keyframes can be modified in the same manner as animation set and camera. To edit a section plane set keyframe, right-click on the desired section plane keyframe in the timeline view; a shortcut menu will be displayed, refer to Figure 6-11. Choose the **Edit** option from the menu; the **Edit Key Frame** dialog box will be displayed, as shown in Figure 6-14. The various options in this dialog box are discussed next.

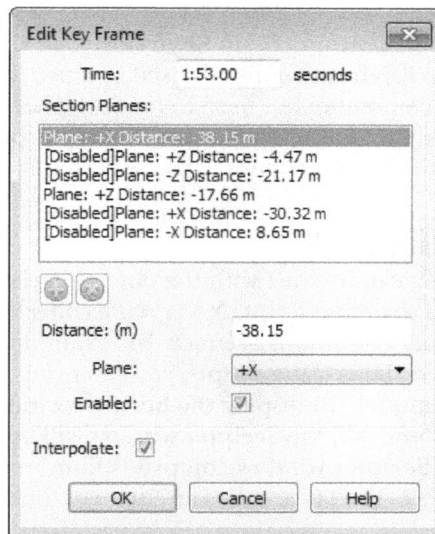

*Figure 6-14 The **Edit Key Frame** dialog box for the section plane set keyframe*

Time

The current time of the selected keyframe will be displayed in the **Time** edit box. To change the time of the selected keyframe, specify the required value (in seconds) in this edit box.

Section Planes

In the **Section Planes** area, the list box contains a list of current section planes. To add a section plane, choose the **Add** button. To delete a section plane from the list, choose the **Delete** button.

Distance

The distance of the selected section plane across the model will be displayed in the **Distance** edit box. To change the distance of the section plane across the model, specify a value in the **Distance** edit box.

Plane

To define the angle of section plane, select the desired up vector from the **Plane** drop-down list. To manually define the up vector, select the **Define Custom** option from the drop-down list; the **Up Vector** dialog box will be displayed, as shown in Figure 6-15. Specify the coordinates in the respective edit boxes, and choose the **OK** button; the **Up Vector** dialog box will be closed.

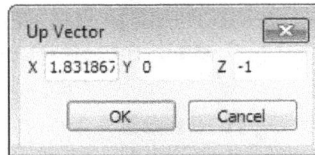

Figure 6-15 *The **Up Vector** dialog box*

Enabled

The **Enabled** check box is used for enabling or disabling the selected section plane in the selected keyframe. To enable the section plane, select the check box.

The other options in the shortcut menu are: **Go to Keyframe**, **Cut**, **Copy**, **Delete**, and **Interpolate**, refer to Figure 6-11. The **Go to Keyframe** option is used to move the Time slider to the selected keyframe. The **Cut** and **Copy** commands are used to move the keyframe to a new position. The **Delete** option is used to delete the selected keyframe. The **Interpolate** option is used to bring a gradual transition between the two keyframes.

THE SCRIPTER WINDOW

Using the **Scripter** window, you can interact with the objects in the model. To interact with the objects in the model, you need to define script. A script is a collection of events and actions that are expected when certain event conditions are met. For example, when the camera reaches a specific spot, the door will open. Thus using scripts, you can bring life to an object which results in better interaction with the model. To display the **Scripter** window, choose the **Scripter** tool from the **Tools** panel in the **Home** tab; the **Scripter** window will be displayed in the Scene View, as shown in Figure 6-16. The **Scripter** window comprises four areas: **Scripts**, **Events**, **Actions**, and **Properties**, refer to Figure 6-16. These areas are discussed next.

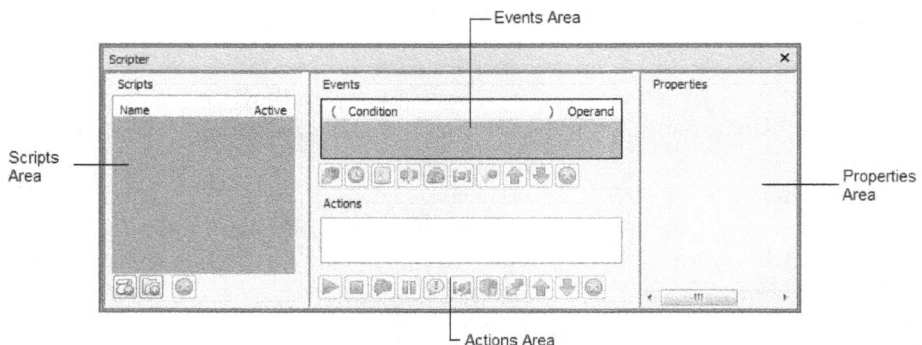

Figure 6-16 *Various areas in the **Scripter** window*

The Scripts Area

The **Scripts** area in the **Scripter** window contains a list of all the added scripts in a hierarchical view. In this area, you can create and manage scripts by using the buttons which are discussed next.

Add New Script

The **Add New Script** button is used to add a new script to the **Scripts** area. To add a script, choose the **Add New Script** button available at the bottom in the **Scripts** area; a script will be added in the **Scripts** area, refer to Figure 6-17. The added script will be displayed in an edit box. To change the name of the added script, specify a new name in the edit box and press ENTER.

Add New Folder

The **Add New Folder** button is used to add a new folder. This folder will contain all the scripts together. To create a folder, choose the **Add New Folder** button available at the bottom to the **Scripts** area; a folder will be created in the **Scripts** area and will be displayed in an edit box, refer to Figure 6-17. To change the name of the created folder, specify the desired name in the edit box and press ENTER.

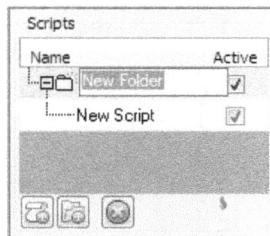

*Figure 6-17 The **Scripts** area in the **Scripter** window*

Delete

The **Delete** button is used to delete a script. To delete a script, select the required script in the **Scripts** area, and then choose the **Delete** button available at the bottom in the window; the selected script will be deleted.

To enable a particular script, select the **Active** check box corresponding to the required script in the **Active** column, refer to Figure 6-17. To enable all the scripts in a particular folder, select the check box corresponding to the required folder in the **Active** column.

You can also right-click and access the above discussed options directly from the shortcut menu displayed, refer to Figure 6-18.

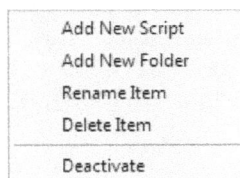

Figure 6-18 The shortcut menu

The Events Area

The **Events** area contains a list of events associated with the selected script. In this area, you can create, manage, and test events. The buttons in this area are used to define and manage the events. These buttons are available at the bottom of the **Events** area, as shown in Figure 6-19. These buttons are discussed next.

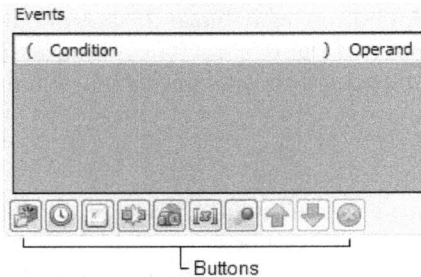

*Figure 6-19 The **Events** area in the **Scripter** window*

On Start

The **On Start** button is used to add start events. The start events are used to set up the initial conditions of the scripts. This will activate a script as soon as it is enabled. On choosing this button, the **On Start** event will be added to the **Events** area, refer to Figure 6-20.

On Timer

The **On Timer** button is used to add timer events. The on timer events activate the script at predefined intervals. On choosing this button, the **On Timer** event will be added to the **Events** area, refer to Figure 6-20.

On Key Press

The **On Key Press** button is used to add key press events. The key press events activate a script with a particular keyboard button. On choosing this button, the **On Key Press** event will be added to the **Events** area, refer to Figure 6-20.

On Collision

The **On Collision** button is used to add collision events. The collision events activate a script when camera collides with a particular object. On choosing this button, the **On Collision** event will be added to the **Events** area, refer to Figure 6-20. The **Collision** option must be selected while adding this event and navigating through a model.

On Hotspot

The **On Hotspot** button is used to add hotspot events. The hotspot events activate a script when camera is within a particular range. Thus, an event will start when the camera approaches a specified area called hotspot. On choosing this button, the **On Hotspot** event will be added to the **Events** area, refer to Figure 6-20.

On Variable

The **On Variable** button is used to add variable events. The variable events activate a script when a variable meets a predefined criterion. On choosing this button, the **On Variable** event will be added to the **Events** area, refer to Figure 6-20.

On Animation

The **On Animation** button is used to add animation events. The animation events are used to activate a script when a selected animation starts or stops. On choosing this button, the **On Animation** event will be added to the **Events** area, refer to Figure 6-20.

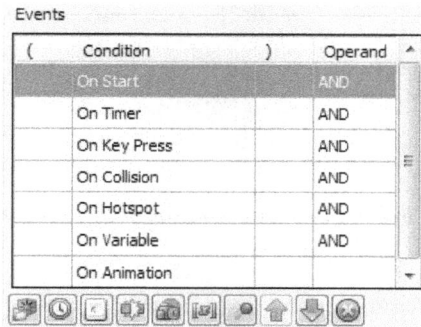

*Figure 6-20 Events added in the **Events** area*

Move Up

The **Move Up** button is used to move a selected event up in the event list. To do so, select the event in the **Events** area and choose the **Move Up** button; the selected event will be shifted up in the list.

Move Down

The **Move Down** button is used to move a selected event down in the event list. To do so, select the event in the **Events** area and choose the **Move Down** button; the selected event will be shifted down in the list.

Delete Event

The **Delete Event** button is used to delete a selected event. To do so, select the event in the **Events** area and choose the **Delete Event** button; the selected event will be removed from the list.

A script can contain more than one event. A condition is to be formed if more than one event is required to start a script. The conditions can be formed with the use of two operands: **AND** and **OR**. To use these operands, click in the **Operand** column corresponding to the selected event condition; a drop-down list will be displayed. Select the required operand from the list. The conditions need to be formed in such a way that it should make a boolean logic sense and should be enclosed in brackets. You can access these brackets from the brackets column in the **Events** area.

You can also perform actions such as adding and renaming events. To do so, right-click in the **Events** area; a shortcut menu will be displayed, as shown in Figure 6-21. To add an event condition, choose **Add Event** from the menu; a cascading menu will be displayed, refer to Figure 6-21. Choose the event types from the cascading menu. To add brackets, choose **Brackets**; a cascading menu will be displayed. Choose the bracket options from the cascading menu. To define the logic operators, choose **Logic**; a cascading menu will be displayed. Choose the logic operators **AND** or **OR** from the cascading menu. To test the validity of an event condition, choose **Test Logic** from the shortcut menu.

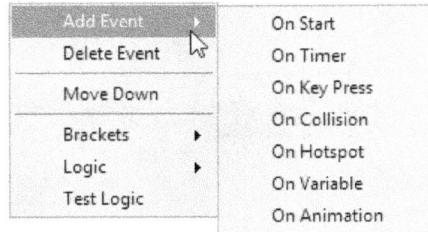

*Figure 6-21 The shortcut menu in the **Events** area*

The Actions Area

The **Actions** area contains a list of actions associated with the selected script. An action is an activity which is carried out when a script is activated by an event. An action can be playing animation such as opening and closing of the doors. After creating events, you need to define actions for these events. You can add multiple actions to a script. An action will start when the event conditions are met. Using the options in the **Actions** area, you can create, manage, and test actions. The buttons used to define actions are available at the bottom in the **Actions** area, as shown in Figure 6-22. These buttons are discussed next.

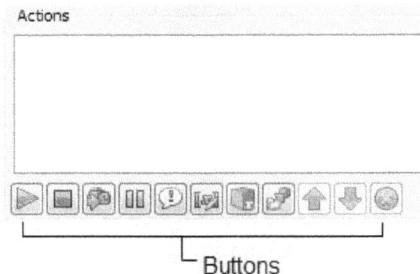

*Figure 6-22 The buttons in the **Actions** area*

Play Animation

The **Play Animation** button is used to add play animation actions. Play animation actions play a selected animation when a script is activated. On choosing this button, the **Play Animation** action will be added to the **Actions** area, refer to Figure 6-23.

Stop Animation

The **Stop Animation** button is used to add stop animation actions. Stop animation actions stop the current playing animation when a script is activated. On choosing this button, the **Stop Animation** action will be added to the **Actions** area, refer to Figure 6-23.

Show Viewpoint

The **Show Viewpoint** button is used to add show viewpoint actions. Show viewpoint actions are used to specify the viewpoint to be used when a script is activated. On choosing this button, the **Show Viewpoint** action will be added to the **Actions** area, refer to Figure 6-23.

Pause

The **Pause** button is used to add the pause actions. Pause actions delay the actions for a specified period of time. On choosing this button, the **Pause** action will be added to the **Actions** area, refer to Figure 6-23.

Send Message

The **Send Message** button is used to add send message actions. Send message actions will create a text file with a specified message when a script is activated. On choosing this button, the **Send Message** action will be added to the **Actions** area.

Set Variable

The **Set Variable** button is used to add set variable actions. Set variable actions are used to set variables when a script is activated. On choosing this button, the **Set Variable** action will be added to the **Actions** area, refer to Figure 6-23.

Store Property

The **Store Property** button is used to add store property action. Store property actions will store object property in a variable. On choosing this button, the **Store Property** action will be added to the **Actions** area, refer to Figure 6-23.

Load Model

The **Load Model** button is used to add load model actions. Load model actions will close the current model and load a new model when a script is activated. On choosing this button, the **Load Model** action will be added to the **Actions** area, refer to Figure 6-23.

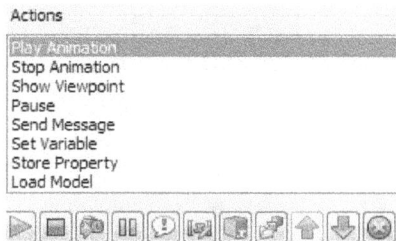

Figure 6-23 Actions added to the Actions area

You can also perform actions such as adding and deleting. To do so, right-click in the **Actions** area; a shortcut menu will be displayed, as shown in Figure 6-24. To add an action, choose **Add Action** from the shortcut menu; a cascading menu will be displayed, refer to Figure 6-24. Choose the desired actions from this cascading menu. Select the **Test Action** option from the menu to execute the selected action. Choose the **Stop Action** option to stop the execution of the selected action.

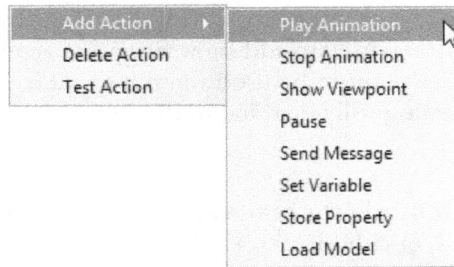

*Figure 6-24 The shortcut menu in the **Actions** area*

The Properties Area

The right pane of the **Scripter** window is the **Properties** area. When you select an event or action, its properties will be displayed in the **Properties** area. These event and action properties can be configured in the **Properties** area which is discussed next.

Event Properties

When you add an event to a script, the **Properties** area displays its properties. In this area, you can configure the properties of all event types except the **On Start** event. Various properties of each event type and their configurations are discussed next.

On Timer Event Property

When you add the **On Timer** event to the **Events** area, its properties will be displayed in the **Properties** area, as shown in Figure 6-25. These properties are discussed next.

In the **Interval (seconds)** edit box, you can specify the time needed before an action starts. To define the frequency of an event, select the required option from the **Regularity** drop-down list. The **Once After** option is used to create an event that starts after a certain period of time. The **Continuous** option is used to repeat an event continuously at specified time intervals.

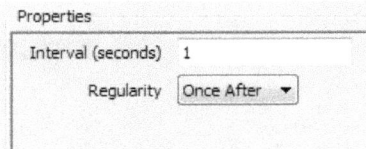

*Figure 6-25 The properties of the **On Timer** event*

On Key Press Event Property

When you add the **On Key Press** event to the **Events** area; its properties will be displayed in the **Properties** area, as shown in Figure 6-26. These properties are discussed next.

Specify a particular key to activate an event in the **Key** edit box of the **Properties** area. To define how the trigger from the keyboard will start, use the options from the **Trigger on** drop-down list. Select the **Key Up** option to trigger an event when you press and release the specified key. Select the **Key Down** option to trigger an event as soon as the specified key is pressed down.

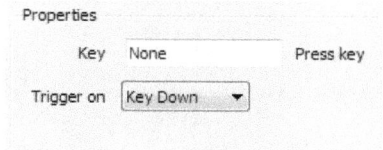

Figure 6-26 The properties of the **On Key Press** *event*

On Collision Event Property

When you add the **On Collision** event to the **Events** area, its properties will be displayed in the **Properties** area, as shown in Figure 6-27. These properties are discussed next.

To define the colliding objects, first you need to select the required object in the model. Next, choose the **Sets** button; a flyout will be displayed. Choose the **Set From Current Selection** option from the flyout to attach a selected object. Choose the **Set From Current Selection Set** option from the flyout to attach a search set or a selection set. Notice that on attaching objects, the **Show Parts(s)** button will be activated. You can use this button to verify the selected object. Select the **Include the effects of gravity** check box to include the effect of gravity in collision.

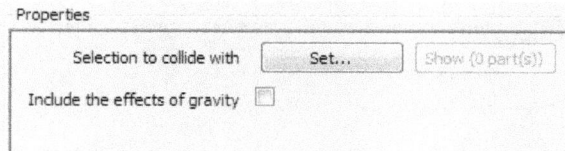

Figure 6-27 The properties of the **On Collision** *event*

On Hotspot Event Property

When you add the **On Hotspot** event to the **Events** area; its properties will be displayed in the **Properties** area, as shown in Figure 6-28. These properties are discussed next.

The **Hotspot** drop-down list contains the options to define the hotspot type. Select the **Sphere** option from the drop-down list to create a sphere around the hotspot. Select **Sphere on selection** to create a sphere around a hotspot defined by a selection. To define how an event will be activated, use the options from the **Trigger when** drop-down list. Select the **Entering** option from the drop-down list to activate an event when you enter the hotspot. Select the **Leaving** option from the list to activate an event when you leave the hotspot. Select the **In Range** option from the list to activate an event when you are inside the hotspot.

If you select **Sphere** from the **Hotspot** drop-down list; the **Position** and **Radius** options will be activated in the **Hotspot Type** area, refer to Figure 6-28. You can specify the position of hotspot by entering the X, Y, and Z coordinates in their respective edit boxes. Alternatively, choose the **Pick** button; a cursor will be displayed. Click at the required location in the model; the coordinates for the hotspot will be displayed in the **Position** edit boxes. Specify the radius around the hotspot in the **Radius** edit box.

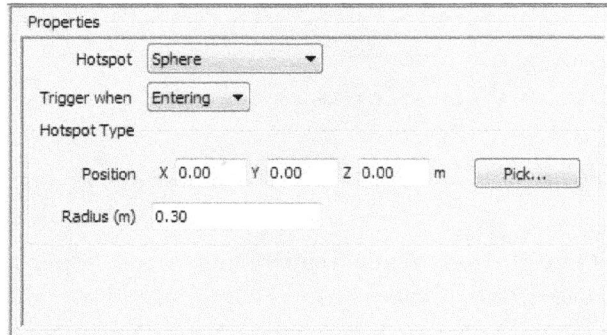

Figure 6-28 *Options displayed on selecting* **Sphere** *from the* **Hotspot** *drop-down list*

If you select **Sphere On selection** from the **Hotspot** drop-down list, the **Selection** and **Radius** options will be activated in the **Hotspot Type** area, as shown in Figure 6-29. Choose the **Set** button to define the hotspot objects; a menu will be displayed. Select the **Set From Current Selection** option to set the current selected object as a hotspot. Select the **Set From Current Selection Set** option to set the current search/selection set as a hotspot. Select the **Clear** option to clear all the current selections. Specify the radius of the hotspot in the **Radius** edit box.

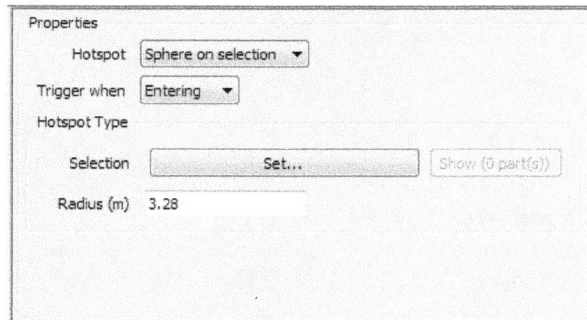

Figure 6-29 *Options displayed on selecting* **Sphere on selection** *from the* **Hotspot** *drop-down list*

On Variable Event Property

When you add the **On Variable** event to the **Events** area, its properties will be displayed in the **Properties** area, as shown in Figure 6-30. These properties are discussed next.

Enter the name of the variable in the **Variable** edit box. Enter the value to be tested against the variable in the **Value** edit box. This value may be of the following types: Number, Float, String, Variable, and Boolean. To define the operator used for variable comparison, select an option from the **Evaluation** drop-down list.

Figure 6-30 *The properties of the* **On Variable** *event*

On Animation Event Property

When you add the **On Animation** event to the **Events** area, its properties will be displayed in the **Properties** area, as shown in Figure 6-31. These properties are discussed next.

The **Animation** drop-down list contains a list of object animations created in the model. Select the required animation from this list to activate the event.

The **Trigger on** drop-down list contains options to define when an event will start. Select the **Starting** option to activate an event when the animation starts. Select the **Ending** option to activate an event when the animation ends.

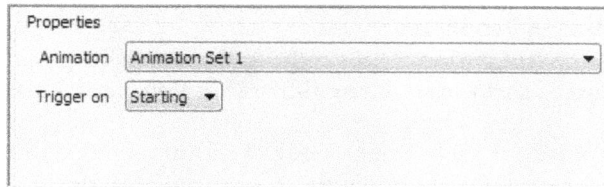

*Figure 6-31   The properties of the **On Animation** event*

Action Properties

When you add an action for an event type, its properties will be displayed in the **Properties** area. The properties of the various action types are discussed next.

Play Animation Action Property

When you add the **Play Animation** action in the **Actions** area, its properties will be displayed in the **Properties** area, as shown in Figure 6-32. These properties are discussed next.

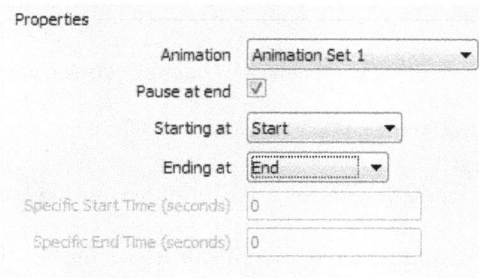

*Figure 6-32   The properties of the **Play Animation** action*

The **Animation** drop-down list contains a list of all the object animations created in the model. Select the animation to be played from this list.

Select the **Pause at end** check box to stop the animation at the end. If this check box is cleared then the selected animation will restart.

You can select an option from the **Starting at** drop-down list to define the starting position of the animation. Select the **Start** option from the drop-down list to start animation from the beginning. Select the **End** option to play the animation backward from the end. Select the **Current Position** option to play the animation from its current position. Select the option

Specified Time to play the animation at a specified time. On selecting the **Specified Time** option, the **Specific Start Time** edit box will be activated. Specify the required time in this edit box.

You can select an option from the **Ending at** drop-down list to define the finishing position of the animation. Select the **Start** option to end the animation at the beginning of an animation sequence. Select the **End** option to end the animation at the end of an animation. Select the **Specified Time** option to end the animation at a specific time. On selecting the **Specified Time** option, the **Specific End Time** edit box will be activated. Specify the required time in this edit box.

Stop Animation Action Property

When you add the **Stop Animation** action in the **Actions** area, its properties will be displayed in the **Properties** area, as shown in Figure 6-33. These properties are discussed next.

The **Animation** drop-down list contains a list of all the object animations created in the model. Select the required object animation to be stopped from this list.

The **Reset to** drop-down list contains the options to define the playback position of the stopped animation. Select the **Default Position** option to reset the animation to its starting position. Select the **Current Position** option to set the animation at its end position.

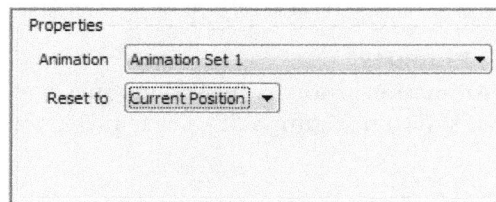

Figure 6-33 The properties of the Stop Animation action

Show Viewpoint Action Property

When you add the **Show Viewpoint** action in the **Actions** area; the **Viewpoint** drop-down list will be displayed in the **Properties** area, as shown in Figure 6-34.

This list contains a list of all the saved viewpoints in the model. Select the required viewpoint to be displayed from this list.

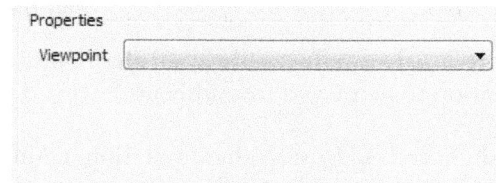

Figure 6-34 The Viewpoint drop-down list

Pause Action Property

When you define the **Pause** action in the **Actions** area, the **Delay** edit box will be displayed in the **Properties** area, as shown in Figure 6-35. This edit box is used to specify the lag time before the next action in the script.

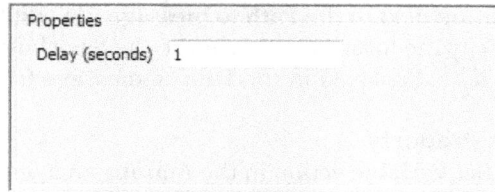

*Figure 6-35 The **Delay** edit box*

Send Message Action Property

When you add the **Send Message** action in the **Actions** area, the **Message** edit box will be displayed in the **Properties** area, as shown in Figure 6-36. In this edit box, you can define the message to be sent to the text file.

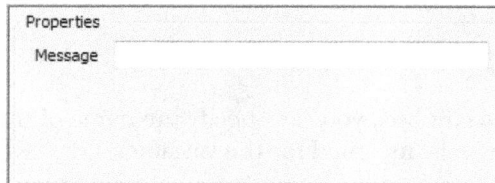

*Figure 6-36 The **Message** edit box*

Before entering the message you have to define the location of the message file. To do so, first invoke the **Options Editor** dialog box by choosing **Application Button > Application Menu > Options**. In the **Options Editor** dialog box, expand the **Tools** node and select **Scripter** under the **Tools** node; the options related to the **Scripter** will be displayed in the right pane of the dialog box, refer to Figure 6-37.

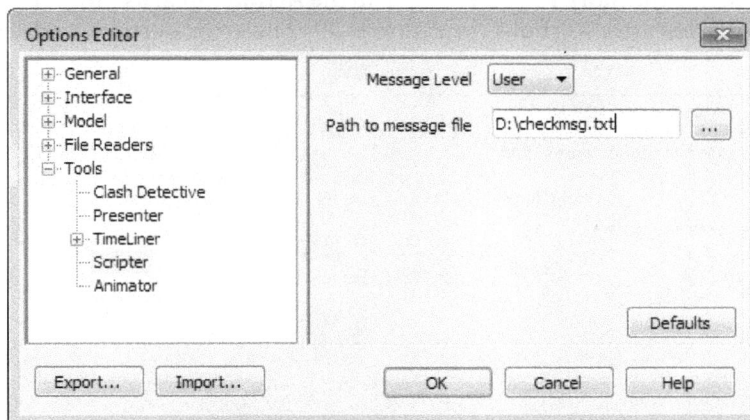

*Figure 6-37 The **Scripter** options in the **Options Editor** dialog box*

The **Message Level** drop-down list contains two options: **User** and **Debug**. On selecting the **User** option, the text file will contain the specified message. On selecting the **Debug** option, the text file will contain the specified message along with the debug messages. The debug messages will describe everything that scripter has done. The text file will include sequence of actions and defined variables with their values. To define location of the message file, choose the **Browse** button next to the **Path to message file** edit box; the **Save As** dialog box will be displayed. Specify the location, and choose the **Save** button to save the file; the path of the message file will be displayed in the **Path to message file** edit box.

Set Variable Action Property

When you define the **Set Variable** action in the **Actions** area, its properties will be displayed in the **Properties** area, as shown in Figure 6-38. These properties are discussed next.

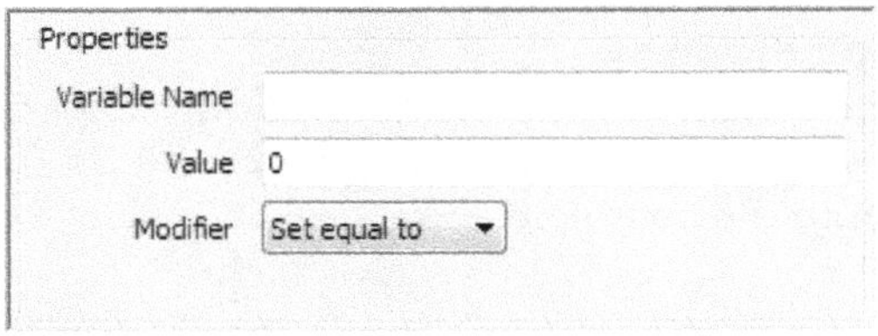

Figure 6-38 The properties of the Set Variable action

In the **Variable Name** edit box, you can specify the name of the variable. In the **Value** edit box, specify the value to be assigned for the variable.

The **Modifier** drop-down list contains operators for the variable. The **Set equal to** option is used to make the variable value equal to the value assigned in the **Value** edit box. The **Increment by** option is used to increase the variable value by the number specified in the **Value** edit box. The **Decrement by** option is used to decrease the variable value by the number specified in the **Value** edit box.

Store Property Action Property

When you define the **Store Property** action in the **Actions** area, its properties will be displayed in the **Properties** area, as shown in Figure 6-39. These properties are discussed next.

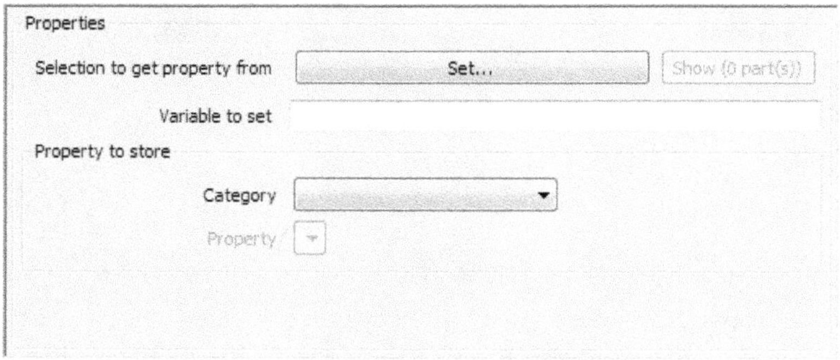

Figure 6-39 The properties of the Store Property action

Select the required object in the model and choose the **Set** button; a flyout will be displayed. This flyout contains three options that are used to define the source objects. Source object is the object whose properties are to be stored. Choose the **Set From Current Selection** option to set the current object selection for storing the property. Choose the **Set From Current Selection Set** to set the objects from the current search/selection set for storing the property. Choose the **Clear** option to clear all the selections.

You can verify the source objects by choosing the **Show (parts)** button next to the **Sets** button; the selected objects will be highlighted in the Scene View.

In the **Variable to set** edit box, specify the name of the variable. In the **Property to store** area, select the property category from the **Category** drop-down list. Select the property type from the **Property** drop-down list.

Load Model Action Property
When you define the **Load Model** action in the **Actions** area, the **File to load** edit box will be displayed in the **Properties** area, as shown in Figure 6-40.

If you want to replace the current model, choose the Browse button next to the **File to load** edit box; the **Open** dialog box will be displayed. Browse to the folder containing the model, and then choose the **Open** button; the path to the new model will be displayed in the **File to load** edit box.

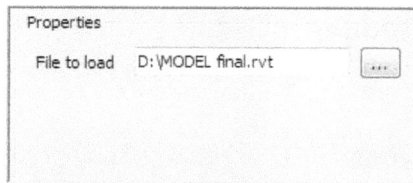

*Figure 6-40 The **File to load** edit box*

ENABLING SCRIPTS

After creating the scripts you need to activate them. To do so, choose the **Enable Scripts** button from the **Script** panel in the **Animation** tab; the **Scripter** window will be disabled and now you can interact with the model. To disable the added scripts, choose the **Enable Scripts** button again; the **Scripter** window will be enabled.

TUTORIALS

General instructions for downloading tutorial files:

1. Download the *c06_nws_2017_tut* zip file for this tutorial from *http://www.cadcim.com*. The path of the file is as follows: *Textbooks > Civil/GIS > Navisworks > Exploring Autodesk Navisworks 2017*.

2. Now, save and extract the downloaded folder at the following location:
   C:\ nws_2017\ c06_nws_2017_tut

Note
*The default unit system used in the tutorials is metric. To change the units to imperial, select the required units from **Options Editor > Interface > Display Units**.*

Tutorial 1   Creating Animation

In this tutorial, you will open the *c06_nws_2017_tut1.nwf* file and create object animation using the options in the **Animator** window.   **(Expected time: 45min)**

The following steps are required to complete this tutorial:

a.   Open the *c06_nws_2017_tut1.nwf* file.
b.   Display the **Saved Viewpoints** window.
c.   Display the **Animator** window.
d.   Create a scene folder.
e.   Add a scene in the scene folder.
f.   Create animation using Animation set.
g.   Play animation.
h.   Save the project.

Opening the Existing Model

In this section, you will open the model *c06_nws_2017_tut1* that was created using the Revit software.

1.   Choose the **Open** button from the Quick Access Toolbar; the **Open** dialog box is displayed on the screen.

2.   In this dialog box, browse to the following location:
     C:\nws_2017\c06_nws_2017_tut

3.   Select the **Navisworks File Set (*.nwf)** file format from the **Files of type** drop-down list.

4.   Next, select *c06_nws_2017_tut1* from the displayed list of files; the file is displayed in the **File name** edit box.

5.   Now, choose the **Open** button; the model is displayed in the Scene View, as shown in Figure 6-41.

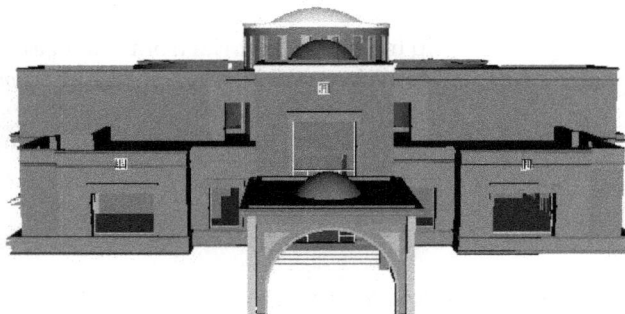

Figure 6-41 *Model displayed in the Scene View*

Displaying the Saved Viewpoints Window

In this section, you will display the **Saved Viewpoints** window and display the view.

1. Choose the **Saved Viewpoints Dialog Launcher** tool from **Viewpoint > Save, Load & Playback > Save Viewpoint drop-down**; the **Saved Viewpoints** window is displayed, as shown in Figure 6-42.

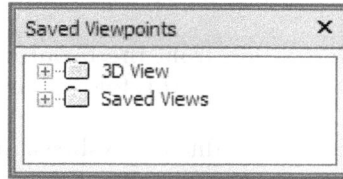

Figure 6-42 The Saved Viewpoints window

2. In this window, expand the **Saved Views** folder and select the **Doors** viewpoint; the door is displayed in the Scene View, as shown in Figure 6-43. Next, close the **Saved Viewpoints** window.

Figure 6-43 The door displayed in the scene view

Displaying the Animator Window

In this section, you will display the **Animator** window.

1. Choose the **Animator** tool from the **Tools** panel in the **Home** tab; the **Animator** window is displayed in the Scene View.

Creating a Scene Folder

In this section, you will create a scene folder and group various scene components in this folder.

1. Click the **Add Scene** button at the lower left corner of the **Animator** window; a flyout is displayed, as shown in Figure 6-44.

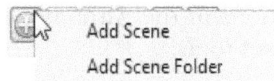

Figure 6-44 The flyout displayed on clicking the Add Scene button

Note
*The options in the **Animator** window will not be enabled if the **Enable Scripts** button is chosen from the **Scripts** panel in the **Animation** tab.*

2. Choose the **Add Scene Folder** option from the flyout; a scene folder is created in the **Animator** tree view. Note that the name of this folder is in editable mode.

3. Specify the name **Animation Set** in the edit box, as shown in Figure 6-45, and press ENTER.

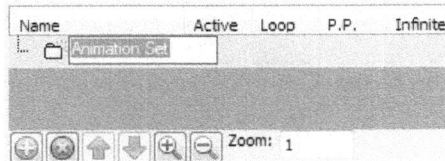

Figure 6-45 The scene folder

Creating a Scene in the Scene Folder

In this section, you will create a scene in the **Animation Set** folder.

1. Choose the **Add Scene** button at the bottom of the **Animator** window; a flyout is displayed refer to Figure 6-44.

2. Choose **Add Scene** from the flyout; a scene with the name **Scene 1** is added to the **Animation Set** folder.

3. Next, double-click on **Scene 1**; an edit box is displayed.

4. Specify the name **Door** in the edit box, refer to Figure 6-46 and press ENTER.

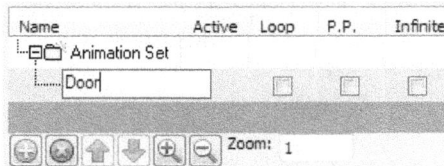

*Figure 6-46 The scene added in the **Animation Set** folder*

Note
You can also create scene folders and their components using the options in the shortcut menu displayed on right-clicking in the tree view.

Creating Animation Using the Animation Set

In this section, you will create animations in the **Animation Set** scene by using the **Translate animation set** and **Rotate animation set** tools.

1. Invoke the **Select** tool from the Quick Access Toolbar and select the door and the door frame in the Scene View by using the CTRL key, as shown in Figure 6-47.

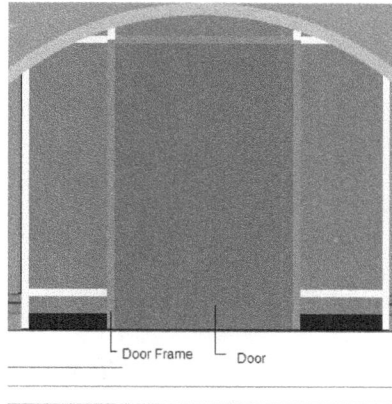

Figure 6-47 *The door and door frame selected in the model*

2.   Next, choose the **Hide Unselected** button from the **Visibility** panel in the **Home** tab; the unselected objects are hidden and the selected objects are visible, refer to Figure 6-48.

Note

*On choosing the **Hide Unselected** button if the **Autodesk Navisworks Manage 2017** message box is displayed then choose the **OK** button.*

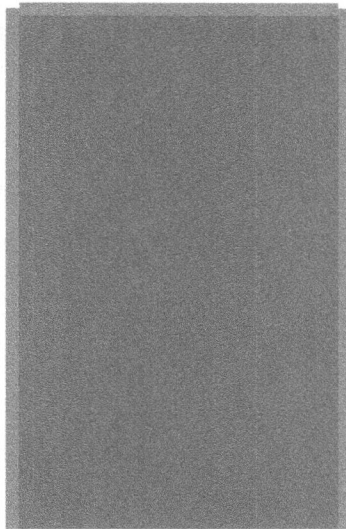

Figure 6-48 *Selected objects in the Scene View*

3.   Deselect the selected objects by clicking in the Scene View and select only the door.

4.   Using the **Orbit** tool, rotate the door, as shown in Figure 6-49.

Figure 6-49 *The rotated door in the Scene View*

5.   Next, right-click on the **Door** in the Animator tree view of the **Animator** window; a shortcut menu is displayed.

6.   In this menu, choose **Add Animation Set > From current selection**, refer to Figure 6-50; **Animation Set 1** is added under **Door**, as shown in Figure 6-51.

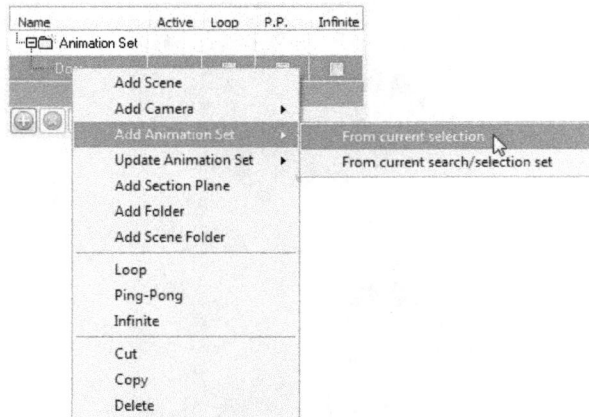

Figure 6-50 *Choosing an option from the shortcut menu*

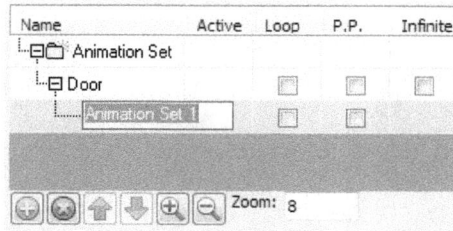

Figure 6-51 The Animation Set created in the Animator tree view

7. Rename **Animation Set 1** as **Opening Door** and press ENTER.

8. Choose the **Capture keyframe** button in the Animator toolbar to capture the initial position of the model.

9. Specify the time as **0:05.00** in the **Time position** edit box.

10. Choose the **Rotate Animation Set** button from the Animator toolbar; the **Rotate** gizmo appears in the Scene View, as shown in Figure 6-52.

*Figure 6-52 The **Rotate** gizmo displayed in the Scene View*

11. Place the cursor on the center ball of the gizmo; the ball is highlighted and a hand cursor appears on the screen, as shown in Figure 6-53.

12. Press and hold the left mouse button and then drag the gizmo and place it, as shown in Figure 6-54.

Figure 6-53 *Cursor at the center ball of the Gizmo*

Figure 6-54 *Gizmo placed at the lower corner*

13. Place the cursor on the green curve; the curve is highlighted and a hand cursor appears, as shown in Figure 6-55.

14. Press and hold the left mouse button keeping the CTRL key pressed and rotate the gizmo, as shown in Figure 6-56. Note that you need to rotate the gizmo not the door.

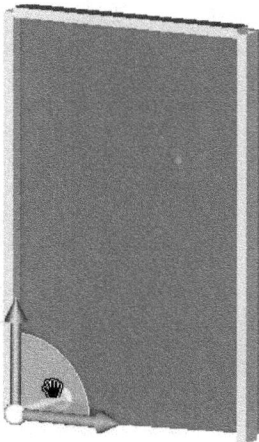

Figure 6-55 *Placing cursor on the green curve*

Figure 6-56 *Gizmo after being rotated*

15. Place the cursor on the blue curve; the curve is highlighted, as shown in Figure 6-57.

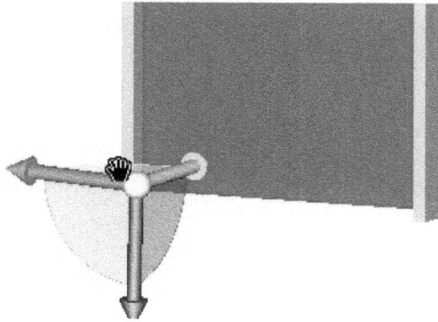

Figure 6-57 Cursor placed on the blue curve

16. Press and hold the left mouse button and drag the gizmo slowly in the left direction to open the door, as shown in Figure 6-58. After opening the door, release the left mouse button.

17. Choose the **Rotate animation set** button in the Animator toolbar; the **Rotate** gizmo disappears.

18. Next, choose the **Capture keyframe** button to capture the current position of the door.

Figure 6-58 Gizmo Moved in left direction to open door

19. Create another animation set, as shown in Figure 6-59, following the procedure explained in the earlier section.

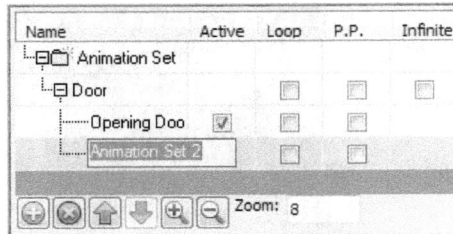

Figure 6-59 *Creating another Animation Set*

20. Rename the Animation Set as **Closing Door**.

21. Specify the time as **0:08.00** in the **Time position** edit box and press ENTER.

22. Choose the **Capture Keyframe** button to capture the current position of the selected door.

23. Specify the time as **0:13.00** in the **Time position** edit box.

24. Next, choose the **Rotate animation set** tool from the Animator toolbar; the **Rotate** gizmo is displayed in the Scene View. Place the gizmo, as shown in Figure 6-60.

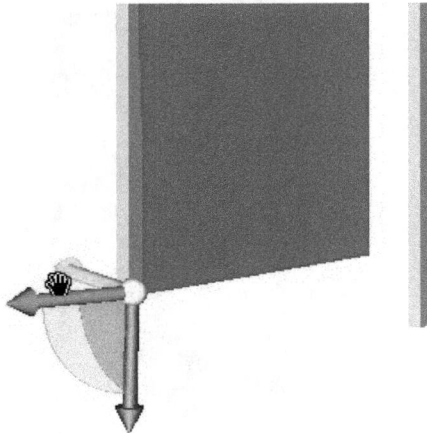

Figure 6-60 *The **Rotate** gizmo displayed in the Scene View*

25. Place the cursor on the blue curve of the gizmo; the curve is highlighted and a hand cursor appears.

26. Press and hold the left mouse button and then drag the mouse in the left direction to close the door, refer to Figure 6-61.

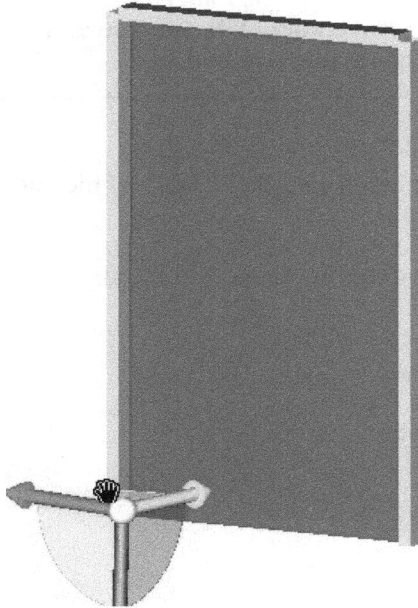

Figure 6-61 *The closed door*

27. Next, choose the **Rotate animation set** tool from the **Animator** toolbar; the **Rotate** gizmo disappears.

28. Choose the **Capture keyframe** button to capture the current position of the door.

29. Next, choose the **Unhide All** option from **Home > Visibility > Unhide All** drop-down; all the objects become visible in the Scene View.

30. Next, deselect the selected objects and display the **Door** viewpoint under the **Saved Views** folder in the **Saved Viewpoints** window.

Playing Animation
In this section, you will play the animation.

1. Select the **Door** animation set in the Animator tree view of the **Animator** window and choose the **Play** button from the Animator toolbar to bring back the animation to the start and play both the animations.

Saving the Project
In this section, you will save the project.

1. Close the **Animator** window and choose **Save As** from the Application Menu; the **Save As** dialog box is displayed.

2. Browse to the *nws_2017/c06_nws_2017_tut* folder and enter **c06_nws_2017_tut01** in the **File name** edit box. Choose the **Navisworks File Set (*.nwf)** file format from the **Save as type** drop-down list, and then choose the **Save** button.

Tutorial 2   Creating Scripts

In this tutorial, you will open the *c06_nws_2017_tut2.nwf* file and create scripts in the **Scripter** window. **(Expected time: 45min)**

The following steps are required to complete this tutorial:

a. Open the *c06_nws_2017_tut2.nwf* file.
b. Display the **Scripter** window.
c. Create the script folder.
d. Add script in the script folder.
e. Add events.
f. Add actions.
g. Adjust the model
h. Enable scripts.
i. Navigate inside the model.
j. Save the project.

Opening the Model
In this section, you will open the model *c06_nws_2017_tut2* which was created in Revit software.

1. To open the existing model, choose the **Open** button from Quick Access Toolbar; the **Open** dialog box is displayed on the screen.

2. In this dialog box, browse to the following location:
   C:\nws_2017\c06_nws_2017_tut

3. Select **Navisworks File Set (*.nwf)** file format from the **Files of type** drop-down list.

4. Next, select the *c06_nws_2017_tut2* from the displayed list of files; the file is displayed in the **File name** edit box.

5. Now choose the **Open** button; the model is displayed in the Scene View.

Displaying the Scripter Window
In this section, you will display the **Scripter** window.

1. Choose the **Scripter** tool from the **Tools** panel in the **Home** tab; the **Scripter** window is displayed in the Scene View.

Creating the Script Folder
In this section, you will create script folder in the **Scripter** window.

1. Choose the **Add New Folder** button available at the lower left corner in the **Scripter** window; a new folder is created. Note that the name of this folder is displayed in editable mode, refer to Figure 6-62.

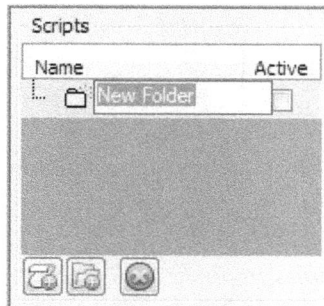

*Figure 6-62 Specifying the folder name in the **Scripts** area*

2. Specify the name **Door_Scripts** in the edit box and press ENTER.

Creating Scripts in the Script Folder

In this section, you will add scripts in the script folder.

1. Choose the **Add New Script** button available at the lower left corner in the **Scripter** window; a new script is added to the **Door_Scripts** folder and is displayed in an edit box, refer to Figure 6-63.

*Figure 6-63 New Script added in the **Door_Scripts** folder*

2. Rename the scripts as **Open & Close Door** and press ENTER.

Adding Events and Configuring Event Properties

In this section, you will add events and configure event properties.

1. Choose the **On Key Press** button [.] available at the bottom of the **Events** area; the **On Key Press** event is added to the **Events** area, as shown in Figure 6-64.

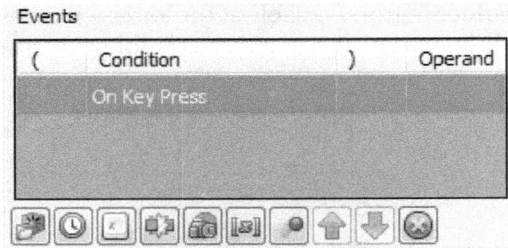

*Figure 6-64 The **On Key Press** event added to the **Events** area*

Note
*On adding the event, its properties are displayed in the **Properties** area, as shown in Figure 6-65.*

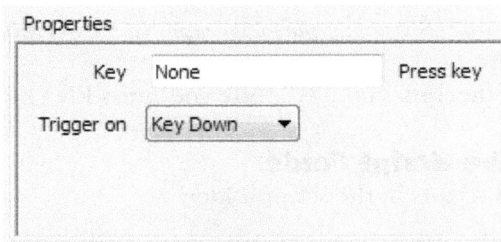

*Figure 6-65 The **On Key Press** event properties displayed to the **Properties** area*

2. Place the cursor and click in the **Key** edit box in the **Properties** area, refer to Figure 6-65.

3. Next, press the **Up** key from the keyboard; the **Up** key is specified in the **Key** edit box in the **Properties** area, refer to Figure 6-66.

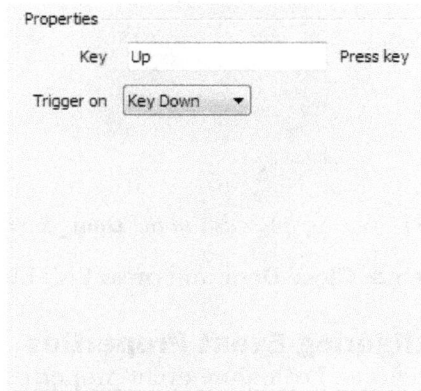

*Figure 6-66 The **Up** key specified in the **Key** edit box*

4. Select the **Key Up** option from the **Trigger on** drop-down list in the **Properties** area, refer to Figure 6-67.

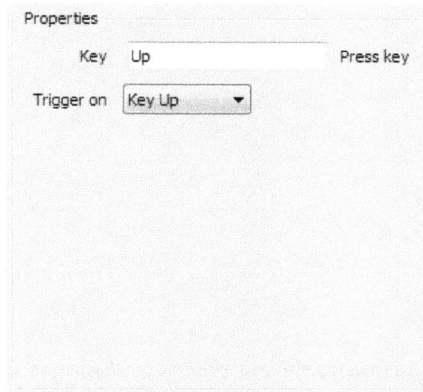

*Figure 6-67 The **Key Up** option selected from the **Trigger on** drop-down list*

5. Choose the **On Animation** button ⦿ available at the bottom in the **Events** area; the **On Animation** event is added to the **Events** area, as shown in Figure 6-68.

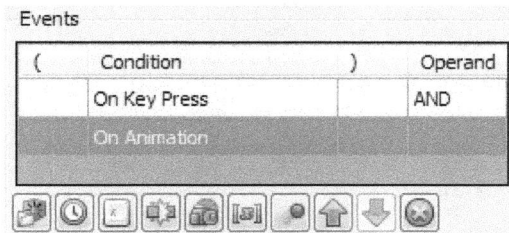

*Figure 6-68 The **On Animation** event added to the **Events** area*

6. Click in the **Operand** column corresponding to the **On Key Press** event in the **Events** area; a drop-down list is displayed, as shown in Figure 6-69.

7. Select the **OR** option from the displayed drop-down list.

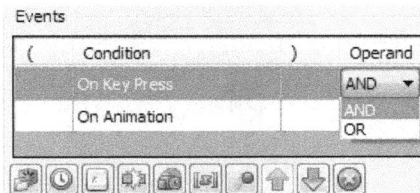

*Figure 6-69 The drop-down list displayed in the **Operand** column*

8. Select the **On Animation** event in the **Events** area; its properties are displayed in the **Properties** area, as shown in Figure 6-70.

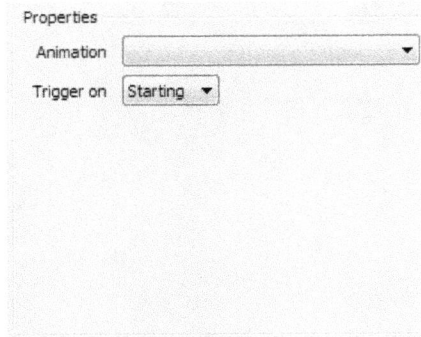

Figure 6-70 *The **On Animation** event properties displayed in the **Properties** area*

9. Select the **Opening Door** animation from the **Animation** drop-down list in the **Properties** area, refer to Figure 6-71.

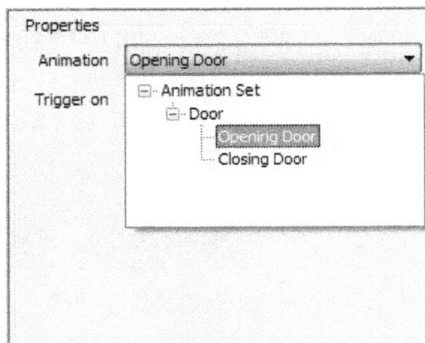

Figure 6-71 *Selecting the **Opening Door** option from the **Animation** drop-down list*

10. Select the **Ending** option from the **Trigger on** drop-down list, as shown in Figure 6-72.

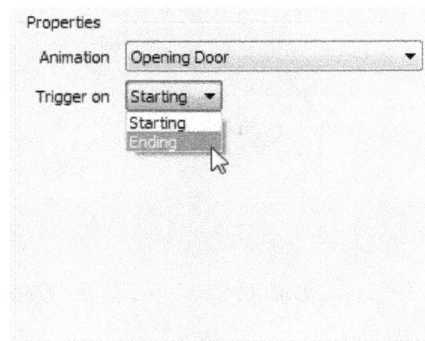

Figure 6-72 *Selecting the **Ending** option from the **Trigger on** drop-down list*

Adding Actions and Configuring Their Properties

In this section, you will add actions and configure their properties.

1. Choose the **Play Animation** button ▷ available at the bottom of the **Actions** area ; the **Play Animation** action is added to the **Actions** area, refer to Figure 6-73. Also, the properties of the **Play Animation** are displayed in the **Properties** area, as shown in Figure 6-74.

Figure 6-73 The **Play Animation** action added to the **Actions** area

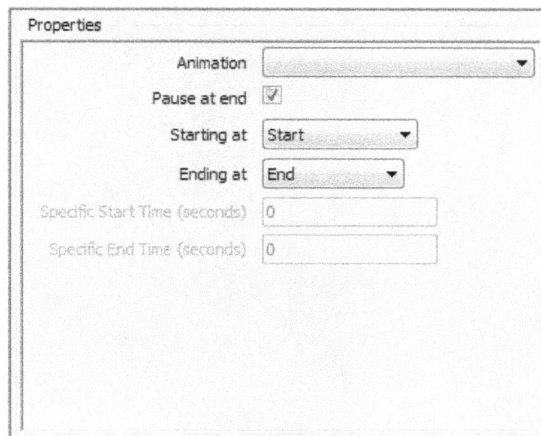

Figure 6-74 The **Play Animation** properties displayed in the **Properties** area

2. Select the **Opening Door** option from the **Animation** drop-down list in the **Properties** area, refer to Figure 6-75.

3. Ensure that the **Start** and **End** options are selected in the **Starting at** and **Ending at** drop-down lists. Also, make sure the **Pause at end** check box is selected, refer to Figure 6-76.

4. Repeat step 1 and add the **Play Animation** action again to the **Actions** area, refer to Figure 6-77. Its properties are displayed in the **Properties** area, as shown in Figure 6-78.

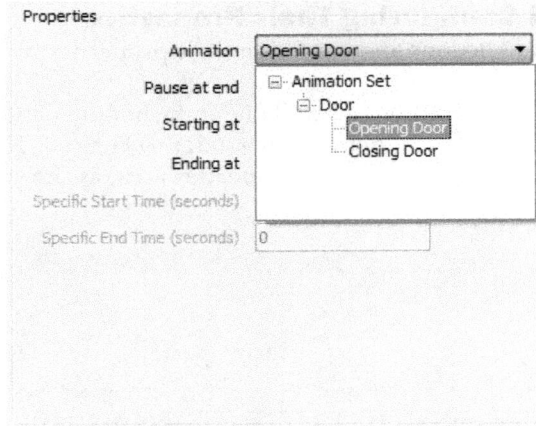

*Figure 6-75 Selecting the **Opening Door** animation from the **Animation** drop-down list*

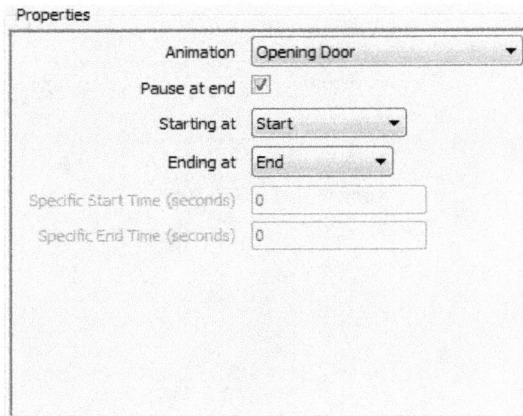

*Figure 6-76 The **Start** and **End** options selected in corresponding drop-down lists*

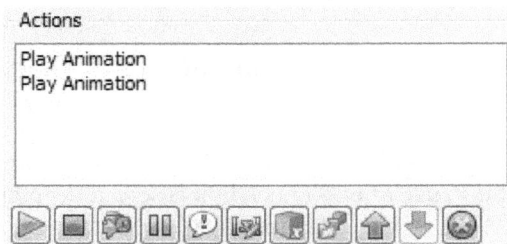

*Figure 6-77 The **Play Animation** actions added in the **Actions** area*

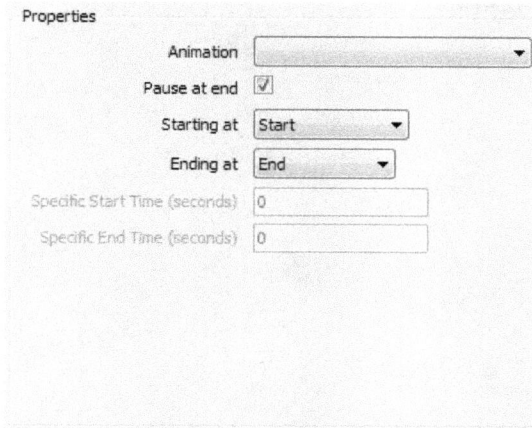

*Figure 6-78 The **Play Animation** properties displayed in the **Properties** area*

5. Select the **Closing Door** option from the **Animation** drop-down list in the **Properties** area, refer to Figure 6-79.

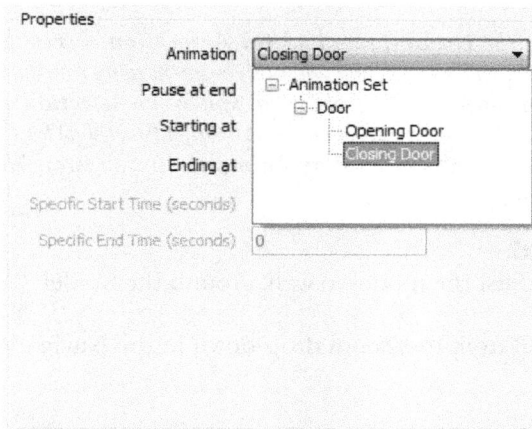

*Figure 6-79 Selecting the **Closing Door** option from the **Animation** drop-down list*

6. Ensure that the **Start** and **End** options are selected in the **Starting at** and **Ending at** drop-down lists, respectively. Also, make sure that the **Pause at end** check box is selected, refer to Figure 6-80.

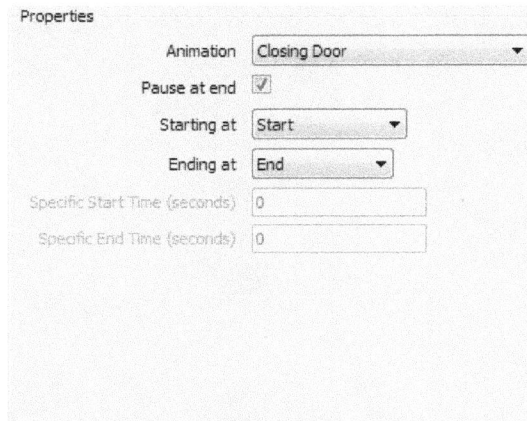

*Figure 6-80 The **Start** and **End** options selected in corresponding drop-down lists*

7. After specifying the properties, click on the **Auto Hide** button in the **Scripter** window to hide the **Scripter** window.

> **Tip**
> *Instead of adding the **On Keypress** and **On Animation** events, you can also add the **On Hotspot** event type. In this case, you will add the event twice and combine them using the **OR** operand. Select the door and then select the **Sphere on selection** option from the **Hotspot** drop-down list and specify the radius value as **0.06** m (0.164 ft) for the first event and **0.50** m (1.64 ft) for the second event. The **Play Animation** action parameters will be same as of the **On Key** press event.*

Adjusting the Model
In this section, you will adjust the model to walk around the model.

1. Choose the **Zoom** tool from the **Zoom** drop-down in the Navigation Bar; the shape of the cursor changes.

2. Press and hold the left mouse button and drag the mouse in the forward direction. Release the mouse button when the model is displayed, as shown in Figure 6-81.

Figure 6-81 Enlarged view of the model

Enabling Scripts
In this section, you will enable the scripts defined in the **Scripter** window.

1. Choose the **Enable Scripts** tool from the **Script** panel in the **Animation** tab; the **Scripter** window is deactivated.

Navigating Inside the Model
In this section, you will walk around the model and use the scripts for interaction.

1. Choose the **Walk** tool from the **Walk/Fly** drop-down in the Navigation Bar; the cursor changes into a feet shaped cursor.

2. Press and hold the left mouse button and drag the mouse in forward direction. Release the button when you reach near the door, as shown in Figure 6-82.

Figure 6-82 Walking towards the door

3. Press and release the **Up** arrow key from the keyboard; the door opens, as shown in Figure 6-83.

*Figure 6-83 Door opens after pressing the **Up** arrow key*

4. Walk inside the model; the **Closing Door** animation starts after completion of the **Opening Door** animation.

5. Using the **Zoom** tool, change the model view, as shown in Figure 6-84.

Figure 6-84 *Model after using the* **Zoom** *tool*

> **Tip**
> *In the* **Hotspot** *event type, use the avatar while navigating. The parameters for the avatar will be same as the parameters used in the Tutorial 1 of Chapter 2.*

Saving the Project
In this section, you will save the project.

1. To save the project with the current view, choose **Save As** from the Application Menu; the **Save As** dialog box is displayed.

2. Browse to the *nws_2017/c06_nws_2017_tut* folder and enter *c06_nws_2017_tut02* in the **File name** edit box, choose **Navisworks File Set (*.nwf)** file format from the **Save as type** drop-down list and then choose the **Save** button.

Self-Evaluation Test

Answer the following questions and then compare them to those given at the end of this chapter:

1. The _____ dialog box is used to modify the keyframes.

2. The _____ event activates a script using a particular key from the keyboard.

3. The _____ actions will play the animation when a script is triggered.

4. The _____ button in the **Scripter** window is used to add show viewpoint actions.

5. The _____ tool is used to activate scripts.

6. The **On Collision** event activates a script when collision occurs. (T/F)

7. Each scene in the **Animator** window can contain more than one camera animation. (T/F)

8. You can modify the scripts defined in the **Scripter** window after choosing the **Enable Scripts** button. (T/F)

9. The **Test Logic** option in the **Scripter** window is used to check the logic of an event condition. (T/F)

10. The **Test Action** option in the **Scripter** window is used to verify the execution of an action. (T/F)

Review Questions

Answer the following questions:

1. Which of the following buttons is used to add an action that closes the current model and loads a new model?

   a) **Play Animation**   b) **Store Property**
   c) **Load Model**   d) **Set Variable**

2. Which of the following events activate a script when the camera is within a particular range?

   a) **On Timer**   b) **On Start**
   c) **On Hotspot**   d) **On Animation**

3. Which of the following actions create a text file with a message?

   a) **Set Variable**   b) **Pause**
   c) **Send Message**   d) **Store Property**

4. Which of the following buttons is used to add the **Pause** action?

   a) **Play**   b) **Stop**
   c) **Pause**   d) **Load Model**

5. Which of the following buttons is used to add the **On Variable** event?

   a) **Add Scene**   b) **Add New Script**
   c) **On Variable**   d) **Move**

6. The **Animator** window contains the list of scripts. (T/F)

7. The **Capture keyframe** button is used to play the animation. (T/F)

8. The Time slider bar shows the end position of an animation. (T/F)

9. The **Ping Pong** option is used to play the animation indefinitely. (T/F)

10. In Navisworks, you cannot modify the keyframes. (T/F)

EXERCISE
Exercise 1   Creating Car Animation

Download and open the *c06_nws_2017_ex1* file from *http://www.cadcim.com*. Create the vehicle animation using the animation set option in the **Animator** window. Create scripts for stopping the vehicle when they reach the same point. Figure 6-85 shows the model to be used for this exercise.   **(Expected time: 1 hr)**

The following steps are required to complete this exercise:

a. Open the *c06_nws_2017_ex1* file.
b. Display the **Saved Viewpoints** window.
c. Display the **Animator** window.
d. Create scene folder.
e. Add scene to the scene folder.
f. Create Animation Set.
g. Create animation.
h. Play animation.
i. Save the project as *c06_navisworks_2017_ex01*.

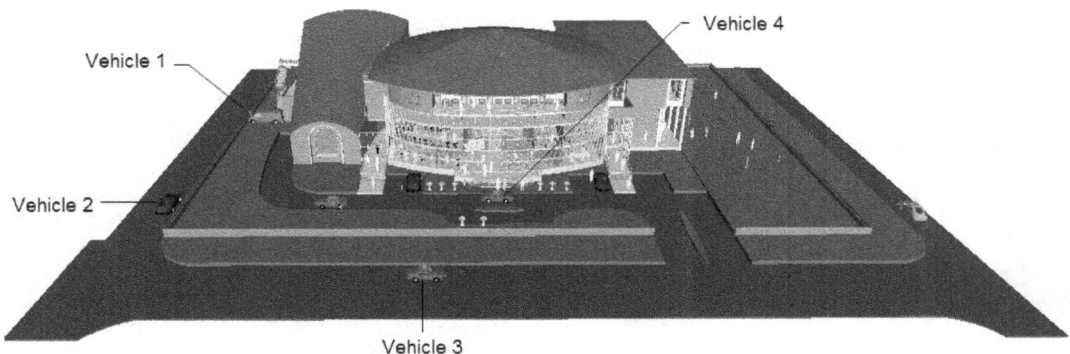

Figure 6-85 *The Building model*

Answers to Self-Evaluation Test
1. Edit Keyframe, 2. On Key Press, 3. Play Animation, 4. Show Viewpoint, 5. Enable Scripts, 6.T, **7.** F, **8.** F, **9.** T, **10.** T

Chapter 7

Quantification

Learning Objectives

After completing this chapter, you will be able to:
- *Understand the concept of Quantification*
- *Use various options in the Quantification Workbook window*
- *Use Item Catalog*
- *Use Resource Catalog*
- *Perform model and 2D takeoff*

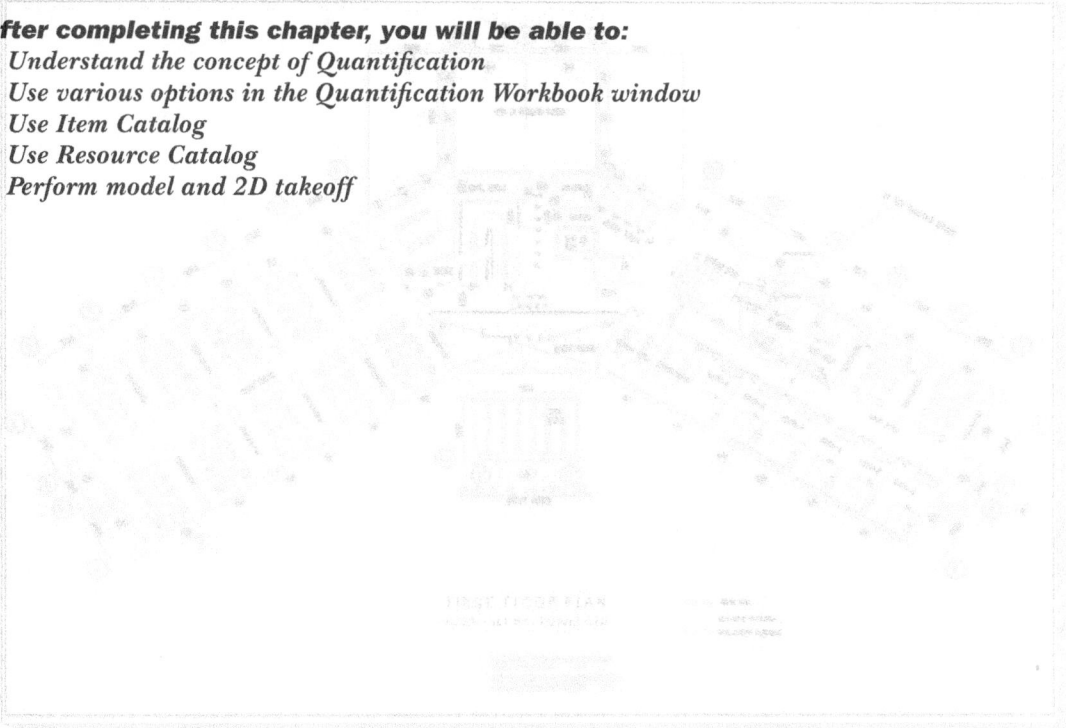

QUANTIFICATION

The material take off is the process of analyzing drawings, and preparing a list of all the materials that are required to complete the design. The list of materials includes the material quantities and material types.

In Navisworks, the **Quantification Workbook** window is used for carrying out the take off process.

QUANTIFICATION WORKBOOK WINDOW

The **Quantification Workbook** window contains various tools that are used for estimating the material quantities and their types, measuring areas, and counting bui   lding components. Instead of calculating the takeoff manually, you can use this feature for the estimation which helps in saving time as well as resources that are used for estimating the material quantities. To perform the take off, first you need to invoke the **Quantification Workbook** window. To do so, choose the **Quantification** tool from the **Tools** panel in the **Home** tab; the **Quantification Workbook** window will be displayed, as shown in Figure 7-1. Alternatively, to display the **Quantification Workbook** window, select the **Quantification Workbook** check box from **View > Workspace > Windows** drop-down. Various components in this window are discussed next.

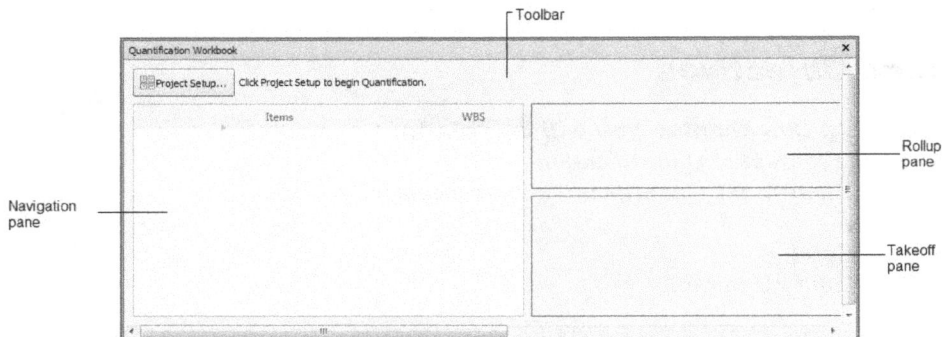

*Figure 7-1 Various components in the **Quantification Workbook** window*

Components of Quantification Workbook Window

The various components in the **Quantification Workbook** window are discussed next.

Navigation

In the **Quantification Workbook** window, the Navigation pane is available in the left side, refer to Figure 7-1. It displays a list of items and their WBS codes.

Rollup

In the **Quantification Workbook** window, the Rollup pane is available at the upper right side, refer to Figure 7-1. This pane displays the summary of takeoff items.

Takeoff

In the **Quantification Workbook** window, the Takeoff pane is available at the bottom right side, refer to Figure 7-1. This pane displays all the takeoff items.

Toolbar

In the **Quantification Workbook** window, the Toolbar is available at the top, refer to Figure 7-1. The Toolbar contains various buttons that are used for the takeoff process, refer to Figure 7-2. These buttons will be available only after creating a project. These buttons and options are discussed next.

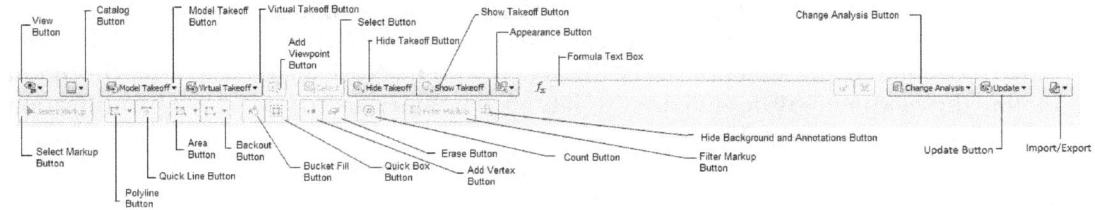

*Figure 7-2 Various buttons of the Toolbar in the **Quantification Workbook** window*

View

The **View** button is used to switch between the Item view and the Resource view. To change the view, choose this button; a drop-down list will be displayed. This drop-down list contains two options: **Item View** and **Resource View**. By default, the **Item View** option is selected. As a result, the Item view is displayed after creating a project in the **Quantification Workbook** window. Select the **Resource View** option to display the Resources view in the window.

Catalog

The **Catalog** button is used to display the Item and Resource catalogs. To display a catalog, choose this button; a drop-down list will be displayed. This drop-down list contains two options: **Item Catalog** and **Resource Catalog**. Select the **Item Catalog** option from the drop-down list to display the **Item Catalog** window. Select the **Resource Catalog** option to display the **Resource Catalog** window. These catalog window will be discussed later in the chapter.

Model Takeoff

The **Model Takeoff** button is used for quantity estimation of the selected model objects. Choose this button; a drop-down list will be displayed. This drop-down list contains two options: **Take off to Selected** and **Take off**. The **Take off to Selected** option is used to display the quantity takeoff in an existing item in the window. To display the takeoff in an existing item in the window, first select the required object and then select the catalog item in the Navigation pane. The **Take off** option is used to display the quantity takeoff in a new item in the **Quantification Workbook** window. Note that these options will be activated only after selecting the required objects.

Virtual Takeoff

The **Virtual Takeoff** button is used to create a new blank takeoff. Choose this button; a drop-down list will be displayed. This drop-down list contains two options: **Create In Selected Catalog Item** and **Create In New Catalog Item**. Select the **Create In Selected Catalog Item** option to create a takeoff in an existing item in the window. Select the **Create In New Catalog Item** option to create a takeoff in a new item in the **Quantification Workbook** window.

Add Viewpoint

The **Add Viewpoint** button is used to add a viewpoint to the takeoff. This button will be activated after taking off items. You can also use this button to update an existing viewpoint. To add a viewpoint to a takeoff, select the required item in the Takeoff pane. Make the required adjustments for the desired viewpoint. Next, choose the **Add Viewpoint** button; the viewpoint will be added in the Takeoff pane and is saved under the **Quantification Views** folder in the **Saved Viewpoints** window.

Select

The **Select** button is used to highlight the model objects associated with the selected take off in the Scene View. To do so, select the required item in the Takeoff pane. Next, choose the **Select** button; the object will be highlighted in the Scene View.

Hide Takeoff

The **Hide Takeoff** button is used to hide the taken off model objects in the Scene View. To do so, choose the **Hide Takeoff** button; the objects will be hidden in the Scene View.

Show Takeoff

The **Show Takeoff** button is used to display the model objects that have been taken off. To do so, choose the **Show Takeoff** button; the objects that have been taken off will be displayed in the Scene View.

Appearance

The **Appearance** button is used to control the model objects appearance. On choosing this button, a drop-down list will be displayed. This drop-down list contains two options: **Reapply Quantification Appearance** and **Reapply Original Model Appearance**. Select the **Reapply Quantification Appearance** option to apply the quantification appearance. On selecting this option, the object will appear transparent in the Scene View. Select the **Reapply Original Model Appearance** option to apply the original model appearance. On selecting this option, the original object will appear in the Scene View.

Formula

The **Formula** text box will display the formula associated with the takeoff such as model length, width, area, and so on. You can also change the formula in this text box. To do so, click on the required value in the Rollup pane; the related formula will be displayed in the **Formula** text box. Specify the new formula and choose the **Apply the formula** button; the formula will be changed.

Change Analysis

The **Change Analysis** button is used to analyze the model for changed or missing items. On choosing this button, a drop-down list will be displayed. This drop-down list contains two options: **Analyze Changes** and **Clear Analysis Results**. Select the **Analyze Changes** option from the list to analyze the changes. Select the **Clear Analysis Results** option to clear the results.

Update

The **Update** button is used to update the takeoff to match with the model. On choosing this button, a drop-down list will be displayed. This list contains four options: **Select Row**, **Update Selected From Model**, **Remove Overrides From Selected**, and **Delete Selected Takeoff**. Select the **Select Row** option to select a row. Select the **Update Selected From Model** option to update takeoff to match model properties. Select the **Remove Overrides From Selected** option to remove an overridden formula. Select the option **Delete Selected Takeoff** to delete the selected takeoff.

Import/Export

The **Import/Export** button is used to import and export the catalogs or quantities. On choosing this button, a drop-down list will be displayed. This drop-down list contains following options: **Import Catalog**, **Export Catalog to XML**, **Export Quantities to Excel**, and **Export Selected Quantities to Excel**. These options are discussed next.

The **Import Catalog** option is used to import a catalog. On selecting this option, the **Import Catalog** dialog box will be displayed. Browse to the location and select the file. Next, choose the **Open** button; the selected catalog will be loaded in the **Quantification Workbook**.

The **Export Catalog to XML** option is used to export the catalog to an XML file. On selecting this option, the **Export Catalog to XML** dialog box will be displayed. In the dialog box, browse to the required location to save the file and choose the **Save** button; the catalog will be saved to the XML file in the specified location.

The **Export Quantities to Excel** option is used to export the takeoff quantities to an excel file. On selecting this option, the **Export Quantities to Excel** dialog box will be displayed. In the dialog box, browse to the required location to save the file and choose the **Save** button; the file will be saved to an excel file at the specified location.

The **Export Selected Quantities to Excel** option is used to export the selected takeoff quantities to an excel file. To do so, first select the required takeoff quantity in the Navigation pane. Next, choose the **Import/Export** button and then select the **Export Selected Quantities to Excel** option; the **Export Quantities to Excel** dialog box will be displayed. In the dialog box, browse to the required location, specify a name, and choose the **Save** button; the file will be saved to an excel file at the specified location.

Next, you will learn about the buttons used for 2D takeoff. These tools will be enabled only in the 2D DWF sheet. You can import a 2D sheet in Navisworks using the **Project Browser** window.

The process of importing 2D sheet has been discussed earlier in the first chapter.

Select Markup

After performing the 2D takeoff in a sheet, the takeoff item will be labeled in different colors which are called markups. The **Select Markup** button is used to select a single as well as multiple takeoff markups in a 2D sheet. You can use this facility only if there are takeoff markups in the sheet. For example, if you have performed the 2D takeoff using the rectangle, polyline, area, and other functions then several markups will be generated. To select a single takeoff markup, choose this button and click on the desired markup. To select multiple takeoff markups, choose this button, press and hold the CTRL key and click on the desired markups. Alternatively, drag a box around the markup to select it. After selecting the markup, when you will hover the cursor on the selected markup its vertices will change to solid color.

Polyline

The **Polyline** button is used to measure the length and perimeter of the geometries which are non-standard in shape. To measure the length and perimeter of the geometries, choose this button; a drop-down list will be displayed. Select the **Polyline (L)** option from the list to perform polyline takeoff. To perform takeoff for a single line such as a single wall, place the cursor on the wall and click to specify the start point. Click for the next point and then right-click to specify this point as the end point. On doing so, the takeoff will be calculated and displayed in the Takeoff pane. Similarly, to calculate takeoff for multiple line segments, you can draw them by moving and clicking at the required places.

You can also draw rectangular markups for measuring room perimeter. To do so, select the **Rectangle Polyline (L)** option from the drop-down list. Click to specify the start point and then drag the cursor to draw a rectangle.

Note
Notice that the green cursor snaps at the vertex while drawing polylines. To draw perfect horizontal and vertical lines, use the SHIFT key while drawing polylines.

Quick Line

The **Quick Line** button is used for creating a linear takeoff. To do so, choose this button and move the cursor in the Scene View; a pick cursor will be displayed. Now, place the cursor on the required model segment; a red line will be displayed. Click on the red line; a marking menu will appear in the Scene View, refer to Figure 7-3. You can select any of the marking options from this menu, the line will be drawn according to the selected option and the takeoff will be displayed in the Takeoff pane of the window.

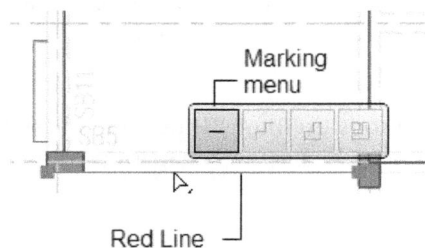

Figure 7-3 Red line along with the Marking menu

Area

The **Area** button is used to perform an area takeoff. On choosing this button, a drop-down list will be displayed which contains two options: **Area (A)** and **Rectangle Area (A)**. To perform the area takeoff using a series of line segments, select the **Area (A)** option from the list. Move the cursor in the Scene View; a plus cursor will be displayed. Click on the 2D sheet to specify the start point and then continue clicking to form an enclosed region. Right-click on the last vertex to exit the drawing; the properties of the created area will be displayed in the Takeoff pane of the **Quantification Workbook** window.

Similarly, to perform the area takeoff by drawing a rectangle or a square, select the **Rectangle Area (A)** option from the list. Specify the start point on the 2D sheet and then drag the cursor to draw the rectangle.

Backout

The **Backout** button is used to cut the portion from the geometry which is to be excluded from the area markup. This button will be activated only after selecting a takeoff markup. Using this button, you can remove a takeoff measurement from the closed polyline area or a rectangle geometry. When you choose this button, a drop-down list will be displayed. Select the **Backout (B)** option from the list to cut a closed polyline area from the area markup. Select the **Rectangle Backout (B)** option to cut a rectangular area from the area markup. Figure 7-4 shows the area markup with a rectangular backout. Here, the area of the created backout will be excluded from the markup area. You will notice that the backout area is displayed in white color.

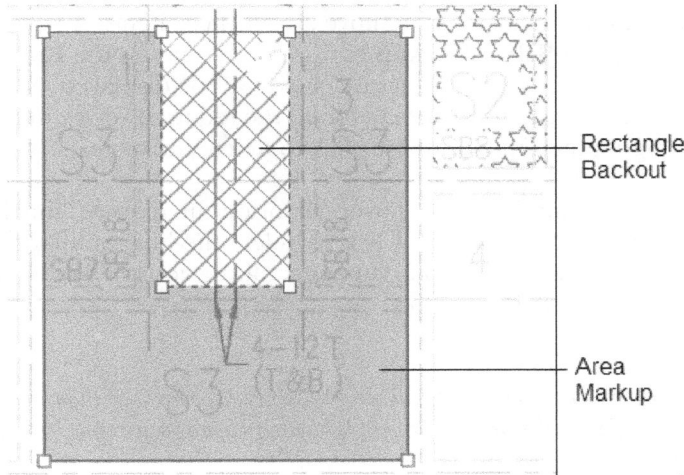

Figure 7-4 Area markup with rectangular backout

Bucket Fill

The **Bucket Fill** button is used to create a linear or area takeoff by clicking within a geometry. To create a linear or area takeoff, choose this button and move the cursor in the 2D sheet; a bucket shaped cursor will be displayed. Now, click in the required enclosed geometry; the region will be highlighted in red. Next, press ENTER; the takeoff will be displayed in the **Takeoff** pane of the **Quantification Workbook** window.

> **Tip**
> *You can switch between the linear and area takeoff by clicking in the geometry boundary. The linear takeoff will be denoted by linear boundary and the area takeoff will be denoted by semi transparent fill.*

Quick Box

The **Quick Box** button is used to perform a quick takeoff by selecting multiple parts of a geometry. To perform a quick takeoff, first select the required item from the Navigation pane and then choose the **Quick Box** button. Next, in the 2D sheet, drag the cursor to create a selection box around the desired geometry; the selected items will be highlighted and a **Quick Box** menu will be displayed. Choose the required option from the menu; the geometry selection will be highlighted and the takeoff will be displayed in the window.

Add Vertex

The **Add Vertex** button is used to add a vertex to the existing markup geometry. This button can also be used for breaking line segments and adding more line segments to a polygon. To add a vertex to an existing markup, choose the **Select Markup** button and select a markup in the 2D sheet and then choose the **Add Vertex** button. Next, place the cursor in the 2D sheet; a plus cursor will be displayed. Now, in the markup area, click at the positions where you want to add the vertex.

Erase

The **Erase** button is used to remove any vertex, line, polygon, and any unwanted geometry in the markup. To remove any of these items, first you need to select the markup area using the **Select Markup** button. Next, choose the **Erase** button and place the cursor in the 2D sheet; the erase cursor will be displayed. Now, to delete a vertex in the markup area, hover the cursor on the required vertex in the markup area; the vertex will be highlighted in blue. Now, click to delete the vertex.

To delete a line in the markup area, hover cursor on the required line in the markup area; the line will be highlighted in blue. Click to delete the line. Similarly, to delete an entire markup, hover the cursor on the markup area; all the vertices will be highlighted in blue. Click on the markup to delete it.

> **Note**
> *If you delete a vertex from a polyline or a rectangular markup, then the polygon will be closed automatically. If you delete a line segment from an area markup or polyline, then the entire markup will be deleted.*

Count

The **Count** button is used to create a count takeoff for the objects in a model. To create a count takeoff, choose the **Count** button and move the cursor in the 2D sheet; the plus cursor will be displayed. Click on the required objects in the 2D sheet; a circular pin will be added to the objects and the takeoff will be displayed in the Takeoff pane of the **Quantification Workbook** window.

Filter Markup

The **Filter Markup** button is used to display the markups of takeoff items selected in the **Quantification Workbook** in the 2D sheet. To filter the markups, first select the takeoff items or group of items in the Navigation pane and then choose the **Filter Markup** button; the selected items will be highlighted in orange color and the takeoff markups of the selected item will be displayed in the 2D sheet.

Hide

The **Hide** button is used to hide the background and annotations. Next, you will learn about creating a project in the **Quantification Workbook** window.

Creating a Project

In Quantification, a project is a collection of files and takeoff items that are used for estimating material quantities. In Navisworks, there are three predefined project templates that you can use for takeoff calculations. These project templates are also known as catalogs. To create a project, first open the model to be used for takeoff. Next, choose the **Project Setup** button from the Toolbar in the **Quantification** window; the **Quantification Setup Wizard** will be displayed, as shown in Figure 7-5. The various pages in the wizard are discussed next.

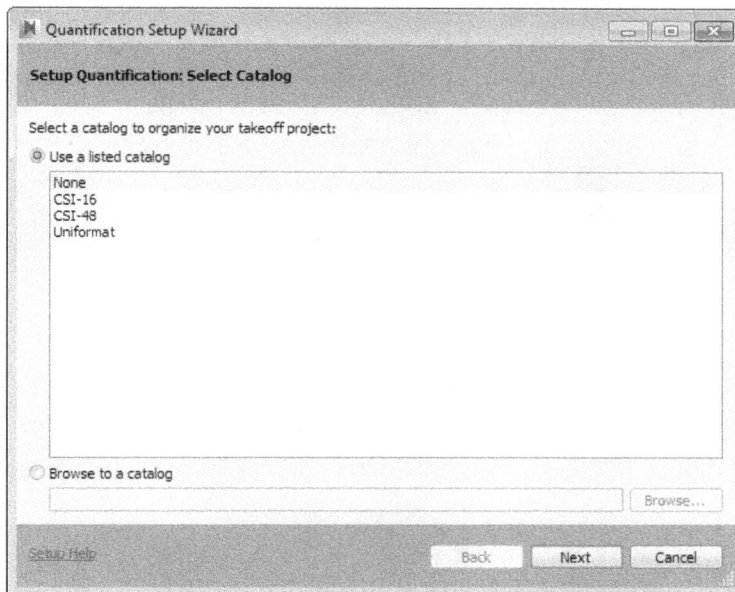

*Figure 7-5 The **Setup Quantification: Select Catalog** page in the Quantification Setup Wizard*

Setup Quantification: Select Catalog Page

In the **Setup Quantification: Select Catalog** page of the wizard, you can select the required catalog. By default, the **Use a listed catalog** radio button is selected. As a result, a list of predefined catalogs will be displayed in the list box. You can select the required catalogs from this list. You can also select a custom catalog. To do so, select the **Browse to a catalog** radio button;

the corresponding **Browse** button will be enabled. Next, choose the **Browse** button; the **Open** dialog box will be displayed. In this dialog box, browse and select the required catalog and then choose the **Open** button; the path to the selected catalog will be displayed in the **Browse to a catalog** edit box. Now, choose the **Next** button; the **Setup Quantification: Select Units** page will be displayed, as shown in Figure 7-6. The options in this page are discussed next.

Setup Quantification: Select Units Page

In the **Setup Quantification: Select Units** page, you can select the required units that you need for measuring quantities. There are three measurement units in this page, refer to Figure 7-6. These units are discussed next.

Imperial

The **Imperial** units are used to convert the model unit values to imperial units such as feet and pounds. Select the **Imperial** radio button for using it in a project.

Metric Units

The **Metric** units are used to convert the model unit values to metric units such as metres and kilograms. Select the **Metric** radio button for using it in a project.

Variable

The **Variable** option is used to use the existing model units. On selecting the **Variable** radio button, you can change the unit of individual takeoff property in the **Setup Quantification: Select Takeoff Properties** page. The various options in this page are discussed next.

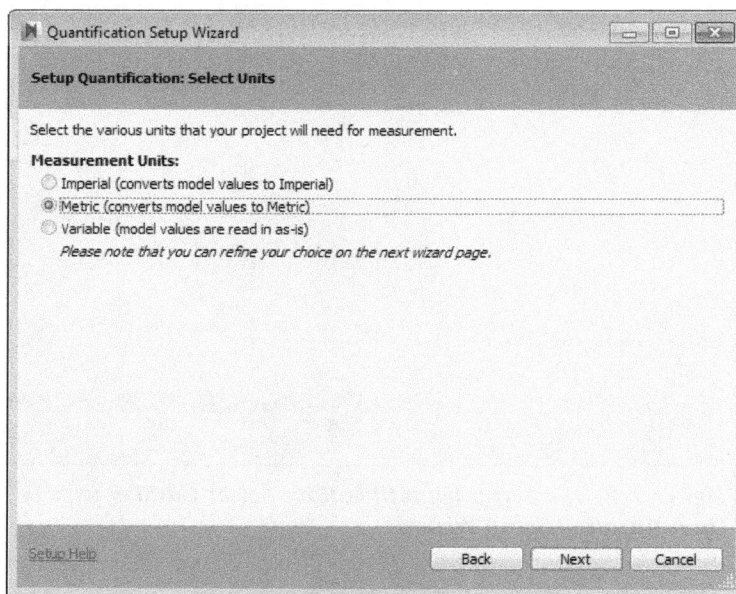

Figure 7-6   *The **Setup Quantification: Select Units** page in the **Quantification Setup Wizard***

Setup Quantification: Select Takeoff Properties Page

In the **Setup Quantification: Select Takeoff Properties** page, you can select the model properties and assign units to the model properties which will be used while calculating the takeoff. In this page, there are two columns **Takeoff Property** and **Units**, as shown in Figure 7-7. These two columns are discussed next.

Takeoff Property Column

The **Takeoff Property** column displays a list of all the general properties of the model such as model length, width, thickness, and so on. To specify a property, select the required property check box in the **Takeoff Property** column.

Units Column

The **Units** column displays the unit set for each model property. You can select the required unit from the corresponding drop-down list. The units in this drop-down list depend on the unit that you have selected from the **Setup Quantification: Select Units** page.

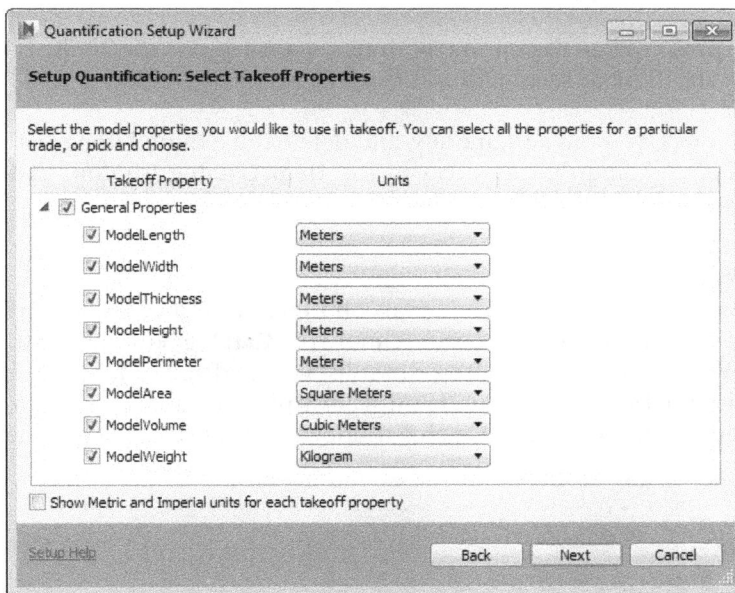

*Figure 7-7 The **Setup Quantification: Select Takeoff Properties** page in the Quantification Setup Wizard*

After specifying the model properties and units, you can select the **Show Metric and Imperial units for each takeoff property** check box to display both the units in the project for each property. Now, choose the **Next** button; the **Setup Quantification: Ready to Create Database** page will be displayed. Next, choose the **Finish** button in this page; the **Quantification Setup Wizard** will be closed and the project will be created. Also, the catalog items will be displayed in the **Quantification Workbook** window, as shown in Figure 7-8.

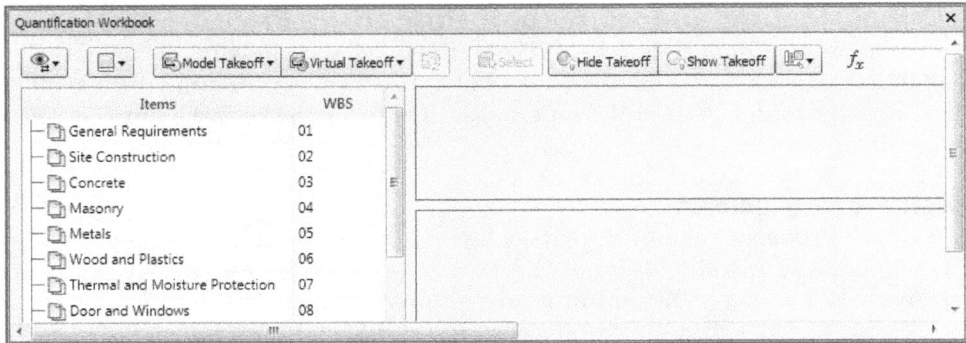

*Figure 7-8 Partial view of the catalog items displayed in the **Quantification Workbook** window*

Note
*The **Show Metric and Imperial units for each takeoff property** check box will be disabled on selecting the **Variable** radio button in the **Setup Quantification: Select Units** page.*

After creating the project, note that various buttons are displayed in the Toolbar that have already been discussed in the Toolbar section of this chapter.

After creating a project, you can add, modify, and delete the required items in the **Item Catalog** window. You can also define the resources in the **Item Catalog** window. The **Item Catalog** window is discussed next.

Item Catalog
The **Item Catalog** window contains a list of takeoff items. You can also define resources in this window. To display the **Item Catalog** window, choose the **Catalog** button from the Toolbar in the **Quantification Workbook** window; a drop-down list will be displayed. Select the **Item Catalog** option from the list; the **Item Catalog** window will be displayed, as shown in Figure 7-9. Figure 7-9 shows various components of the **Item Catalog** window and these components are discussed next.

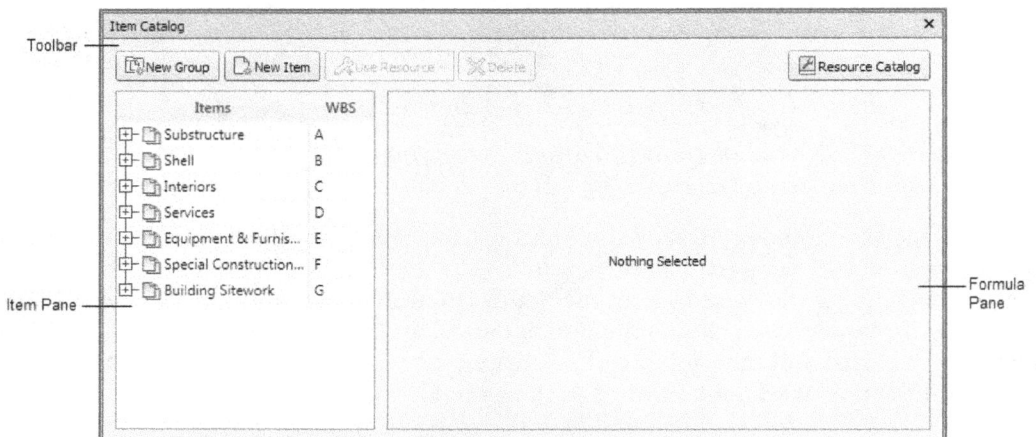

*Figure 7-9 Various components of the **Item Catalog** window*

Item Pane

The Item pane, as shown in Figure 7-9, contains a list of all takeoff items. This list is same as the **Item View** list displayed in the **Quantification Workbook** window. You can add more groups and items in the Item pane by using the buttons available in the Toolbar.

Toolbar

The Toolbar in the **Item Catalog** window contains buttons which are used for creating groups, adding resources, and adding items in the groups. These buttons are discussed next.

New Group

The **New Group** button is used to create a new group in the Item pane. On choosing this button, a new group will be created and displayed with a default name **New Group** in the Item pane of the window. To rename the group, specify the required name in the edit box and press ENTER.

New Item

The **New Item** button is used to add a new item in a group. To do so, first select the required group and then choose the **New Item** button; a new item will be created and displayed with a default name **New Item** in the Item pane of the window. To rename this item, specify the required name in the edit box and press ENTER.

Use Resource

The **Use Resource** button enables you to use a master resource for the catalog item. First select the required item in the Item pane. Next, choose the **Use Resource** button; a drop-down list will be displayed. This list contains two options: **Use New Master Resource** and **Use Existing Master Resource**.

The **Use New Master Resource** option is used to create a new resource. On selecting this option, the **New Master Resource** dialog box will be displayed, as shown in Figure 7-10.

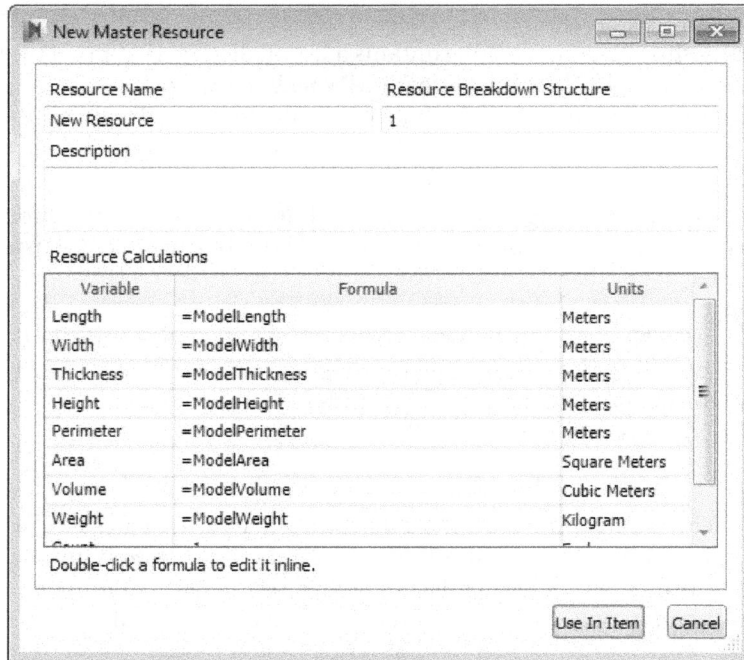

*Figure 7-10 The **New Master Resource** dialog box*

In this dialog box, specify the name of the resource in the **Resource Name** text box. The resource breakdown structure number will be displayed in the **Resource Breakdown Structure** edit box. Resource breakdown structure is a hierarchical list of resources. Next, specify the resource description in the **Description** area. The **Resource Calculations** area displays a list of variables, related formulas, and their units. You can change the formulas and their units by clicking in their corresponding columns. After specifying the required parameters, choose the **Use In Item** button; the defined resource will be assigned to the selected item in the Item pane of the **Item Catalog** window.

The **Use Existing Master Resource** option helps you to use the existing resources. On selecting this option, the **Master Resource List** dialog box will be displayed, as shown in Figure 7-11. This dialog box displays a list of all the existing resources. Select the required resource from the list displayed in this dialog box and then choose the **Use In Item** button; the selected resource will be added in the Item pane of the catalog.

Note

*The **Use Existing Master Resource** option will be available only if you have earlier added resources to the **Resource Catalog**. You will learn more about it later in the chapter.*

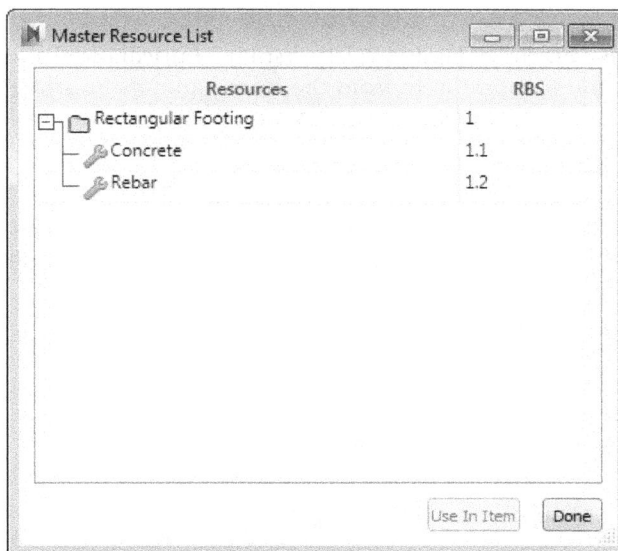

*Figure 7-11 The **Master Resource List** dialog box*

Delete

The **Delete** button is used to delete the selected group or item from the Item pane. To do so, first select the group or item to be removed from the Item pane and then choose the **Delete** button; the selected item or group will be removed.

Property Mapping

The **Property Mapping** button is used for mapping the takeoff properties to the model properties. The mapping will be applied to the entire 2D sheet and 3D model takeoff calculation. On choosing this button, the **Property Mapping** dialog box will be displayed. In this dialog box, choose the **Add Mapping Rule** button; a row will be added to the dialog box. Double-click in the **Takeoff Property** column; a drop-down list will be displayed. Select the takeoff property to be mapped. Next, select the corresponding category to be mapped from the drop-down list available in the **Category** column. Similarly, select the property from the drop-down list available in the **Property** column. The properties displayed in this drop-down list will be the model properties. To delete any of the takeoff properties or categories, select that property or category and press DELETE.

To remove a row from the dialog box, select the required row and choose the **Remove Mapping Rule** button. Choose the **OK** button to close the dialog box and apply the mapping rule.

Resource Catalog

Choose the **Resource Catalog** button to open the **Resource Catalog** window. The **Resource Catalog** window is discussed later in this chapter.

Formula Pane

The Formula pane, as shown in Figure 7-12, displays information and formula related to different items. Select the required item from the Item pane; its information will be displayed in the Formula pane, as shown in Figure 7-12. Now, choose the Close button; the dialog box will be closed.

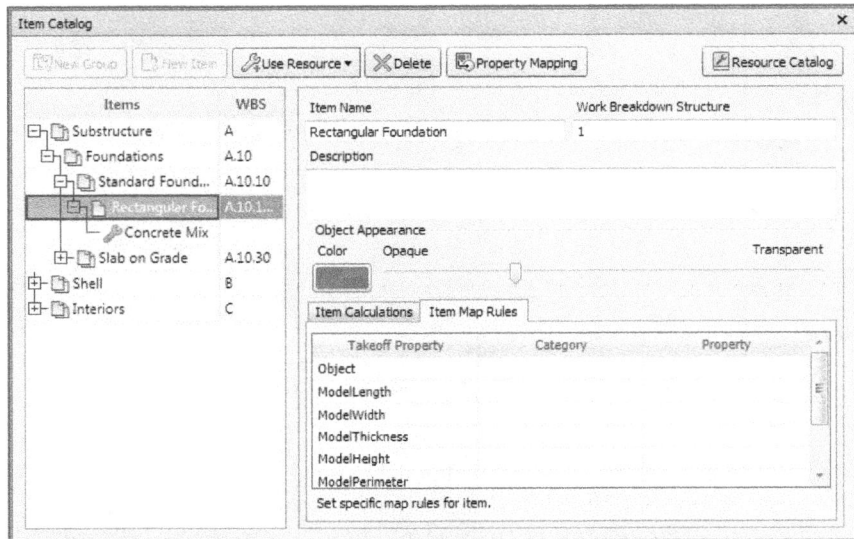

*Figure 7-12 The Formula pane in the **Item Catalog** window*

In the Formula pane, the item name and its Work Breakdown Structure will be displayed in the **Item Name** and **Work Breakdown Structure** text boxes, refer to Figure 7-12.

In the **Object Appearance** area, you can define the appearance of the object after quantification. To change the color of the takeoff object, choose the **Color** button; the **Color** dialog box will be displayed. In the **Color** dialog box, select the desired color and choose the **OK** button; the color will be changed. You can also change the transparency by dragging the slider.

The **Item Calculations** tab displays a list of variables, related formulas, and units in a tabular format. To change a formula, click in the formula cell corresponding to the required variable; the cell will become editable. Specify the new formula in the edit box and press ENTER. Similarly, to change the unit, click in the required cell; a drop-down list will be displayed. Select the desired unit from the list.

In the **Item Map Rules** tab, you can map individual takeoff properties. For example, to map model length property, click in the **Model Length** cell. Next, specify the category by clicking in the **Category** column; a drop-down list will be displayed. Select the required option from the list. Similarly, select an appropriate option in the **Property** column. Now, you can save the newly mapped property for an item by saving the file.

Resource Catalog

The **Resource Catalog** contains a list of all the resources for a project. This resource list may include materials, equipments, or structural components. To display the **Resource Catalog** window, choose the **Catalog** button from the Toolbar in the **Quantification Workbook** window; a drop-down list will be displayed. Select the **Resource Catalog** option from the list; the **Resource Catalog** window will be displayed, as shown in Figure 7-13. Alternatively, to invoke the **Resource Catalog** window, choose the **Resource Catalog** button from the **Item Catalog** window. Figure 7-13 shows various components of the **Resource Catalog** window, which are discussed next.

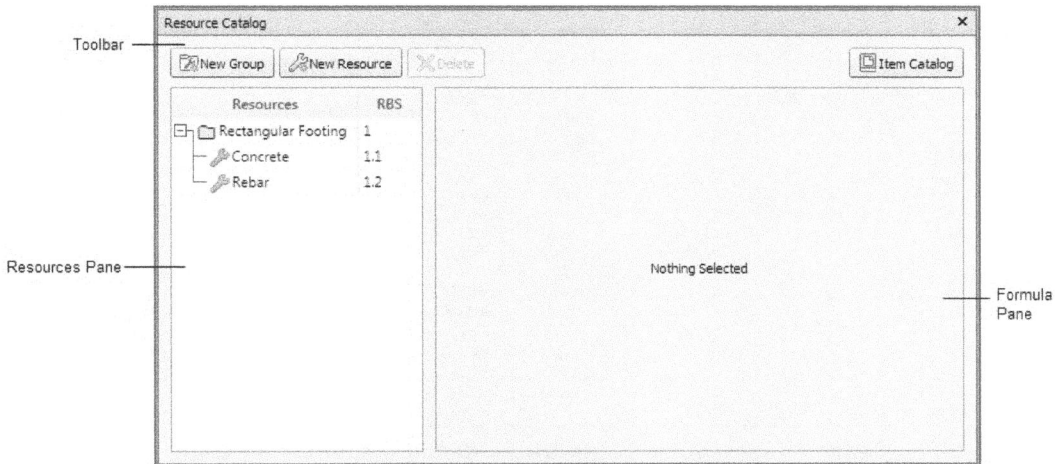

*Figure 7-13 Various components of the **Resource Catalog** window*

Toolbar

The Toolbar contains the buttons that are used for creating resource groups and resources. These buttons are discussed next.

New Group

The **New Group** button is used to create a new group in the **Resource Catalog** window. This group will contain the defined resources. To create a group, choose the **New Group** button; a new group will be created and displayed with a default name in the edit box in the Resource pane. To rename the added group, specify the new name in the edit box and press ENTER.

New Resource

The **New Resource** button is used to create a new resource in the **Resource Catalog** window. To create a resource, first select the created group in the Resource pane. Next, choose the **New Resource** button; the new resource will be created and displayed with a default name in the edit box in the Resource pane. To rename the added resource, specify the desired resource name in the edit box and press ENTER.

Item Catalog

The **Item Catalog** button is used to display the **Item Catalog** window.

Resource Pane

The Resource pane displays a list of all the added resources, refer to Figure 7-14.

Formula Pane

The Formula pane displays the resource information and formulas related to it. Select the required resource from the Resource pane; its information will be displayed in the Formula pane, as shown in Figure 7-14.

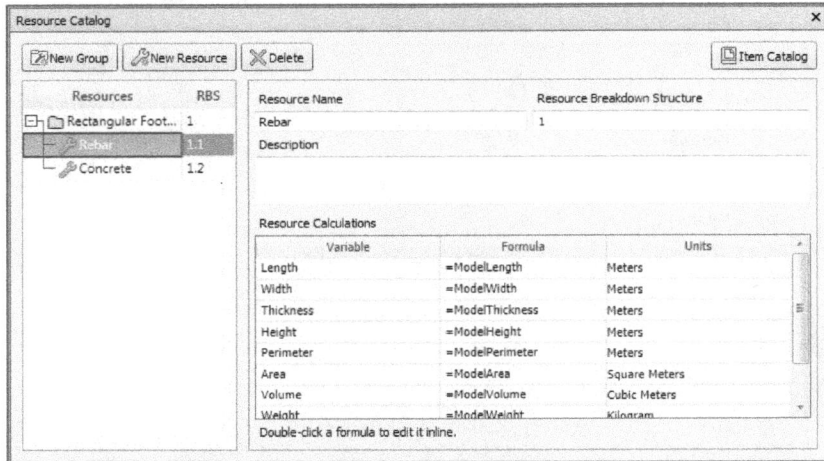

*Figure 7-14 The Formula pane in the **Resource Catalog** window*

In the Formula pane, the resource name and its RBS will be displayed in the **Resource Name** and **Resource Breakdown Structure** text boxes, respectively, refer to Figure 7-14.

The **Resource Calculations** area displays a list of variables, related formulas, and units in a tabular format. To change the formula, click in the formula cell corresponding to the required variable; the cell will become editable. Specify the new formula in the edit box and press ENTER. Similarly, to change the unit, click in the required cell; a drop-down list will be displayed. Select the desired units from the list.

TAKEOFF METHODS

In Quantification, there are two takeoff methods: Model Takeoff and Virtual Takeoff. These two methods are also known as Automatic Takeoff and Manual Takeoff, respectively. These methods are discussed next.

Model Takeoff

In the Model Takeoff method, the objects included in the files are estimated. Using this method, you can perform the takeoff for the objects that have properties mapped in Navisworks. The Model Takeoff method is discussed next.

In the Model Takeoff method, first you need to select the required object(s) in the Scene View or in the **Selection Tree** window. Next, choose the **Model Takeoff** button in the **Quantification**

Workbook toolbar; a drop-down list will be displayed. Now, select the **Take off to New Catalog Item** option from the drop-down list to take off the selected object(s) to a new catalog item. On selecting this option, a new item will be added to the Navigation pane of the **Quantification Workbook** window and the take off results will be displayed in the Takeoff pane. The summary will be displayed in the Rollup pane, as shown in Figure 7-15.

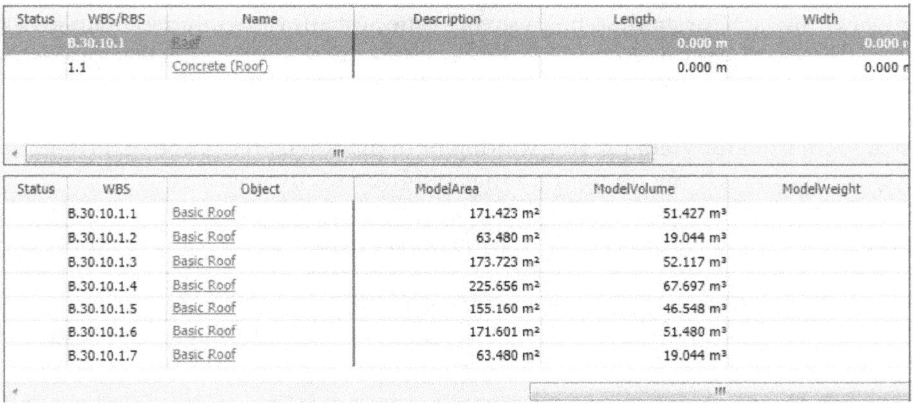

Status	WBS/RBS	Name	Description	Length	Width
	B.30.10.1	Roof		0.000 m	0.000 r
	1.1	Concrete (Roof)		0.000 m	0.000 r

Status	WBS	Object	ModelArea	ModelVolume	ModelWeight
	B.30.10.1.1	Basic Roof	171.423 m²	51.427 m³	
	B.30.10.1.2	Basic Roof	63.480 m²	19.044 m³	
	B.30.10.1.3	Basic Roof	173.723 m²	52.117 m³	
	B.30.10.1.4	Basic Roof	225.656 m²	67.697 m³	
	B.30.10.1.5	Basic Roof	155.160 m²	46.548 m³	
	B.30.10.1.6	Basic Roof	171.601 m²	51.480 m³	
	B.30.10.1.7	Basic Roof	63.480 m²	19.044 m³	

Figure 7-15 The takeoff results displayed in the Takeoff and Rollup pane

The **Take off to Selected Catalog Item** option is used to take off the selected objects to the item selected in the Navigation pane. This option will be available only when you choose an item in the Navigation pane of the **Quantification Workbook** window. To do so, first define an item in the **Item Catalog** window. Next, select the item in the Navigation pane of the **Quantification Workbook** window. Next, select the required object in the Scene View or in the **Selection Tree** window. Now, select the **Take off to Selected Catalog Item** option from the drop-down list; the take off results and its summary will be displayed in the Takeoff pane and Rollup pane.

Virtual Takeoff

The Virtual Takeoff method is used to takeoff those objects that are not linked to a model object or the objects which contain no properties. You can use this method to takeoff the objects that have geometry but no properties.

To takeoff the objects that have geometry but no properties, first you need to add a desired viewpoint so that you can refer to it whenever required. Next, select the item in the Navigation pane and choose the **Virtual Takeoff** button; a drop-down list will be displayed. Select the **Create In Selected Catalog Item** option from the list; the takeoff will be displayed in the Takeoff and Rollup panes.

You can select the **Create In New Catalog Item** option from the drop-down list to takeoff to a new catalog item. You will notice that in the Takeoff pane, the takeoff values are 0. Now, measure the objects by using any of the measurement tools. Specify these measured values such as **Model Length**, **Model Width**, **Area**, and so on in the takeoff pane.

You can save the measurement views by choosing the **Convert to Redline** tool from the **Measure** panel in the **Review** tab. You can also save these views by choosing the **Add Viewpoint** button, for later references.

2D Takeoff

The 2D Takeoff method is used to perform takeoffs such as calculation of area, length, or count of objects in a 2d sheet. This method helps saving time and effort required in manual calculation. To perform 2D takeoff, you need to load the required 2D plan. For this you can use native or scanned DWF files. The workflow involved in performing a 2D Takeoff is given below:

1. Select an item from the **Item Catalog** window or in the Navigation pane of the **Quantification Workbook** window. Create an item if it has not been defined earlier.

2. Invoke any of the takeoff tools such as **Area**, **Polyline**, or **Quick Line** from the toolbar.

3. In the 2D sheet, create the required markup using the selected markup tool; the takeoff will be performed and displayed in the Takeoff pane of the **Quantification Workbook** window.

TUTORIALS

General instructions for downloading tutorial files:

1. Download the *c07_nws_2017_tut* zip file for this tutorial from *http://www.cadcim.com*. The path of the file is as follows: *Textbooks > Civil/GIS > Navisworks > Exploring Autodesk Navisworks 2017*.

2. Now, save and extract the downloaded folder at the following location: *C:\ nws_2017\c06_nws_2017_tut*

> **Note**
> *The default unit system used in the tutorials is metric. To change the unit to imperial, select the required unit from **Options Editor > Interface > Display Units**.*

Tutorial 1   Model Takeoff

In this tutorial, you will open *c07_navisworks_2017_tut1* file and perform quantity takeoff of various building components by using the options available in the **Quantification Workbook** window.   **(Expected time: 45min)**

The following steps are required to complete this tutorial:

a. Open the *c07_navisworks_2017_tut1* file.
b. Display the **Quantification Workbook** window.
c. Set up the project.
d. Perform the takeoff.
e. Save the project.

Opening the File

In this section, you will open the file to be used in this tutorial.

1. Choose the **Open** button from the Quick Access Toolbar; the **Open** dialog box is displayed.

2. In this dialog box, browse to the following location:
    C:\nws_2017\c07_nws_2017_tut

3. Select **Navisworks File Set (*.nwf)** from the **Files of type** drop-down list.

4. Select the file **c07_navisworks_2017_tut1** from the displayed list of files; the file name appears in the **File name** edit box.

5. Next, choose the **Open** button in the dialog box; the model is displayed in the Scene View, as shown in Figure 7-16.

Figure 7-16 *Model displayed in the Scene View*

Displaying the Quantification Workbook Window

In this section, you will display the **Quantification Workbook** window.

1. Choose the **Quantification** tool from the **Tools** panel in the **Home** tab; the **Quantification Workbook** window is displayed, as shown in Figure 7-17.

Note

*If you are starting Quantification for the first time, the **Quantification Getting Started Wizard** message box is displayed. In this message box, you can choose the **Never** button.*

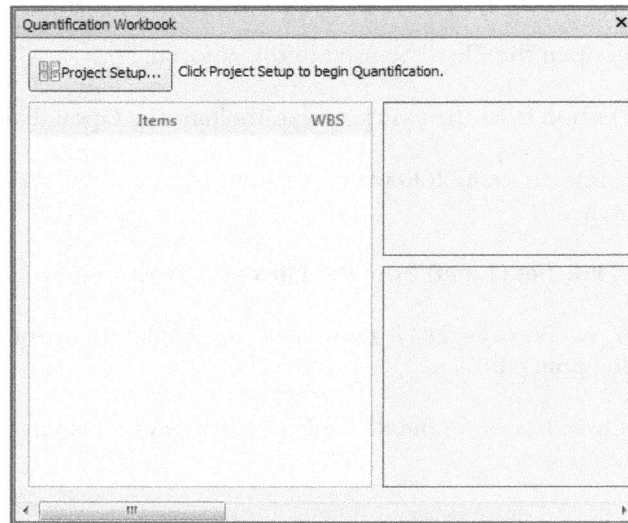

Figure 7-17 *The **Quantification Workbook** window*

Setting up the Project

In this section, you will browse to the catalog and set up a new Quantification session.

1. Choose the **Project Setup** button available at the top of the **Quantification Workbook** window, refer to Figure 7-17; the **Quantification Setup Wizard** dialog box is displayed, as shown in Figure 7-18.

2. In the **Setup Quantification: Select Catalog** page of the wizard, select the **Browse to a catalog** radio button; the **Browse** button is enabled.

3. Next, choose the **Browse** button; the **Open** dialog box is displayed.

4. In the **Open** dialog box, browse to the following location:
    C:\nws_2017\c07_nws_2017_tut

5. Select the **Quantification** xml file displayed in the file list.

6. Next, choose the **Open** button; the path of the selected catalog is displayed in the **Browse** edit box.

7. Now, choose the **Next** button in the **Setup Quantification: Select Catalog** of the wizard; the **Setup Quantification: Ready to Create Database** page is displayed, as shown in Figure 7-19.

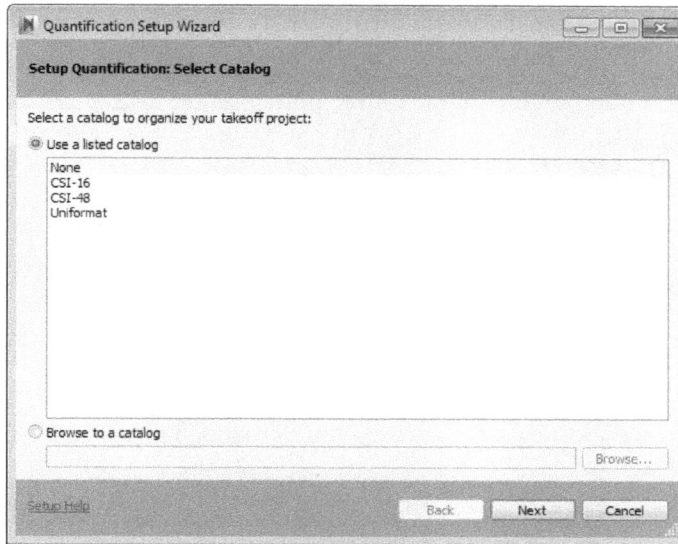

Figure 7-18 The **Quantification Setup Wizard** *dialog box*

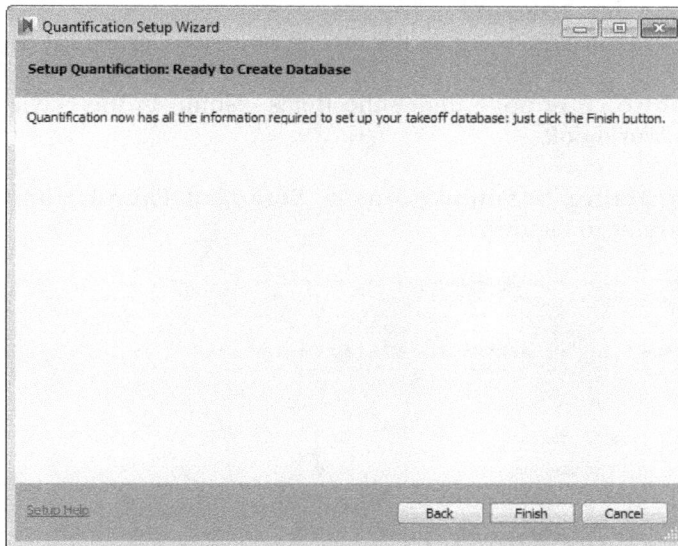

Figure 7-19 The **Setup Quantification: Ready to Create Database** *page in the wizard*

8. Choose the **Finish** button; the catalog is added to the Navigation pane of the **Quantification Workbook** window, refer to Figure 7-20.

Note

For calculating the takeoff in Imperial units, use the Quantification_Imperial xml file from the C:\nws_2017\c07_nws_2017_tut location.

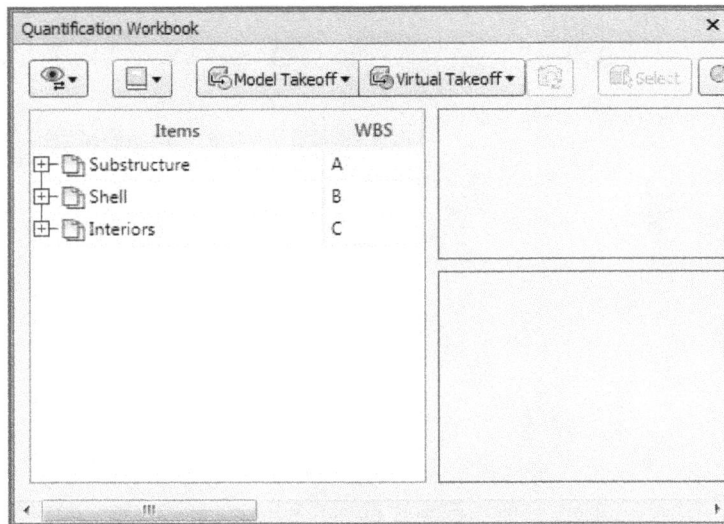

*Figure 7-20 The catalog added to the Navigation pane of the **Quantification Workbook** window*

Performing Quantity Takeoff

In this section, you will calculate the quantity takeoff of various structural components.

1. Expand the **Substructure** node under the Items column in the Navigation pane of the **Quantification Workbook**.

2. Expand **Substructure > Foundations > Standard Foundations > Rectangular Foundation(0)**, refer to Figure 7-21.

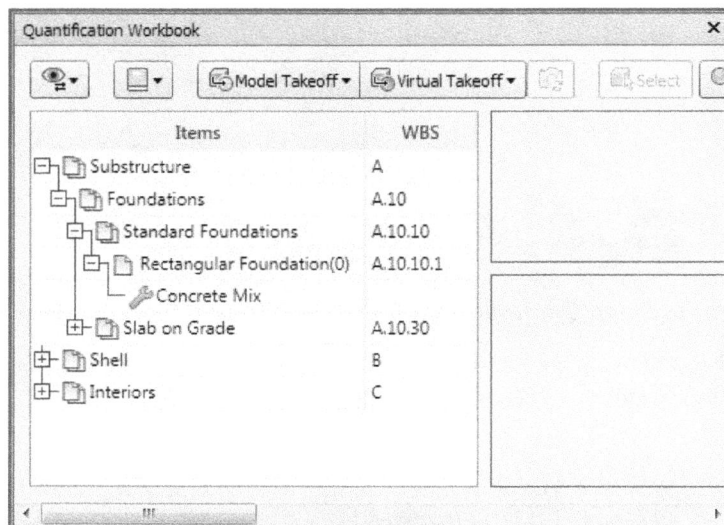

Figure 7-21 The folders expanded in the Navigation pane of the window

3. Now, choose the **Selection Tree** button from the **Select & Search** panel of the **Home** tab; the **Selection Tree** window is displayed.

4. Expand the **tut 2.nwc** folder in the **Selection Tree** window.

5. Next, expand the **Level 3 > Structural Foundations > Footing-Rectangular > 1800x1200x450** in the **Selection Tree** window, refer to Figure 7-22.

6. Now, select the desired foundations in the **Selection Tree** window, as shown in Figure 7-22.

7. Select the **Rectangular Foundation** item in the Navigation pane of the **Quantification Workbook**; its summary is displayed in the Rollup pane of the window.

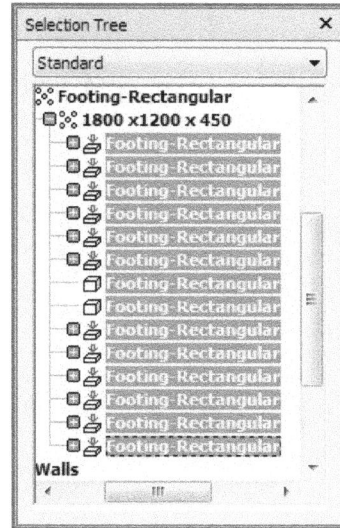

Figure 7-22 Selecting foundations in the Selection Tree window

8. Next, choose the **Model Takeoff** button in the Toolbar of the **Quantification Workbook**; a drop-down list is displayed.

9. Select the **Take off to Selected** option from the list; the take off results are displayed in the Takeoff pane of the **Quantification Workbook**, as shown in Figure 7-23.

> **Note**
> *The appearance of the selected objects changes after the takeoff.*

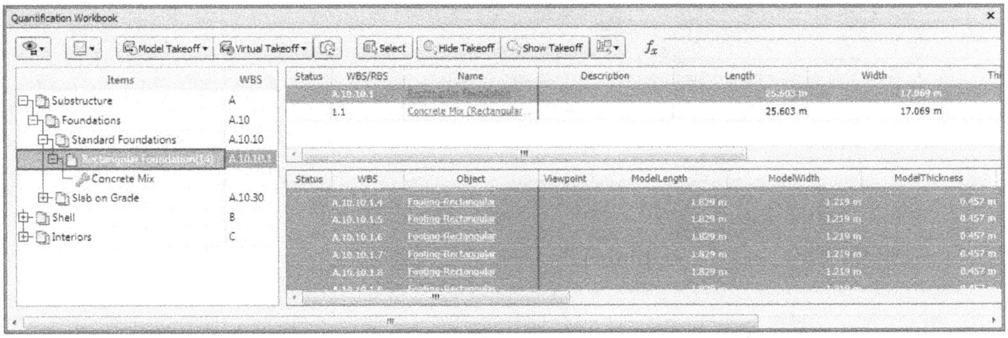

Figure 7-23 Take off results displayed in the Takeoff pane of the window

10. Rotate the model and select the foundation slab using the **Select** tool; the slab is highlighted in blue, as shown in Figure 7-24.

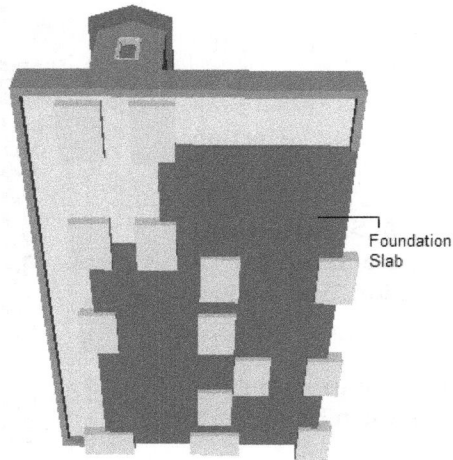

Figure 7-24 *Foundation slab selected in the Scene View*

11. Next, choose the **Select** tool from **Home > Sect&Search > Select** drop-down and then select the **Foundation Slab** item under the **Slab on Grade** node in the Navigation pane of the **Quantification Workbook** window, as shown in Figure 7-25.

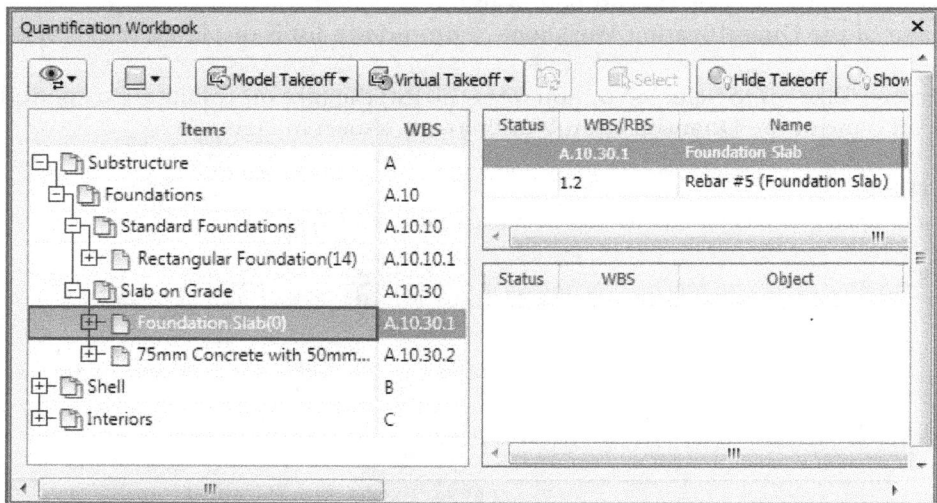

Figure 7-25 *Selecting the **Foundation Slab** item in the Navigation pane of the window*

12. Repeat the procedure followed in steps 8 and 9 to calculate the model takeoff of the selected object; the takeoff result is displayed in the Takeoff pane of the window.

13. Now, select the **Floor** slab in the Scene View; the slab is highlighted in blue, as shown in Figure 7-26.

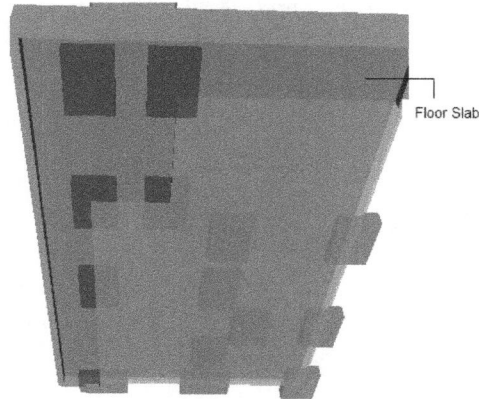

Figure 7-26 *The slab selected in the Scene View*

14. Select the **75mm Concrete with 50mm Metal Deck** item in the Navigation pane of the **Quantification Workbook** window, as shown in Figure 7-27.

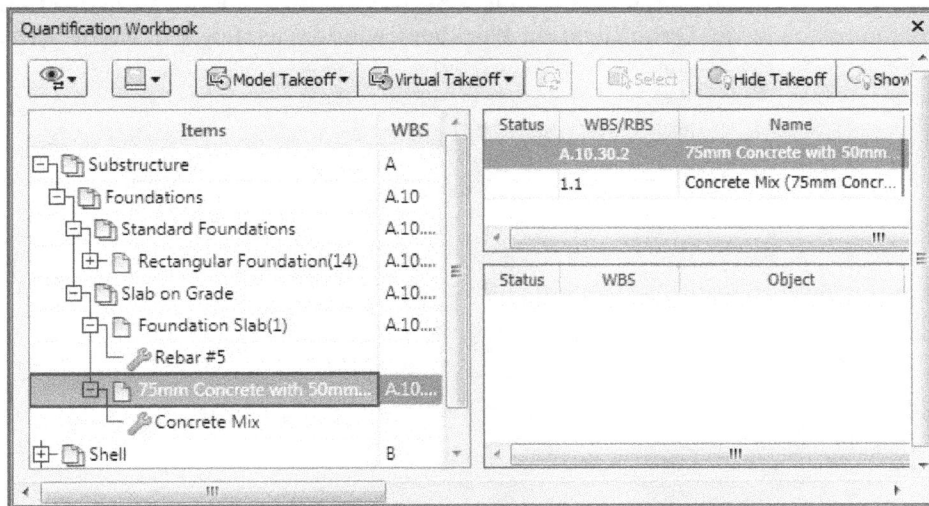

Figure 7-27 *Selecting **75mm Concrete with 50mm Metal Deck** in the window*

15. Repeat the procedure followed in steps 8 and 9 to calculate the model takeoff of the selected object; the takeoff result is displayed in the Takeoff pane of the window.

16. Select the structural columns available under the **Level 3** and **Level 5** nodes in the **Selection Tree** window, as shown in Figure 7-28.

Note
*You can use the CTRL key to select multiple items in the **Selection Tree** window.*

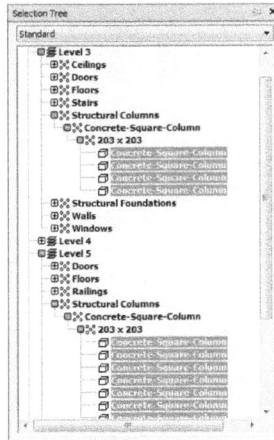

*Figure 7-28 Selecting columns in the **Selection Tree** window*

17. Next, select the **Columns** item from **Shell > Superstructure > Floor Construction** in the Navigation pane of the **Quantification Workbook** window, as shown in Figure 7-29.

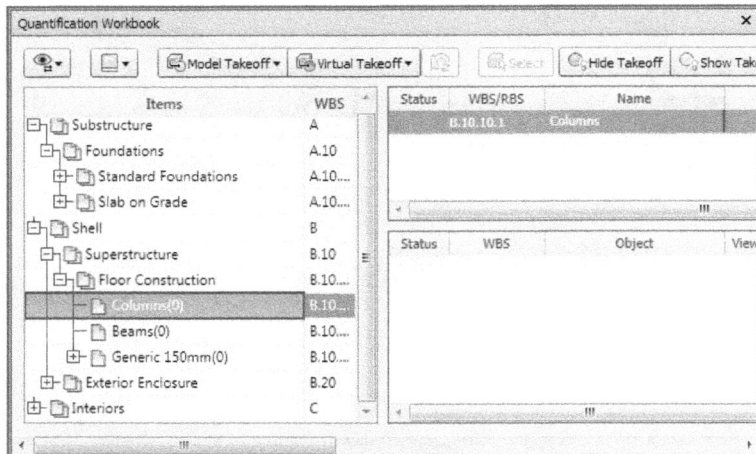

Figure 7-29 Selecting columns in the Navigation pane of the window

18. Repeat the procedure followed in steps 8 and 9 to calculate the model takeoff for the selected object; the takeoff result is displayed in the Takeoff pane of the window.

19. Next, in the **Selection Tree** window, select all the members by pressing the CTRL key under **Level 5 > Structural Framing > Concrete-Rectangular Beam > 203*304** node, as shown in Figure 7-30.

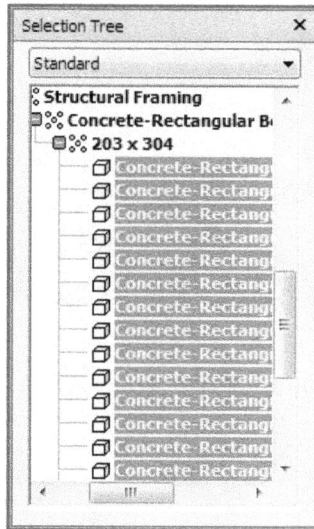

Figure 7-30 *Selecting beams in the **Selection Tree** window*

20. Select the **Beams** item under the **Shell** group in the Navigation pane of the **Quantification Workbook** window, as shown in Figure 7-31.

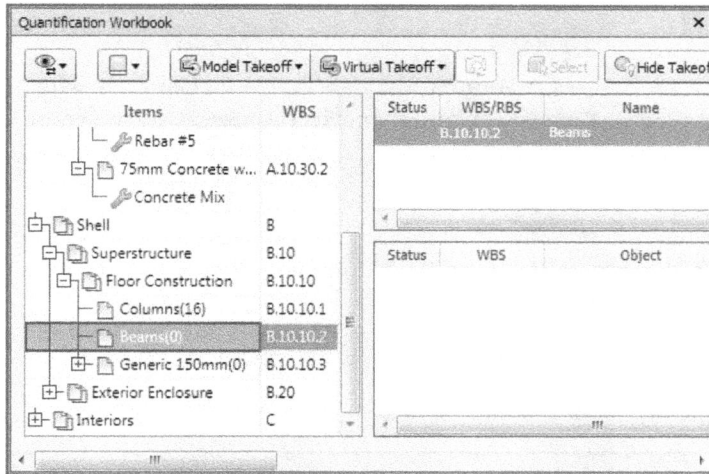

Figure 7-31 *Selecting **Beams** item in the Navigation pane of the window*

21. Repeat the procedure followed in steps 8 and 9 to calculate the model takeoff for the selected object.

22. Select the floor of the model in the Scene View; the selected object is highlighted in blue color, as shown in Figure 7-32.

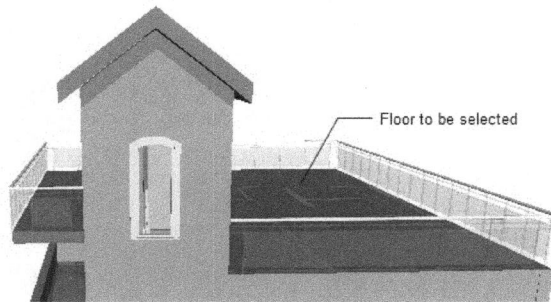

Figure 7-32 *Selecting floor in the model*

23. Select the **Generic 150mm** item from **Shell > Superstructure > Floor Construction** in the Navigation pane of the workbook.

24. Repeat the procedure followed in steps 8 and 9 to calculate the model takeoff the selected object.

Exporting Quantities to Excel

In this section, you will export the takeoff quantities to an excel file.

1. In the Navigation pane of the **Quantification Workbook** window, select **Rectangular Foundations(14)**.

2. Press the CTRL key and select **Foundation Slab(1), 75mm Concrete with 50mm Metal Deck slab(1), Columns(16), Beams(21)**, and **Generic 150mm(1)**, as shown in Figure 7-33.

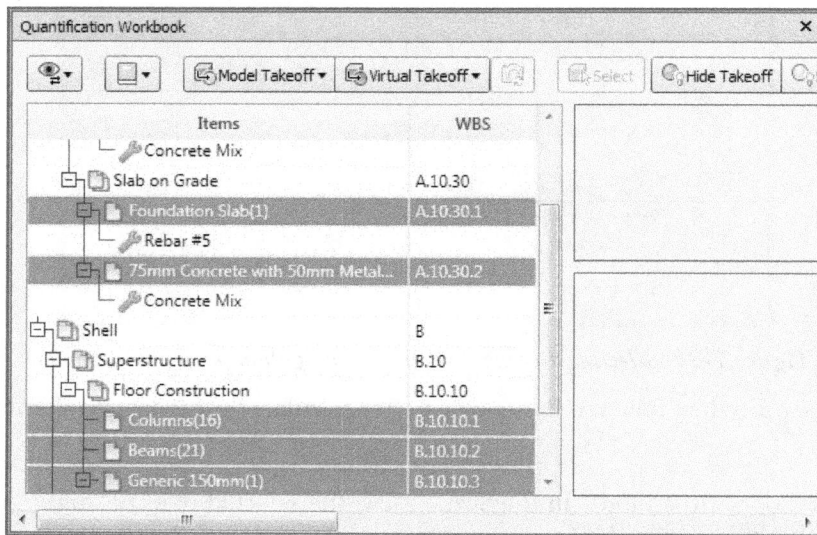

Figure 7-33 *Selecting items in the Navigation pane of the **Quantification Workbook** window*

3. Next, select the **Export Selected Quantities to Excel** option from the **Import/export Catalogs and export Quantities** drop-down list located at the extreme right in the **Quantification Workbook** window; the **Export Quantities to Excel** dialog box is displayed.

4. In this dialog box, browse to the following location: *C:\Navisworks_2017\c07_tutorial*.

5. Specify the name **Takeoff Quantities** in the **File name** edit box, as shown in Figure 7-34 and choose the **Save** button; the file is saved at the specified location and the **Export Quantities to Excel** message box is displayed.

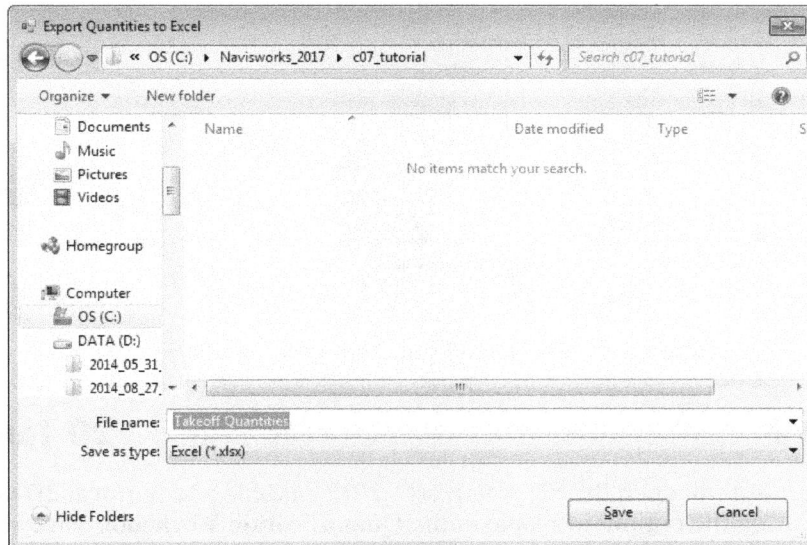

Figure 7-34 The Export Quantities to Excel dialog box

6. In this message box, choose the **Yes** button; the takeoff quantities are exported and opened in an excel file, as shown in Figure 7-35.

Saving the Project

In this section, you will save the project.

1. Before saving, choose the **Appearance** button in the Toolbar of the **Quantification Workbook** window; a drop-down list is displayed.

2. Select the **Reapply Original Model Appearance** option from the list; the original appearance is applied to the model.

3. To save the project with the current view, choose **Save As** from the Application Menu; the **Save As** dialog box is displayed.

4. Browse to the *Navisworks_2017* folder and enter **c07_navisworks_2017_tut01** in the **File name** edit box. Now, select the **Navisworks File Set (*.nwf)** file format from the **Save as type** drop-down list, and choose the **Save** button.

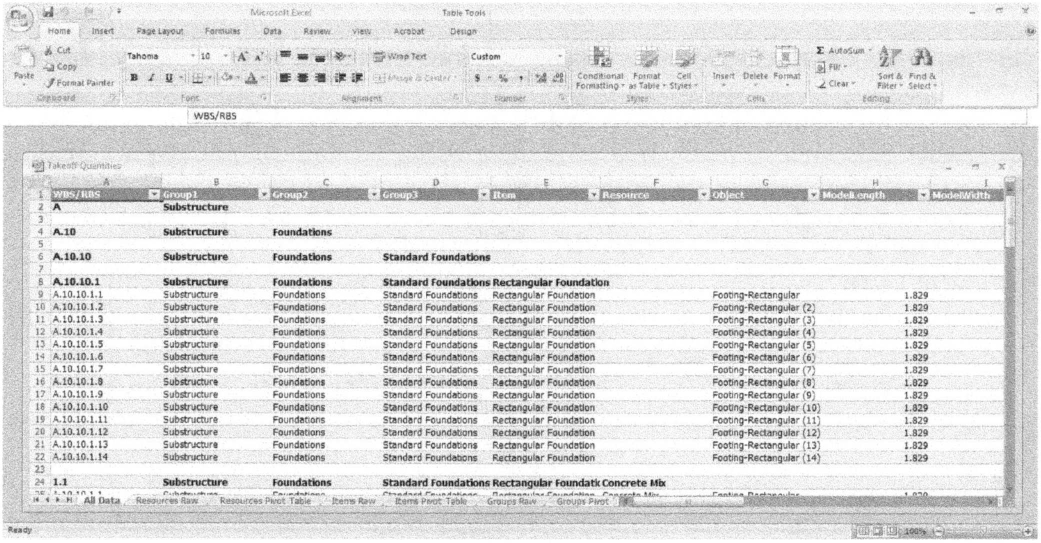

Figure 7-35 The Takeoff quantities opened in an excel file

Tutorial 2   2D Takeoff

In this tutorial, you will open the *c07_navisworks_2017_tut2* file and perform 2D takeoff on the floor plan by using the options available in the **Quantification Workbook** window.

(Expected time: 45min)

The following steps are required to complete this tutorial:

a. Open the *c07_navisworks_2017_tut2* file.
b. Display the **Project Browser** window and import the 2D sheet.
c. Set up the project in the **Quantification Workbook** window.
d. Perform the takeoff.
e. Save the project.

Opening the File
In this section, you will open the file to be used in this tutorial.

1. Choose the **Open** button from the Quick Access Toolbar; the **Open** dialog box is displayed.

2. In this dialog box, browse to the following location:
C:\Navisworks_2017\c07_tutorial

3. Select **Navisworks File Set (*.nwf)** from the **Files of type** drop-down list.

4. Select the file **c07_navisworks_2017_tut2** from the displayed list of files; the file name appears in the **File name** edit box.

5. Next, choose the **Open** button in the dialog box; the model is displayed in the Scene View, as shown in Figure 7-36.

Figure 7-36 *Model displayed in the Scene View*

Importing the 2D Sheet

In this section, you will learn to import the floor plan of the model using the **Sheet Browser** window.

1. Choose the **Sheet Browser** button available in the Status Bar; the **Sheet Browser** window is displayed.

2. In this window, choose the **Import Sheets & Models** button; the **Insert From File** dialog box is displayed. In this dialog box, select the DWF option

3. In this dialog box, browse to the following location:
    C:\Navisworks_2017\c07_tutorial

4. Select the *c07_tut2_floor plan.dwf* file from the displayed list of files. Choose the **Open** button; the file gets added and is displayed with the name **Model** in the **Sheet Browser** window.

5. Choose the **Next Sheet** arrow in the Status Bar; the 2D sheet is displayed in the Scene View, refer to Figure 7-37.

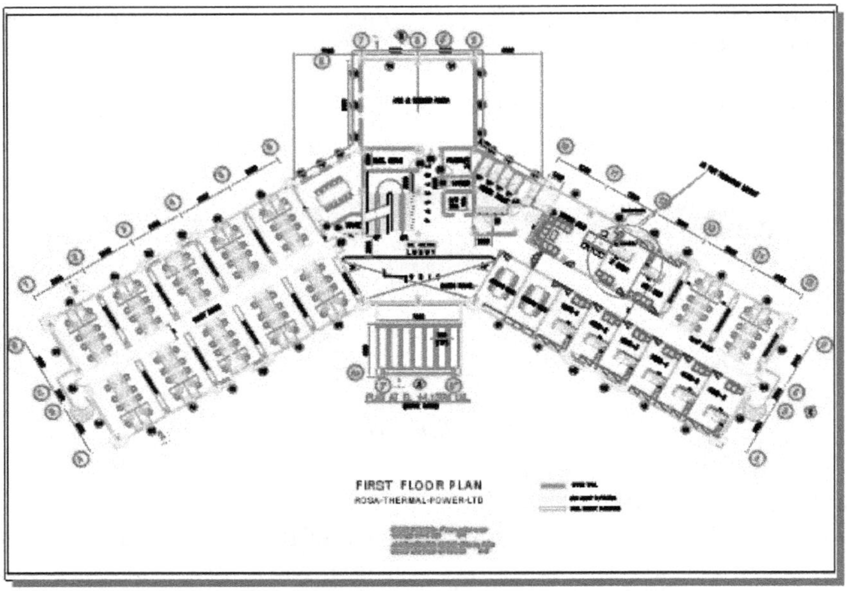

Figure 7-37 2D Sheet displayed in the Scene View

Setting Up the Project in the Quantification Workbook Window

In this section, you will invoke the **Quantification Workbook** window and import a quantification catalog to perform the takeoff.

1. Choose the **Quantification** tool from the **Quantification** panel of the **Home** tab; the **Quantification Workbook** window is displayed.

2. Next, choose the **Project Setup** button; the **Quantification Setup Wizard** is displayed. Select the **Browse to a catalog** radio button and choose the **Browse** button; the **Open** dialog box is displayed.

3. In the **Open** dialog box, browse to the following location:
    C:\Navisworks_2017\c07_tutorial

4. Select the *2D Takeoff.xml* file from the file lists and choose the **Open** button; the **Open** dialog box closes and then choose the **Next** button in the **Quantification Setup Wizard** dialog box; the **Setup Quantification: Ready to Create Database** page is displayed.

5. Choose the **Finish** button; the catalog is imported to the **Quantification Workbook** window, refer to Figure 7-38.

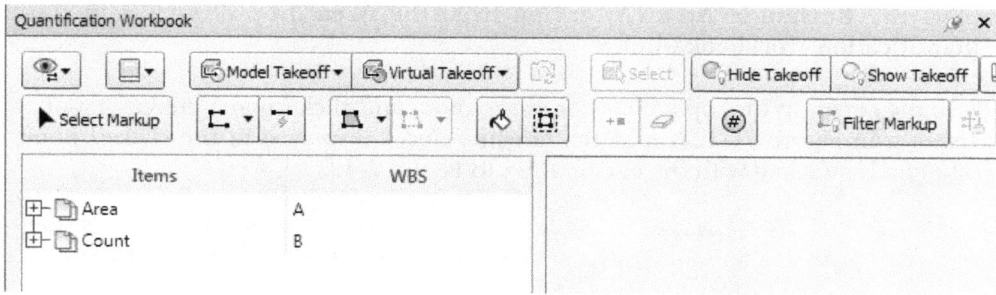

Figure 7-38 The 2D takeoff imported to the **Quantification Workbook** window

Performing the Takeoff

In this section, you will learn to perform the area and count takeoff on the 2D sheet.

1. In the Navigation pane of the **Quantification Workbook** window, expand the **Area** group and select the **Area of Chiller Room** item, refer to Figure 7-39.

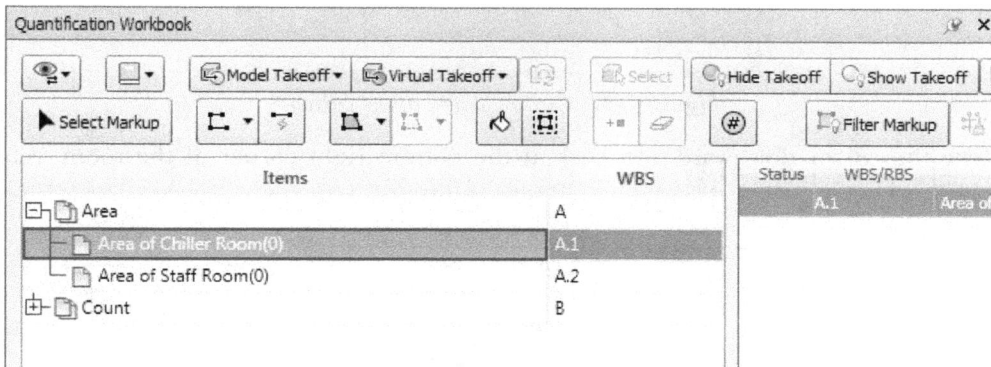

Figure 7-39 The **Area of Chiller Room** item selected in the window

2. Click in the Scene View and zoom in using the middle mouse button to view the chiller room, refer to Figure 7-40.

Figure 7-40 The chiller room

3. Select the **Rectangle Area (A)** option from the **Area** drop-down list in the
   Quantification Workbook window.

4. Place the cursor at the top left corner of the room and click when a green symbol appears,
   as shown in Figure 7-41. Notice that a white dot is displayed at the clicked point and a
   rectangle is attached with the cursor, refer to Figure 7-42.

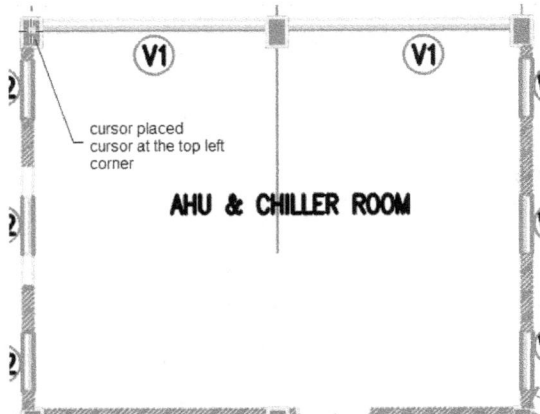

Figure 7-41 *Cursor placed at the corner*

5. Move the cursor downward and click at the bottom right corner of the room, refer to
   Figure 7-43.

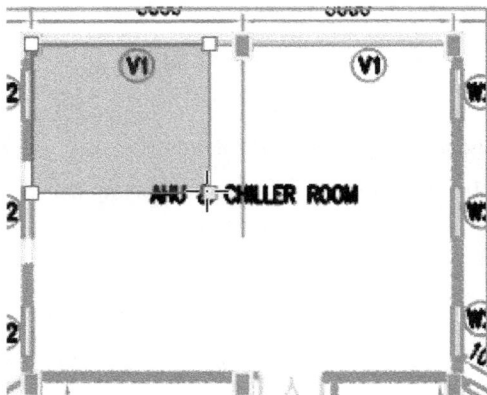

Figure 7-42 *Drawing area markup*

Figure 7-43 *Drawing area markup*

6. The area takeoff of the chiller room is displayed in the Takeoff pane of the **Quantification
   Workbook** window.

Now, you will perform the area takeoff for the staff room located in the left side of the plan.

7. Invoke the **Zoom** tool from the Navigation Bar and zoom in to view the staff room area,
   refer to Figure 7-44.

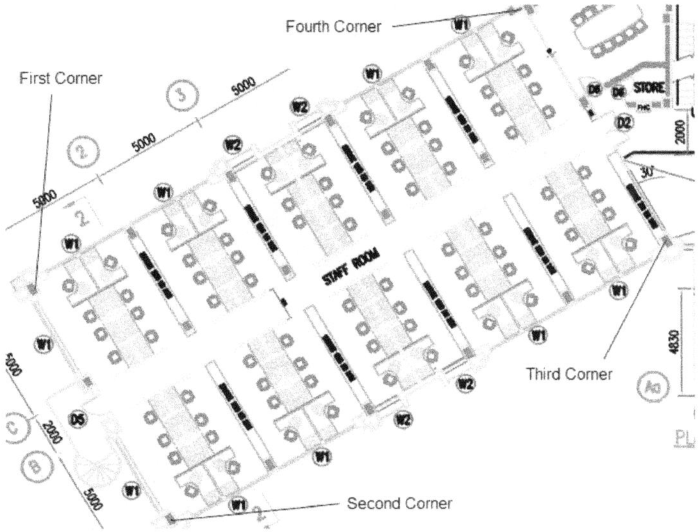

Figure 7-44 *Staff room in the plan*

8. Select the **Area of Staff Room** item in the Navigation pane of the **Quantification Workbook** window and select the **Area (A)** option from the **Area** drop-down list.

9. Place the cursor at the first left corner, refer to Figure 7-44. Click at that point and then move the cursor downward; a blue line is attached to the cursor, refer to Figure 7-45.

10. Click at the second corner and move the cursor in the right direction and click at the third corner, refer to Figure 7-46.

11. Move the cursor upward and click at the fourth corner. Next, right-click to complete the area markup, refer to Figure 7-47. The area takeoff is displayed in the Takeoff pane of the **Quantification Workbook** window.

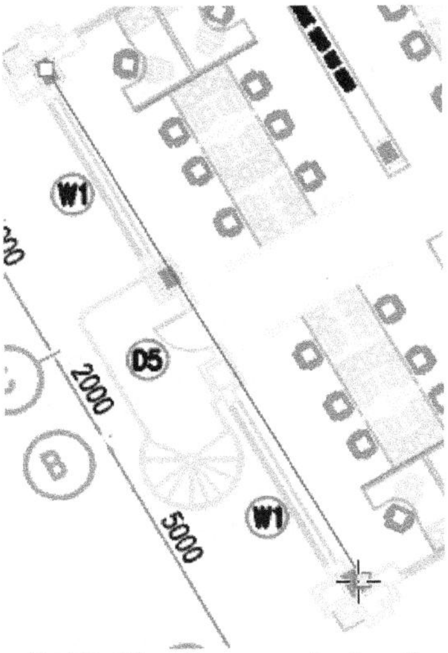

Figure 7-45 *Blue line attached to the cursor*

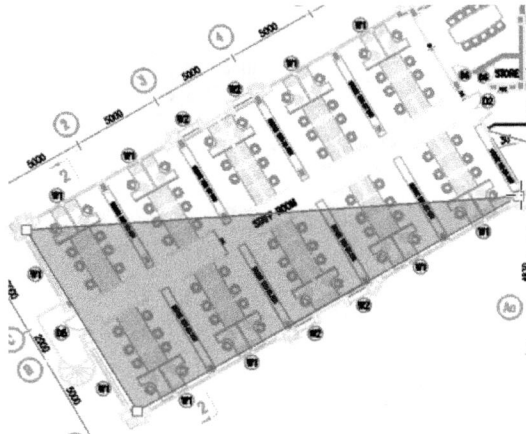

Figure 7-46 *Drawing markup*

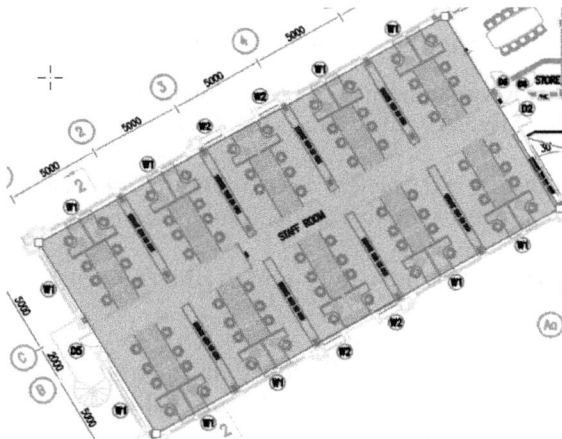

Figure 7-47 *Complete markup*

12. Choose the **Backout** button; a drop-down list is displayed. Now, select the **Backout (B)** option from the list.

13. Place the cursor at the point marked as A, refer to Figure 7-48. Click when a green symbol appears.

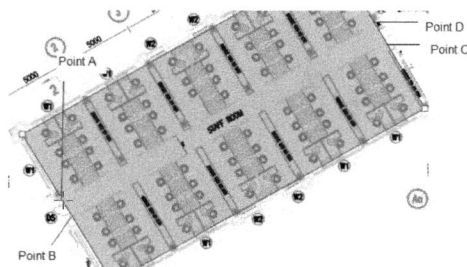

Figure 7-48 *Cursor placed at point A*

14. Move the cursor downward and click at point B. Similarly, click at points C and D and right-click to complete; the backout markup is created, refer to Figure 7-49 and the resulting area is displayed in the Takeoff pane of the **Quantification Workbook** window.

Note
Click in the Scene View to deselect the markup

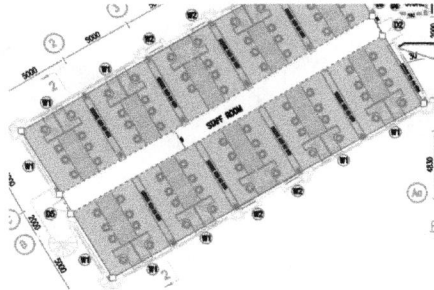

Figure 7-49 Backout Created

15. Choose the **Catalog** button from the Quantification toolbar; a drop-down list is displayed. Select the **Item Catalog** option from the list; the **Item Catalog** window is displayed.

16. In this window, expand the **Count** group and select the **Chairs** item; the properties are displayed in the Formula pane.

17. In the **Object Appearance** area of the Formula pane, choose the **Color** button; the **Color** palette is displayed. Select the pink color and choose the **OK** button and then close the **Item Catalog** window.

18. Expand the **Count** group and select the **Chair** item in the Navigation pane of the **Quantification Workbook** window.

19. Choose the **Count** button from the toolbar in the **Quantification Workbook** window.

20. Click at the chairs in the staff room one by one to count the objects; the count takeoff is displayed in the Takeoff pane of the **Quantification Workbook** window and the number of chairs are displayed in the Navigation pane. Figure 7-50 shows the chairs marked with pink dots as count takeoff.

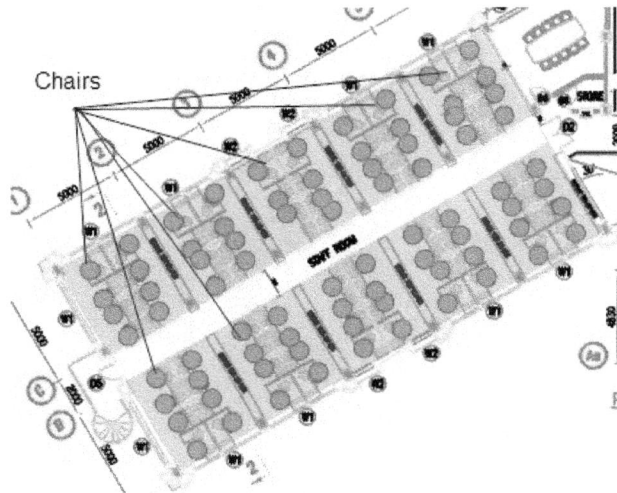

Figure 7-50 Chairs marked as count takeoff

Saving the Project

In this section, you will save the project.

1. To save the project with the current view, choose **Save As** from the Application Menu; the **Save As** dialog box is displayed.

2. Browse to the *nws_2017\ c07_nws_2017_tut* folder and enter **c07_navisworks_2017_tut02** in the **File name** edit box. Now, select the **Navisworks File Set (*.nwf)** file format from the **Save as type** drop-down list, and choose the **Save** button.

Self-Evaluation Test

Answer the following questions and then compare them to those given at the end of this chapter:

1. The **Quantification Workbook** window can be invoked from the _____ panel of the _____ tab.

2. The _____ option is used to apply the original appearance of the model to the objects after performing the takeoff.

3. The _____ button is used to select the markups in the sheet.

4. The catalog items can be imported/exported from the _____ and _____ files.

5. The **Polyline (L)** option is used to draw markups for the area which are non-standard in shape. (T/F)

6. Viewpoints cannot be added to the takeoff objects. (T/F)

7.   Model Takeoff is a method of quantifying model objects. (T/F)

8.   Units of the takeoff results cannot be changed. (T/F)

9.   The **Item Catalog** window contains a list of all the resources for the project. (T/F)

10. The **Master Resource List** dialog box contains a list of existing resources in a project.(T/F)

Review Questions

Answer the following questions:

1.   Which of the following options is used to filter the markup of the takeoff items?

     a) **Select Markup**   b) **Filter Markup**
     c) **Count**   d) None of these

2.   Which of the following options is used for excluding an area from the area markup?

     a) **Polyline (L)**   b) **Backout**
     c) **Quick Line**   d) **Model Takeoff**

3.   Which of the following options is used for creating a linear takeoff?

     a) **Quick Box**   b) **Quick Line**
     c) **Erase**   d) None of these

4.   Which of the following options is used to create a blank takeoff?

     a) **Model Takeoff**   b) **Virtual Takeoff**
     c) **Update**   d) **Change Analysis**

5.   Which of the following options is used to display all the items in the Quantification Workbook window?

     a) **Item View**   b) **Resource View**
     c) **2D Takeoff**   d) None of these

6.   Navisworks does not allow you to takeoff objects from the 2D sheet. (T/F)

7.   You cannot change the formula for any of the takeoff methods. (T/F)

8.   The **Count** button is used to create the count takeoff of the objects. (T/F)

9.   You cannot remove any unwanted geometry from the markup. (T/F)

10. The color of the takeoff objects cannot be changed. (T/F)

EXERCISE
Exercise 1   Model Takeoff

Download and open the c07_navisworks_2017_ex2 file from http://www.cadcim.com. Calculate the Quantity takeoff of the Rectangular-Footings, Wall Foundation, Columns, Beams, and Roof Slab. Figure 7-51 shows the model to be used for this exercise.   **(Expected time: 45min)**

The following steps are required to complete this exercise:

a) Open the c07_navisworks_2017_ex2 file.
b) Use the catalog Quantification_Exer for metric and Quantification_ Ex_imp for imperial unit.
c) Perform the takeoff of beams, columns, foundation, footing, and roof slab.
d) Export the takeoff quantities to an excel file.
e) Save the project as c07_navisworks_2017_ex02.

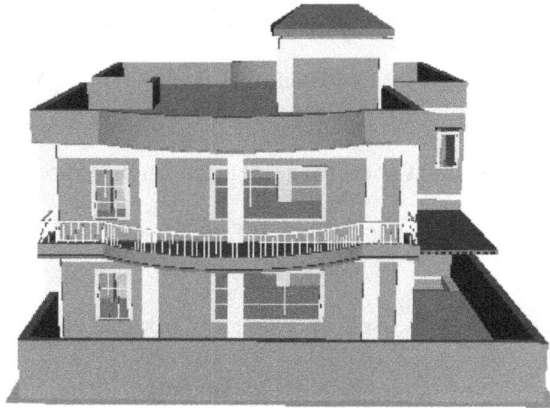

Figure 7-51 The Residence Building

Exercise 2   2D Takeoff

Download and open the c07_navisworks_2017_ex3 file from http://www.cadcim.com. Calculate the Quantity takeoff of the Common Room and Bed Room. Figures 7-52 and 7-53 show the model and its floor plan to be used for this exercise.   **(Expected time: 30min)**

The following steps are required to complete this exercise:

a) Open the c07_navisworks_2017_ex2 file.
b) Import the c07_residence.dwf file using the **Project Browser** window.
c) Open the loaded dwf file using the **Next Sheet** button from the Status Bar.
d) Invoke the **Quantification Workbook** window and import the 2D Takeoff_ex2 xml file.
e) Perform the area takeoff for the common room and bed room using the area takeoff tool.

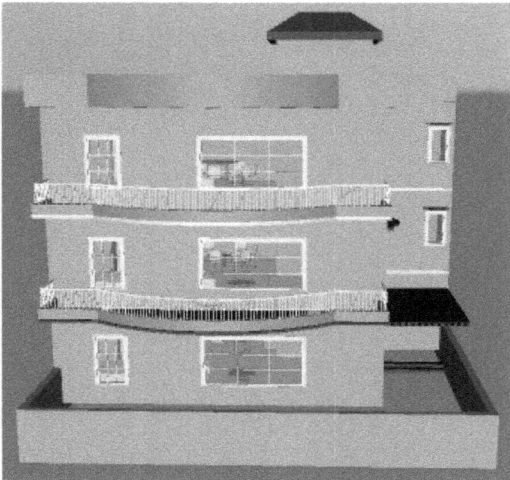

Figure 7-52 *The Residence building*

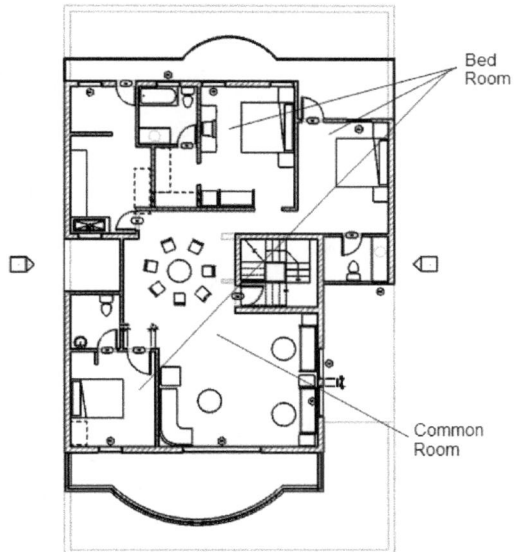

Figure 7-53 *Floor plan of the residence building*

Answers to Self-Evaluation Test

1. Tools, Home, 2. Reapply Original Model Appearance, 3. Select Markup, 4. Excel, XML,
5. T, 6. F, 7. T, 8. F, 9. F, 10. T

Chapter 8

Clash Detection

Learning Objectives

After completing this chapter, you will be able to:
- *Understand the concept of clash detection*
- *Create a test*
- *Select objects for clash test*
- *Perform the clash test*
- *View clash results*
- *Generate a clash report*

INTRODUCTION

Clash detection and coordination, though is not an entirely new concept has gained an impetus in the recent years. It has invariably played an important role in lowering down the construction costs. This concept had helped various project managers to control the huge cost overruns and limits the overall budget of construction. Clash detection has also helped to avert the unnecessary delays in the project due to lack of coordination between the stakeholders.

Before using BIM, the clashes were determined with the help of charts and drawings that were overlaid on a light table. This process was time consuming and was not accurate. The use of BIM for clash detection has helped the project managers to examine the intrinsic detection of clashes, at the earliest stages.

Moreover, clash detection and coordination help one to find the incompatible and inconsistencies much before starting the actual construction. Hence, the clash detection process plays a crucial role in avoiding delays and saving time and resources.

In a Building Information Modeling project, a clash is referred as the spatial conflict that occurs between different components. Clashes are characterized on the basis of their existence. Generally, the clash detection process is divided into three types and each has its own significance. The three types of clash detection are: Hard Clash, Soft Clash, and Time Clash. A hard clash occurs when one building component penetrates another building component. For example, when a hydronic pipe penetrates into a structural column, a hard clash occurs. A soft clash also known as clearance clash that occurs when the two or more components are at a distance lesser than the minimum specified clearance. A time clash is a clash that is anticipated at the time of construction between temporary components such as scaffolds and pipes. Time clashes are useful for maintaining adequate clearance between the permanent and the temporary components in a building project.

In a Building Information Modeling project, clash detection is an important process. It is the process of identifying interferences between the model objects. These interferences may be between Structure and the MEP models such as walls and pipes, beams and ducts, or between a single discipline itself such as pipes and ducts, columns and floors, and so on. Using clash detection, you can identify interferences and fix them on time thereby saving time and cost. A user can perform clash detection just after the completion of the design process to find out the errors and remove those errors virtually from the project. It reduces the risk of human error during model inspection. In Navisworks, there are four types of clashes: Hard, Clearance, Duplicate, and 4D Workflow. In hard clashes, two objects occupy the same space thus creating interferences. For example, a pipe running through a wall. Clearance clashes need to be detected for the objects that require certain geometric tolerances for access, insulation, or safety. Duplicate clashes need to be detected for the objects that are identical in type and position. 4D workflow clashes require to be performed for resolving scheduling clashes and project timeline issues.

In Navisworks, you can also link clash detection with other features. For example, you can combine clash detection with an existing animation scene to automatically check clashes in the animation sequence. Similarly, clash detection can be linked with timeliner schedules. You can create custom clash tests through export clash tests. All these features will be discussed in detail in this chapter. The **Clash Detective** window which contains several options used for performing the clash tests is discussed next.

CLASH DETECTIVE WINDOW

The **Clash Detective** window is a dockable window in which you can specify rules and options for performing clash tests. In this window, you can view and arrange results and can also generate clash reports. To display this window, choose the **Clash Detective** tool from the **Tools** panel in the **Home** tab; the **Clash Detective** window will be displayed, as shown in Figure 8-1. Alternatively, to display the **Clash Detective** window, select the **Clash Detective** check box from **View > Workspace > Windows** drop-down.

The **Clash Detective** window is divided into two areas: top and bottom. The top area consists of Test panel and the bottom area contains four tabs. The Test panel in the top area is discussed next.

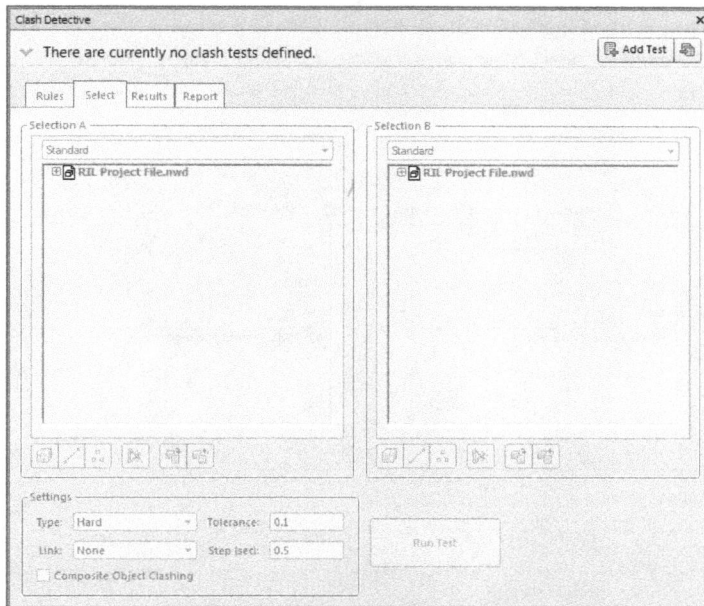

Figure 8-1 *The* **Clash Detective** *window*

Test Panel

The Test panel comprises of all the clash tests listed in a tabular format. In the **Clash Detective** window, the Test panel is not displayed by default. To display it, choose the **Add Test** button located at the top right corner in the window; the Test panel will be displayed with a name **Test1** under the **Name** column, as shown in Figure 8-2. You can also hide the display of the Test panel by clicking on the arrow button located at the top left corner of the window. The Test panel displays the summary of the performed clash test. The columns available in this panel are discussed next.

The **Name** column displays the name of the clash test performed in the project. The **Status** column displays the status of the test. If **New** is displayed in the **Status** column, it means that the test has been added recently and has not yet been executed. If **Done** is displayed in the **Status** column, it means that the test has been performed. The **Clashes** column will display the number of clashes detected in the project. Here the number of clashes represents the total number of clashes excluding the clashes which are grouped.

The other five columns display the clash status. The **New** column displays the clash test that is performed for the first time. The **Active** column displays the clash test that is found in previous test and is not resolved. The **Reviewed** column displays the clash test that has been reviewed in the previous test. The **Approved** column displays the clash that has been approved in the previous test. The **Resolved** column displays the clash that has been resolved in the previous test and is not found in the current test. Note that every clash is represented using a color code that is defined in the column name of the **Test** panel. For example, the newly identified clashes will be displayed in red color in the **Clash Detective** window and in the Scene View.

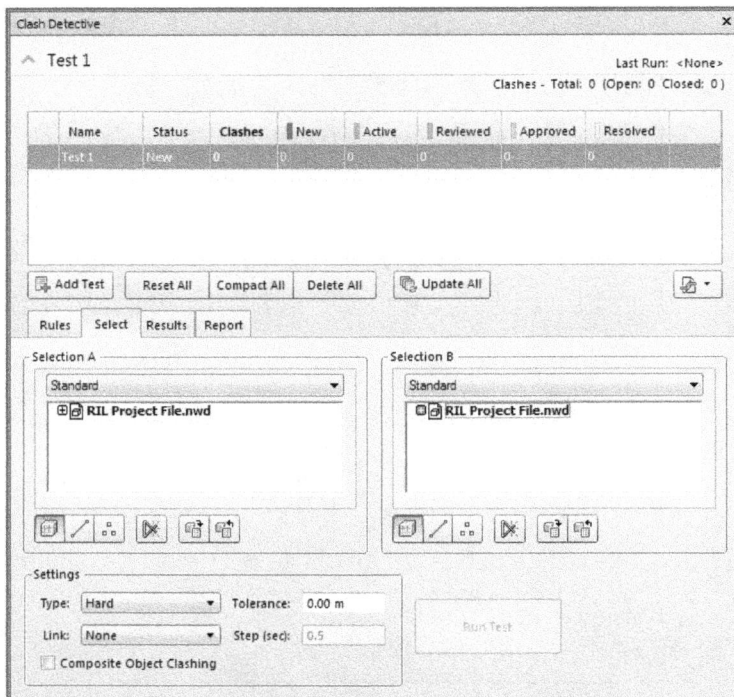

*Figure 8-2 The **Clash Detective** window with the Test panel displayed*

The buttons located at the bottom of the Test panel are used to manage the clash tests and are discussed next.

Add Test

The **Add Test** button is used for adding a new test. On choosing this button, a new clash test will be added to the **Test** panel, refer to Figure 8-3. To change the name of the added test, double-click on the field corresponding to the **Name** column; the field will change into an edit box, as shown in Figure 8-3. Enter a new name in the edit box and press ENTER.

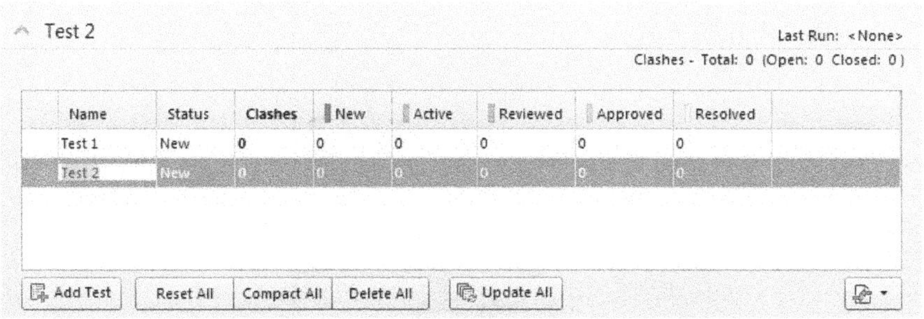

*Figure 8-3 Clash test added to the **Test** panel*

Reset All

Reset All   The **Reset All** button is used for resetting the status of all the tests to **New**.

Compact All

Compact All   The **Compact All** button is used for removing all the resolved clash tests from the clash results and from the summary displayed in the Test panel.

Delete All

Delete All   The **Delete All** button is used for removing all the clash tests. On choosing this button, all tests will be removed from the Test panel in the **Clash Detective** window.

Update All

Update All   The **Update All** button is used for updating all the clash tests. When you choose this button, all clashes will be updated with the current settings.

Import/Export Clash Tests

The **Import/Export Clash Tests** button is used for importing or exporting clash tests as an .xml file. To import or export clash tests, choose this button; a drop-down list will be displayed. This drop-down list contains two options: **Import Clash Tests** and **Export Clash Tests**. To export the clash tests, select the **Export Clash Tests** option from the **Import/Export Clash Tests** drop-down list; the **Export** dialog box will be displayed. In the **Export** dialog box, specify the file location for the file to be exported. Next, specify the file name in the **File name** edit box and then choose the **Save** button; the file will be exported and saved at the specified location.

Similarly, to import the clash test, select the **Import Clash Tests** option from the drop-down list; the **Import** dialog box will be displayed. In this dialog box, browse and select the required file and then choose the **Open** button; the test will be imported and added to the Test panel.

If you right-click in the empty area in the Test panel in the **Clash Detective** window, a shortcut menu will be displayed, as shown in Figure 8-4. The options in this menu will work the same way as the buttons discussed above.

Add Test

Update All
Reset All Tests
Compact All Tests
Delete All Tests

Import Clash Tests...
Export Clash Tests...

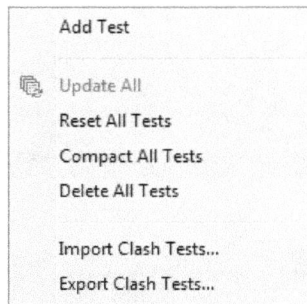

Figure 8-4 *Shortcut menu displayed on right-clicking in the empty area*

You can also manage a currently selected clash test by using the options from the shortcut menu. To do so, right-click in the required test column; a shortcut menu will be displayed, as shown in Figure 8-5. Choose the required option from the shortcut menu.

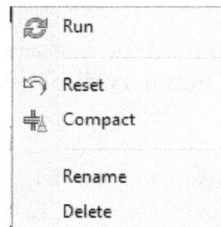

Run

Reset
Compact

Rename
Delete

Figure 8-5 *The shortcut menu displayed by right-clicking on a test*

The parameters of the clash test are defined using the options provided in the tabs below the **Test** panel. These tabs are discussed next.

Rules Tab

When you choose the **Rules** tab in the **Clash Detective** window; a list of rules will be displayed, as shown in Figure 8-6. You can use these rules to define the situations in which Navisworks ignores clash between two objects.

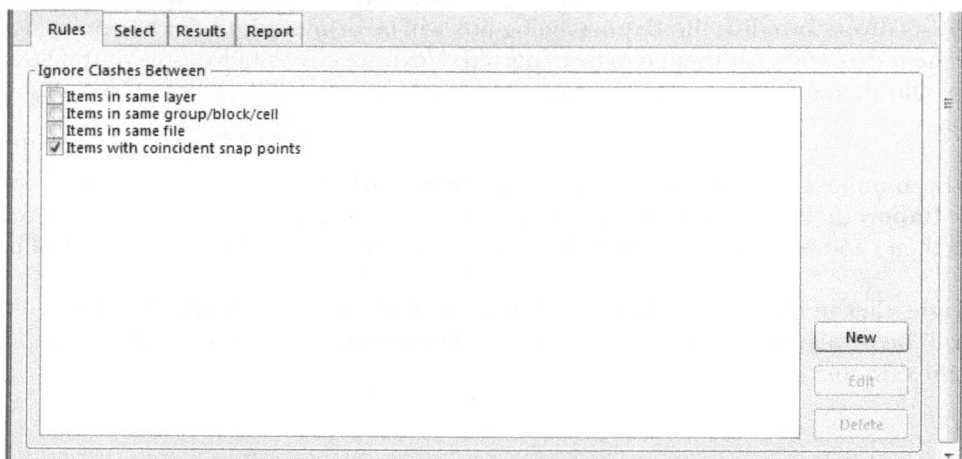

| Rules | Select | Results | Report |

Ignore Clashes Between

☐ Items in same layer
☐ Items in same group/block/cell
☐ Items in same file
☑ Items with coincident snap points

New
Edit
Delete

Figure 8-6 *The **Rules** tab in the **Clash Detective** window*

There are four pre-defined rules and five rule templates. The four pre-defined rules are displayed in the **Rules** tab, refer to Figure 8-6. You cannot customize these default rules. These rules are discussed next.

Items in same layer

The Items in same layer rule is used to ignore clashes between objects in the same layer. To apply this rule, select the **Items in same layer** check box in the **Rules** tab.

Items in same group/block/cell

The Items in same group/block/cell rule is used to ignore clashes between objects in the same group, block, or cell. To apply this rule, select the **Items in same group/block/cell** check box in the **Rules** tab.

Items in same file

The Items in same file rule is used to ignore clashes between items within the same file. To apply this rule, select the **Items in same file** check box in the **Rules** tab. This rule can also be used when you want to check clashes between different files.

Items with coincident snap points

The Items with coincident snap points rule is used to ignore clashes between items that have coinciding snap points. To apply this rule, select the **Items with coincident snap points** check box in the **Rules** tab. For example, this rule can be used for pipe runs and fittings that have snap points at the end of centre lines.

The other rule templates are displayed in the **Rules Editor** dialog box along with the default rules. These rule templates are discussed next.

Insulation Thickness

You can use the Insulation Thickness rule template for creating rule to ignore the clashes between the objects that have the clearance value greater than the specified insulation thickness. To create a rule based on it, choose the **New** button in the **Rules** tab; the **Rules Editor** dialog box will be displayed. Select the **Insulation Thickness** option from the **Rule Templates** list; its description will be displayed in the **Rule description** area, as shown in Figure 8-7. Also the rule name will be displayed in the **Rule name** text box. You can change the name if required.

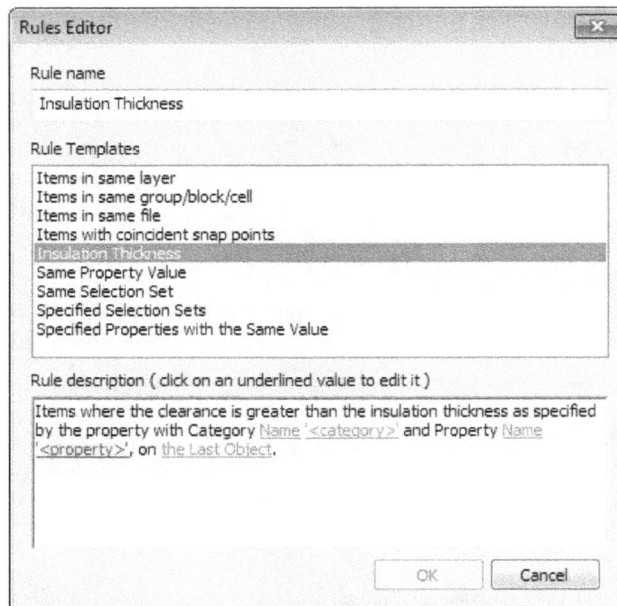

Figure 8-7 *The* **Insulation Thickness** *option selected from the* **Rule Templates** *list*

In the **Rule description** area, click on the underlined values to specify the parameters. For example, to define the category name, click on **Name**; the **Rules Editor** dialog box will be displayed, as shown in Figure 8-8. Select the **Name** option from the **Enter Value** drop-down list. To define a category, click on the '**<category>**' in the **Rule description** area; the **Rules Editor** dialog box will be displayed. Select the category from the **Enter Value** drop-down list and choose the **OK** button to close it. Similarly, define the property name and category by clicking on the **Name** and '**<property>**'. To search for the defined property at the specified location, click on **Last Object**; the **Rules Editor** dialog box will be displayed. Select the required value from the **Enter Value** drop-down list and choose the **OK** button to close it. Next, choose the **OK** button in the **Rules Editor** dialog box; the new rule will be added in the **Rules** tab. You can use this rule with the Clearance clash test.

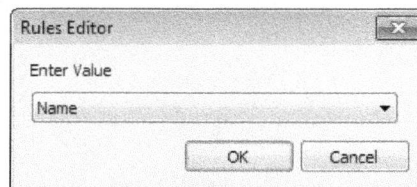

Figure 8-8 *The* **Rules Editor** *dialog box*

Same Property Value

The Same Property Value rule template is used to ignore clashes between the objects that share the same specified properties. To create a custom rule based on it, select the **Same Property Value** from the **Rule Templates** list, refer to Figure 8-7. Next, specify its parameters using the options in the **Rules description** area, as done in the **Insulation Thickness**.

Same Selection Set

The Same Selection Set rule template is used to ignore clashes between objects that are in the same selection sets.

Specified Selection Sets

The Specified Selection Sets rule template is used to ignore clashes between objects that are in two different selection sets.

Specified Properties with the Same Value

The Specified Properties with the Same Value rule template is used to ignore clashes between objects that have same value for different properties.

You can also modify a clash rule. To do so, select a rule in the **Rules** tab. Choose the **Edit** button; the **Rules Editor** dialog box will be displayed, refer to Figure 8-7. Make the required changes and choose the **OK** button; the changes will be applied to the rule. To delete a rule, select the rule in the **Rules** tab, refer to Figure 8-6. Choose the **Delete** button; the rule will be deleted.

Select Tab

The **Select** tab is chosen by default in the **Clash Detective** window. In this tab, you can define the clash test by selecting multiple sets of items at a time. Instead of testing the whole model against itself, a particular set of objects in the model will be tested. The **Select** tab comprises the **Selection A** and **Selection B** areas at the top and the **Settings** area at the bottom, refer to Figure 8-9. These areas are discussed next.

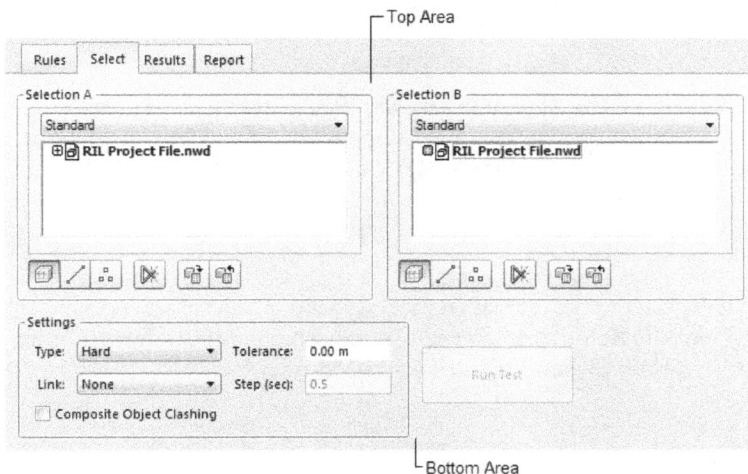

Figure 8-9 The Select tab in the Clash Detective window

Selection A and Selection B Areas

The **Selection A** and **Selection B** areas display all the items in a hierarchical structure. These two areas replicate the **Selection Tree** window. You can select objects from these areas which will be tested against each other during a clash test. The drop-down list available at the top in these areas contains the following options: **Standard**, **Compact**, **Properties**, and **Sets**.

The buttons available in these areas are discussed next.

Surfaces

The **Surfaces** button is used for including surface geometry in the clash tests.

Lines

The **Lines** button is used for including line geometry in the clash tests.

Points

The **Points** button is used for including point geometry in the clash tests.

Self-Intersect

The **Self-Intersect** button is used to test the selected object against itself for clashes.

Use Current Selection

The **Use Current Selection** button is used to select objects for the clash test directly in the Scene View. To select an object for the clash test, select the required object in the Scene View or in the **Selection Tree** window. Next, choose the **Use Current Selection** button in the required area; the object corresponding to the area from which the **Use Current Selection** button is chosen will be highlighted.

Select in Scene

The **Select in Scene** button is used to verify the selected objects for the clash test in the Scene View. To verify an object for the clash test, select the object in any of the selection areas. Next, choose the **Select in Scene** button in the required area; the corresponding object will be highlighted in the Scene View.

> **Tip**
> *You can also perform the above two operations by using the options from the shortcut menu. To do so, right-click anywhere in the selection area; a shortcut menu is displayed. In this menu, there are two options: **Select** and **Import Current Selection**. The **Select** option is used for highlighting the objects selected for the clash test in the Scene View. The function of the **Import Current Selection** option is same as that of the **Use Current Selection** button.*

Settings

In the **Settings** area of the **Select** tab, you can specify various clash detection parameters before running a clash test. These parameters are discussed next.

Type

You can define the clash type by selecting an option from the **Type** drop-down list. Select the **Hard** option for detecting a clash when two objects will actually intersect with each other. Select the **Hard (Conservative)** option for identifying a clash when the two objects will be assumed as intersecting. Select the **Clearance** option for detecting a clash when two objects will be assumed as intersecting when they come within a specified distance to each other. Select the option **Duplicates** for detecting a clash where two objects identical in type and position will intersect with each other.

Tolerance
The **Tolerance** edit box is used to specify a value that filters out the negligible clashes.

Link
You can also link the clash test to the timeliner or object animation by using the options available in the **Link** drop-down list. To link the clash test with the timeliner simulation, select the **TimeLiner** option from the list. To link the clash test with the object animation created in the **Animator** window, select the required animation scene from the **Link** drop-down list.

> **Note**
> *Note that the object animation options will be available only if you have created animations by using the **Animator** window.*

Step
In the **Step (sec)** edit box, you can specify the time interval used while looking for clashes in a simulation. This edit box will be enabled only when the **TimeLiner** option will be selected in the **Link** drop-down list.

Run Test

Run Test

The **Run Test** button is used for running a clash test. After running a test, you will be able to view clash results in the **Result** tab.

Results Tab
In the **Results** tab, you will be able to view the clash results. This tab is divided into three areas: **Results** View, **Display Settings** panel, and **Items** panel. These areas are discussed next.

Result View
The Results View area displays a list of clash results in a tabular format, as shown in Figure 8-10. The **Name** column in this area displays a list of names of all the clash results. The **Status** column displays the status of the clash results. The **Found** column displays the date and time of the performed clash test. The **Description** column displays the clash type used for performing the clash test. If the clashes have saved viewpoints then the viewpoint icon will be displayed in the **Viewpoint** column. You can customize the display of the Result View area which is discussed later in this chapter. The buttons in the Results View area are used for managing results are discussed next.

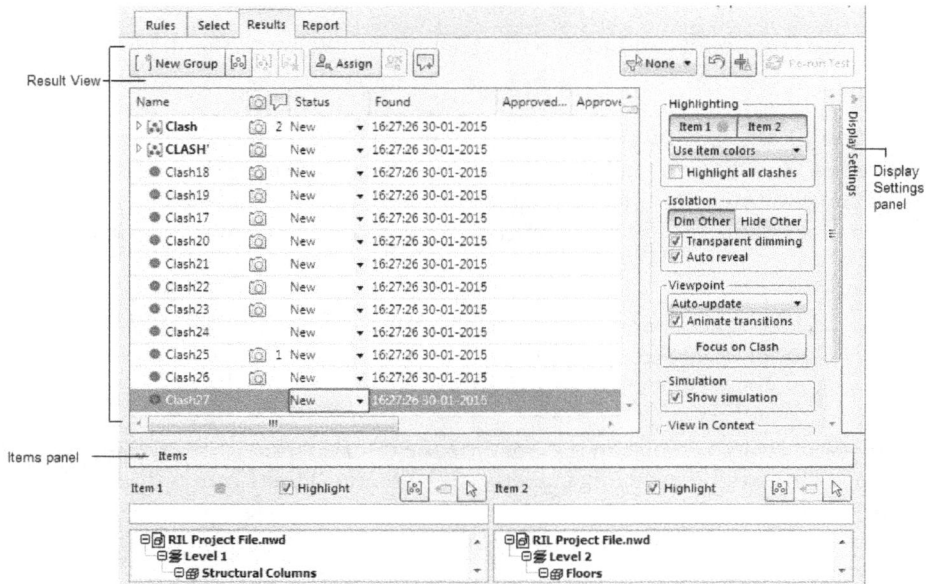

Figure 8-10 *The* **Results** *tab in the* **Clash Detective** *window*

New Group

[⌄]   The **New Group** button is used to create a new empty clash group. On choosing this
button, a new group will be created in the table. Next, drag and place the clashes in
this group.

Group Selected Clashes

[⁸⁰]   The **Group** button is used to group all the selected clash results. To do so, first select
the required clashes and then choose the **Group Selected Clashes** button; the selected
clashes will be grouped in the table.

Remove from Group

[⁰⁸]   The **Remove from Group** button is used to remove the selected clash results from the
group. To do so, select the required clashes in the group and then choose this button;
the selected clashes will be removed from the group.

Explode Group Button

[⁰⁸]   The **Explode Group** button is used to ungroup the selected clash results. To do so,
select the required group and then choose the **Explode Group** button; the selected
group will be ungrouped.

Assign Button

[Assign]   The **Assign** button is used to assign a clash or a clash group to any person or
trade. To do so, first select the required clash. Next, choose this button; the **Assign
Clash** dialog box will be displayed, as shown in Figure 8-11. In this dialog box, specify the
name of the person/trade in the **Assign To** text box. You can also add notes if required in
the **Notes** area. Next, choose the **OK** button; the assigned person/trade will be displayed in
the **Assigned** column corresponding to the selected clash in the Results View area.

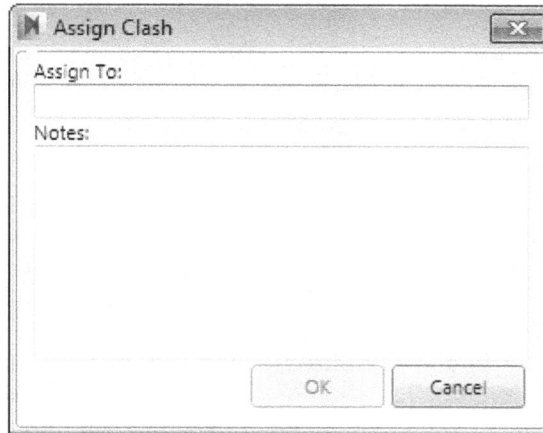

Figure 8-11 *The **Assign Clash** dialog box*

Unassign

The **Unassign** button is used to clear the assignment of selected clashes. To do so, first select the required clashes. Next, choose the **Unassign** button; the assigned person or group will be unassigned from the selected clashes.

Add Comment

The **Add Comment** button is used to add a comment to the selected clashes. To do so, select the required clash. Next, choose the **Add Comment** button; the **Add Comment** dialog box will be displayed. Enter the comment in the dialog box and then choose the **OK** button; the comment will be added to the selected clash. You can view the comments in the **Comments** window.

Filter by Selection

The **Filter by Selection** button is used to filter the clash results based on the objects selected in the Scene View or in the **Selection Tree** window. On choosing this button, a drop-down list will be displayed. Select the **None** option from the list to display all the clashes. Select the **Exclusive** option to display the clashes where both the clashing objects are selected in the Scene View. Select the **Inclusive** option where at least one clashing object is selected in the Scene View.

Reset

The **Reset** button is used to clear all the clash test results.

Compact

The **Compact** button is used to clear all the resolved clashes from the test.

Re-run Test

The **Re-run Test** button is used to re-run the test and also update the clash results.

Display Settings Panel

The right pane of the **Results** tab contains the **Display Settings** panel. It is an expandable panel. You can display or hide this panel by clicking on the **Show/Hide** button. The five areas in this panel are discussed next.

Highlighting

In the **Highlighting** area, there are two buttons which are used to highlight the clash objects in the Scene View. On choosing the **Item 1** and **Item 2** buttons; the corresponding clashing objects will be highlighted in the Scene View. Note that the clashes highlighted in the Scene View will depend upon the button you choose. If you will choose the **Item 1** button, the item 1 involved in the clash will be highlighted. If you choose the **Item 2** button, the item 2 involved in the clash will be highlighted. And if both the buttons are chosen then both the objects will be highlighted. The color of the highlighted object depends on the options selected in the drop-down list below these buttons. If you select the **Use item colors** option from the list, then the objects will be highlighted with the colors assigned to the clashing objects. By default, the colors assigned to the clashed objects are red and green. If you have selected the option **Use status color** from the list, then the objects will be highlighted with the color assigned to the clash status. Select the **Highlight all clashes** check box to highlight all the clashes in the Scene View.

> **Note**
> *You can also change the color of the clashing objects highlighted in the Scene View by selecting required colors from the **Custom Highlight Colors** area in **Options Editor > Tools > Clash Detective**.*

Isolation

In the **Isolation** area, you can use various options to customize the display of clashing objects in the Scene View. Choose the **Dim Other** button to fade the objects that are not involved in the selected clash. Choose the **Hide Other** button to hide all the objects except those which are involved in the selected clash. The **Transparent dimming** check box is activated only when the **Dim Other** button is chosen. Select the **Transparent dimming** check box to make all those items dim that are not involved in the selected clash. Select the **Auto reveal** check box to temporarily hide obstructing item if any to clearly see the selected clash.

Viewpoint

In the **Viewpoint** area, you can specify the method of choosing viewpoints in the Scene View while viewing the clash results. To do so, you can use the options from the **Viewpoint** drop-down list. Select the **Auto-update** option to automatically update the viewpoint for the selected clash in the Scene View. Select the **Auto-Load** option to zoom the camera for displaying all objects in the clash in the Scene View. Select the **Manual** option for not changing the model view to the clash viewpoint. Select the **Animate transitions** check box to show the transition while viewing the clashes one by one. Choose the **Focus on Clash** button to focus on the original clash point.

Simulation

In the **Simulation** area, the **Show simulation** check box is used for the time based and soft clashing tests. On selecting this check box, the playback slider in the timeliner simulation or while playing the animation will move to the exact point at which the clash has occurred.

View in Context

In the **View in Context** area, you can use the options in the **View in Context** drop-down list to temporarily change the zooming reference in the model. Select the **All** option from the list to zoom out the view, which will make the whole view visible in the Scene View. Select the **File** option, to zoom out in such a way that the extent of the file containing the items involved in the selected clash is visible in the Scene View. Select the **Home** option to view the previously defined home view. Choose the **View** button to show the selected context view in the drop-down list in the Scene View.

Items Panel

The bottom area of the **Results** tab contains the **Items** panel. It is an expandable panel. By default, this panel is hidden. To display it, click on the **Show/Hide** button, the **Items** panel will be expanded, as shown in Figure 8-12. This panel contains the information related to both the clashing objects in the selected clash in the **Item1** and **Item2** areas. The clash information displayed in this panel depends on the clash result selected in the Results View area of the **Results** tab. The buttons and options located at the top of the panel are discussed next.

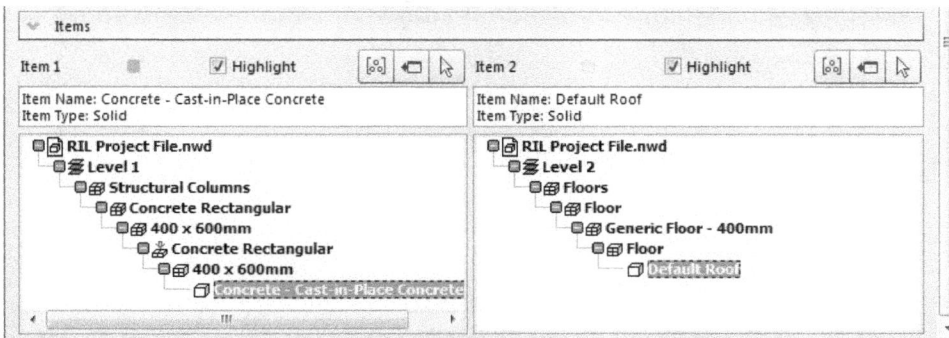

*Figure 8-12 The **Items** panel in the **Results** tab*

Highlight

The **Highlight** check box is used to highlight the clashing objects in the Scene View.

Group Clashes Involving Item

The **Group Clashes Involving Item** button is used to group the objects involved in the selected clash result. When you choose this button, a new group will be created in the table in the Results View.

SwitchBack

The **SwitchBack** button is used to switchback to the design application for the selected clash result. You can choose the **SwitchBack** button to send the current selected object to the original CAD package such as AutoCAD and Revit.

Select in Scene

The **Select in Scene** button is used to select the objects involved in the clash in the Scene View. To do so, choose the **Select in Scene** button; the corresponding objects will be selected in the Scene View as well as in the **Selection Tree** window.

Report Tab

In the **Report** tab, there are several options that are used for generating a report. The generated report will contain the details of all the clash results of the selected clash test. The **Report** tab comprises of three areas: **Contents**, **Include Clashes**, and **Output Settings**, refer to Figure 8-13.

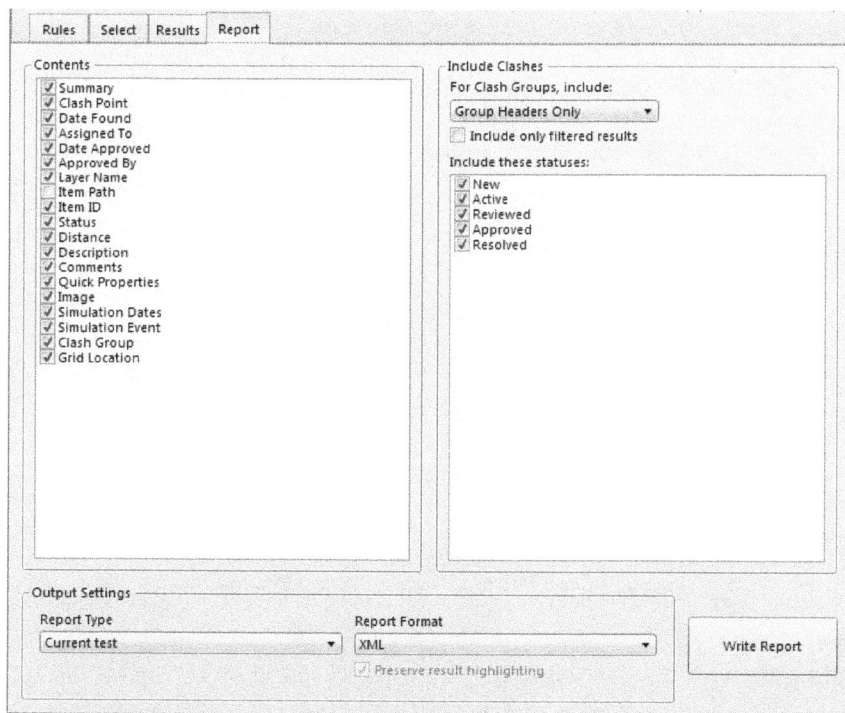

Figure 8-13 *The* **Report** *tab in the* **Clash Detective** *window*

Contents

The **Contents** area in the **Report** tab contains a list of all the contents to be included in the report. In this area, select the check boxes for the contents that you want to include in the report.

Include Clashes

In the **Include Clashes** area of the **Report** tab, you can define the display of clash groups in the report by using the options from the **For Clash Groups, include** drop-down list. Select the

Group Headers Only option to include the summary of clash groups and individual clashes in the report. Select the **Individual Clashes Only** option to include the individual clash results in the report. Select the **Everything** option to display both clash groups and individual clash results in the report. Select the **Include only filtered results** check box to include only filtered clash results in the report. In the **Include these statuses** area, select the required check boxes to define the clashes to be included in the report.

Output Settings

In the **Output Settings** area of the **Report** tab, you can define the report type and its format. You can define the type of report by selecting options from the **Report Type** drop-down list. Select the **Current Test** option to create a single report for the current test. Select the **All Tests (Combined)** option to create a single report for all tests. Select the **All Tests (Separate)** option to create a separate report for each test.

You can specify the format of the report by selecting the options from the **Report Format** drop-down list. Select the **XML** option to create report in an xml file. Select the **HTML** option to create a report in which clashes will be listed sequentially. Select the **HTML (Tabular)** option to create a report in which clashes will be listed in a tabular format. Select the **Text** option to create a report in a text file. Select the **As viewpoints** option to create a folder with the name Test_Name in the **Saved Viewpoints** window. This folder will contain each clash as a viewpoint with an attached comment. When you select the **As viewpoints** option, the **Preserve result highlighting** check box will be activated. Select this check box to maintain the transparency for each clash viewpoint and also to highlight it.

After defining all the parameters, choose the **Write Report** button to create a report. On choosing this button; the **Save As** dialog box will be displayed. In this dialog box, browse to the location and save the report by choosing the **Save** button.

MANAGING CLASH TESTS

In this section, you will learn about merging and managing clash tests. You will learn about merging clash tests from different files, importing and exporting clash tests, and creating custom clash tests.

Merging Clash Tests

In Navisworks, when you merge different files into a single Navisworks file, all the clash data from different files will be combined together. To do so, choose **Open > Merge** from the **Application Menu**; the **Merge** dialog box will be displayed. In this dialog box, browse to the file location to choose the required file and then choose the **Open** button; the selected file will be merged with the current file. Next, invoke the **Clash Detective** window; all the clash data will be combined and displayed in this window.

The procedure of importing and exporting clash tests is discussed earlier in this chapter.

Creating Custom Clash Tests

If you are using a common set of clash tests in the multiple projects, then a custom clash test can be created which will be used everywhere. To create a custom clash test, you need to export the clash test in the .xml file format (as discussed above). After exporting the clash test, copy the exported file to the **custom_clash_tests** folder in the Autodesk Navisworks Manage 2017 search directory. Now, restart the Navisworks session. Choose the **Select** tab in the **Clash Detective** window. Next, choose the required custom test from the **Type** drop-down list of the **Settings** area.

LINKING CLASH TEST WITH TIMELINER AND OBJECT ANIMATION

In Navisworks, you can link the clash test with timeliner simulation and object animation. When you link the clash test with the timeliner, you are able to identify the clashes for the entire project. Similarly, by linking the clash test with the object animation, you can identify the clashes between moving objects. You can also link the clash tests with the animated timeliner schedule in which some tasks are linked to animation scenes. When a clash test is linked with the timeliner simulation then it is called Time-Based Clash Test. When a clash test is linked with the object animation then it is called Soft Clashing. When it is linked with the animated timeliner schedule then it is called Time-Based Soft Clashing. These are discussed in detail next.

Time-Based Clash Test

By linking clash detection with the timeliner simulation, you can identify the clashes at each step in the timeliner sequence. You can link the tasks which are scheduled as temporary, demolished or even construct task types to the clash detection such as cranes, bulldozer, and so on. This feature will help you in monitoring a project on a timely basis. To link a timeliner schedule to a clash test, choose the **Select** tab in the **Clash Detective** window. Select the objects that you want to test in the **Selection A** and **Selection B** areas. Next, select the **TimeLiner** option from the **Link** drop-down list. Specify the time interval required for checking the clashes in the **Step (sec)** edit box. Next, choose the **Run Test** button. If there are any clashes present, clash detective will check them at each interval and display them in the **Clash Detective** window. Next, in the **Results** tab, select the **Show simulation** check box in the **Simulation** area of the **Display Settings** panel. In the **TimeLiner** window, choose the **Simulate** tab, and in the **Clash Detective** window, select the required clash result in the Results View area; the slider in the **TimeLiner** window will move to the exact point at which the clash has occurred. This slider can be used for checking the events which are occurring before and after the clash.

Soft Clashing

In Navisworks, you can also link an object animation scene created in the **Animator** window to a clash test to identify the clashes between moving objects. To do so, choose the **Select** tab in the **Clash Detective** window. Select the objects in the **Selection A** and **Selection B** areas. Next, select the required animation scene from the **Link** drop-down list. Specify the time interval size to be used when looking for clashes in the animation in the **Step (sec)** edit box. Now, choose the **Run Test** button. If there is any collision between the moving objects in the animation, clash detective will check it and display it in the window. Next, choose the **Results** tab in the **Clash Detective** window. Select the **Show simulation** check box in the **Simulation** area of the **Display Settings** panel. Now, select the required result in the Results View area of the **Results** tab and choose the **Animation** tab in the ribbon; the playback slider in the **Playback** panel will move to the exact point at which the clash has occurred.

Time-Based Soft Clashing

Before linking the clash test with the animated timeliner schedule, ensure that at least one animation scene has been linked to one of the tasks in the **TimeLiner** window. Next, in the **Select** tab of the **Clash Detective** window, define the clash test and select the **TimeLiner** option from the **Link** drop-down list. Specify the interval size in the **Step (sec)** edit box, and choose the **Run Test** button; the clash detective will check the clashes at each interval and the results will be displayed in the window. Now, view the clash results in the same way as for the Time-Based Clash Test.

CUSTOMIZING CLASH DETECTIVE WINDOW

You can also control the display of columns in the **Result** tab. To do so, right-click on the column header in the Results View area; a shortcut menu will be displayed. Choose the **Choose Columns** from the menu; the **Choose Columns** dialog box will be displayed, as shown in Figure 8-14. In the dialog box, select the check box corresponding to the column name that you want to display in the result table. Next, choose the **OK** button; the selected column will be displayed in the column table. Note that clearing the check box will remove the column from the Results View area of the **Results** tab.

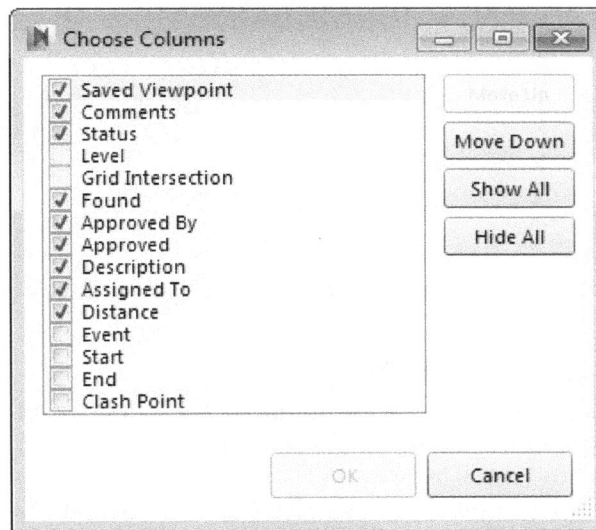

*Figure 8-14 The **Choose Columns** dialog box*

As discussed earlier in this chapter, you can use the options in the **Display Settings** panel to control the display of clash results. Alternatively, you can control the clash results display from the **Options Editor** dialog box. To do so, first invoke the **Options Editor** dialog box from the Application Menu. In this dialog box, expand the **Tools** node in the left pane and then select the **Clash Detective** option from the left pane; several options will be displayed in the right pane of the dialog box, refer to Figure 8-15.

In the **View in Context Zoom Duration (seconds)** edit box, specify the duration that will be used for an animated transition. Specify the duration during which the view will remain zoomed out in the **View in Context Pause (seconds)** edit box. Specify the duration to be used for moving between the views in the **Animated Transition Duration (seconds)** edit box. Specify the dimming level of transparency for the items not involved in the clash tests in the **Dimming Transparency** edit box. Select the **Use Wireframes for Transparent Dimming** check box to display the items not involved in clashes in a wireframe structure. Specify the factor of zoom to be applied to the clash in the Scene View in the **Auto-Zoom Distance Factor** edit box.

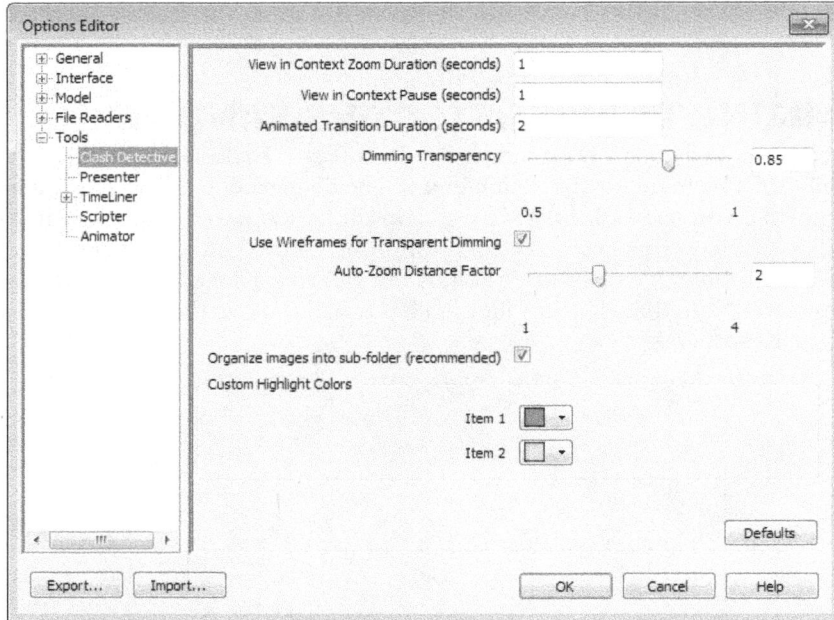

Figure 8-15 *The* *Clash Detective* *options in the* *Options Editor* *dialog box*

CREATING CLASH REPORT

In Navisworks, several types of reports can be produced. You can share these reports with others. To create a clash report after running the clash test, choose the **Report** tab. In the **Contents** area of the **Report** tab, select the check boxes to be included in the report. In the **Include Clashes** area, select the required option from the **For Clash Groups, include** drop-down list for displaying the clash groups in the report. For example, if you select the **Group Headers Only** option then the clash report will include a summary of the clash groups and individual clashes which are not grouped. Select the status check boxes in the **Include these statuses** area for including the status of clash results. Select the type of report from the **Report Type** drop-down list. Select the report format from the **Report Format** drop-down list. Choose the **Write Report** button to write the report. When you choose this button; the **Save As** dialog box will be displayed. Browse to the folder and save the file in the desired report format such as .xml, .html. If you select the HTML report format, then the report will be created, as shown in Figure 8-16.

Name	Clash40
Distance	m
Status	New
Clash Point	-12.88m, -6.48m, 3.60m
Date Created	2015/1/30 10:57:26

Item 1

Element ID	140904
Layer	Level 1
Path	File ->RIL Project File.nwd ->Level 1 ->Structural Columns ->Concrete Rectangular ->400 x 600:
->Concrete - Cast-in-Place Concrete	
TimeLiner Attached to	
Task:1	
TimeLiner Attached to	
Task Start (Planned):1	
TimeLiner Contained in	
Task End (Planned):1	
TimeLiner Attached to	
Task Start (Planned):2	

Figure 8-16 Report created in the HTML format

TUTORIALS
General instructions for downloading tutorial files:

1. Download the *c08_nws_2017_tut* zip file for this tutorial from *http://www.cadcim.com*. The path of the file is as follows: *Textbooks > Civil/GIS > Navisworks > Exploring Autodesk Navisworks 2017*.

2. Now, save and extract the downloaded folder at the following location:
   C:\ nws_2017\ c08_nws_2017_tut

Note
*The default unit system used in the tutorials is metric. To change the units to imperial, select the required units from **Options Editor > Interface > Display Units**.*

Tutorial 1   Clash Detection Between Ducts and Beams

In this tutorial, you will perform clash detection to find the interferences between the ducts and beams in a model. To perform this tutorial, you will use the *c08_navisworks_2017_tut1.nwf* file.
                                                                                              (Expected time: 30 min)
The following steps are required to complete this tutorial:

a. Open the *c08_nws_2017_tut1.nwf* file.
b. Create selection sets.
c. Add test and select test items.
d. Specify rules for clash detection.
e. Perform the clash test.
f. Review clash results.
g. Save the clashed views.
h. Create the clash report.
i. Save the project.

Opening the File

In this section, you will open the file.

1. Choose the **Open** button from the Quick Access Toolbar; the **Open** dialog box is displayed.

2. In this dialog box, browse to the following location:
    C:\nws_2017\c08_nws_2017_tut

3. Select the **Navisworks File Set (*.nwf)** file format from the **Files of type** drop-down list.

4. Next, select the *c08_nws_2017_tut1* from the file list, the file name is displayed in the **File name** drop-down.

5. Choose the **Open** button; the model is displayed, as shown in Figure 8-17.

Note
*Sometimes on choosing the **Open** button, the **Resolve** message box is displayed prompting you to resolve external references. In such a case, choose the **Ignore All** button to ignore the message.*

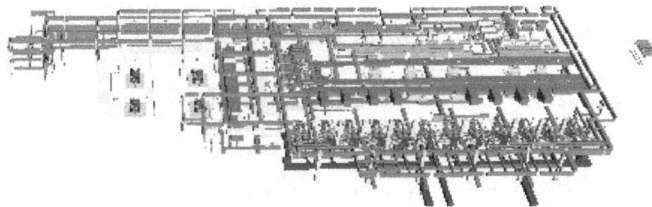

Figure 8-17 *Model displayed in Scene View*

Creating Selection Sets

In this section, you will define two selection sets: Ducts and Beams. Further, you will use this selection set to perform the clash detection test.

1. Choose the **Selection Tree** tool from the **Select & Search** panel of the **Home** tab; the **Selection Tree** window is displayed.

2. Select the **Round Duct** sub node from **file.nwc > Ground Floor > Ducts** node, refer to Figure 8-18; the ducts are highlighted in the Scene View.

3. Choose the **Manage Sets** tool from **Home > Select & Search > Sets** drop-down; the **Sets** window is displayed.

4. In the **Sets** window, choose the **Save Selection** button; a selection set is created [⚙].

5. Enter the name **Ducts** in the selection set and press ENTER.

6. Next, select the **Structural Framing** sub-node in the **No Level** node of the **Selection Tree** window.

7.   Repeat step 4 and save the selection with the name **Beams**. Figure 8-19 shows the **Sets** window with the created selection sets.

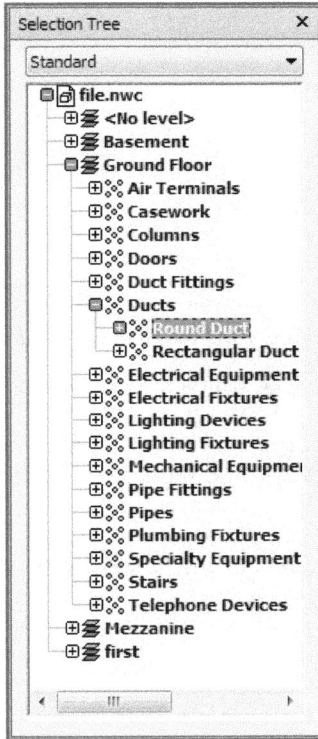

Figure 8-18 *The Round Duct selected in the Selection Tree window*

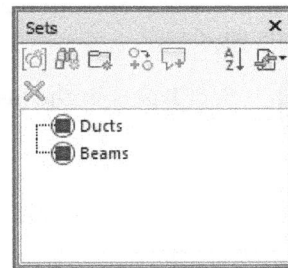

Figure 8-19 *The Sets window*

Adding the Test and Selecting Test Items

In this section, you will add a test in the **Clash Detective** window and then select the test items used for clash testing.

1.   Choose the **Clash Detective** tool from the **Tools** panel in the **Home** tab; the **Clash Detective** window is displayed.

2.   Choose the **Add Test** button in the **Clash Detective** window; a new test named **Test 1** is added in the **Test** panel.

3.   Next, rename **Test1** as **Ducts vs Beams** by double-clicking on it and press ENTER, refer to Figure 8-20.

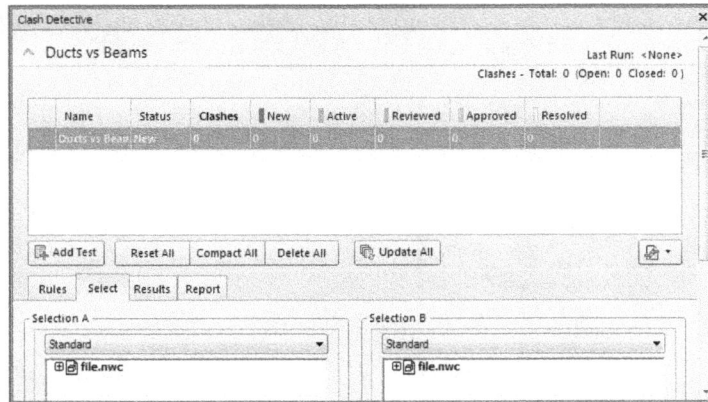

Figure 8-20 Test added in the Test panel of the Clash Detective window

4. Choose the **Select** tab in the **Clash Detective** window, if it is not chosen by default.

5. In the **Select** tab, select the **Sets** option from the drop-down lists available at the top in the **Selection A** and **Selection B** areas; the created selection sets are displayed in both the areas of the window.

6. Next, select **Ducts** in the **Selection A** area and **Beams** in the **Selection B** area, as shown in Figure 8-21.

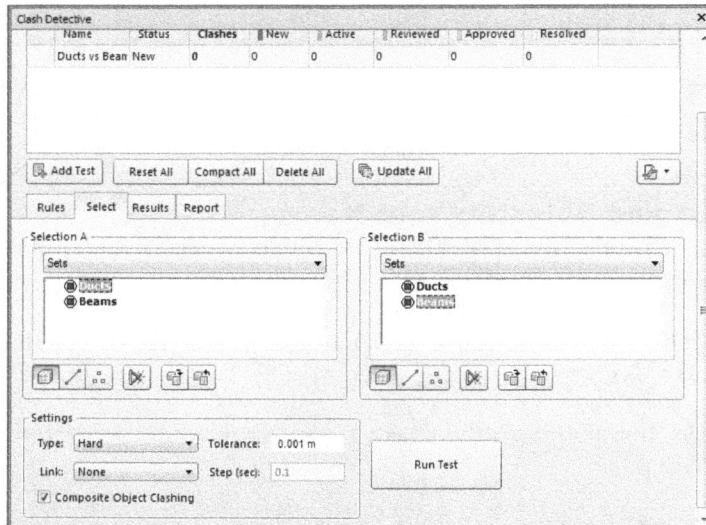

Figure 8-21 Test items selected in the Selection A and Selection B areas

Specifying Rules for Clash Detection

In this section, you will specify the rules to be used for detecting clashes.

1. Choose the **Rules** tab in the **Clash Detective** window.

2. Ensure that the **Items in same layer** check box is selected from the predefined rules displayed in the window, as shown in Figure 8-22.

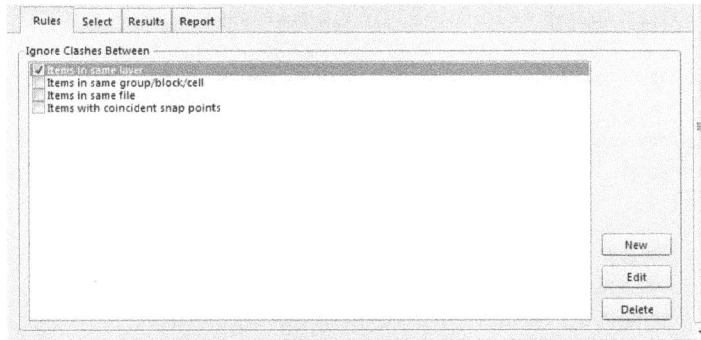

*Figure 8-22 Rule selected in the **Rules** tab*

Performing Clash Test

In this section, you will perform the clash test.

1. Choose the **Select** tab and ensure that the **Hard** option is selected in the **Type** drop-down in the **Settings** area.

2. Next, choose the **Run Test** button in the **Clash Detective** window; the **Results** tab is displayed with the clash results and the summary is displayed in the **Test** panel, as shown in Figure 8-23

.

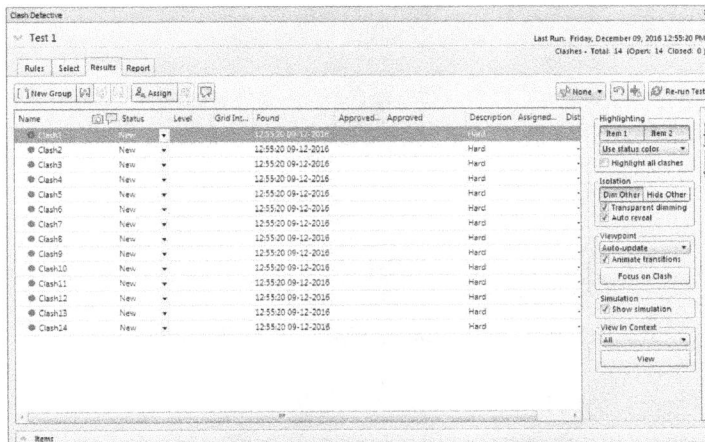

*Figure 8-23 Partial view of the clash test results displayed in the **Results** tab*

Reviewing Clash Results

In this section, you will view the clash results.

1. In the **Results** tab, ensure that the **Item 1** and **Item 2** buttons are chosen in the **Highlighting** area of the **Display Settings** panel.

2. Next, ensure that the **Dim Other** button is chosen in the **Isolation** area of the **Display Settings** panel.

3. Next, select the **Transparent dimming** and **Auto reveal** check boxes in the **Isolation** area.

4. Ensure that the **Auto-update** option is selected in the **Viewpoint** drop-down list.

5. Select the **Animate transitions** check box in the **Viewpoint** area of the **Display Settings** panel, if not selected.

Note
Deselect the clashing objects if selected in the Scene View.

6. Next, select the **Clash1** result in the result table; the corresponding clashed items are displayed in the Scene View, as shown in Figure 8-24.

7. Click in the **Status** column corresponding to the **Clash1** result; a drop-down list is displayed.

8. Select the **Reviewed** option from this list. Next, select the **Use status color** option from the drop-down list in the **Highlighting** area of the **Display Settings** panel; the clashed items get highlighted in the color assigned for the clash status, as shown in Figure 8-25.

9. Repeat the procedure followed in steps 1 to 7 and view the clash results Clash2 to Clash14.

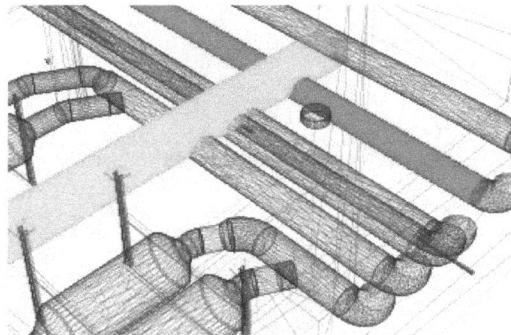

Figure 8-24 Clashing objects in the Scene View

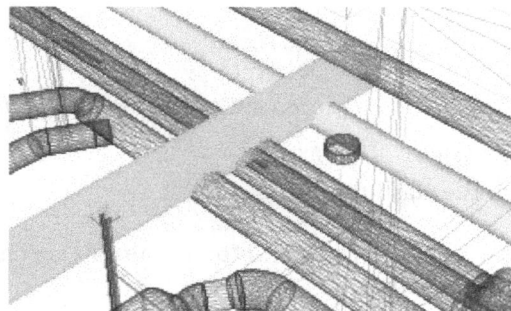

Figure 8-25 Reviewed clash objects

Saving Clashed Views

In this section, you will highlight the clashed item and add text to it.

1. Choose the **Cloud** tool from **Review > Redline > Draw** drop-down; the cursor changes into a pencil cursor.

2. Draw the cloud around the clashed items, as shown in Figure 8-26. Here, Clash1 has been shown.

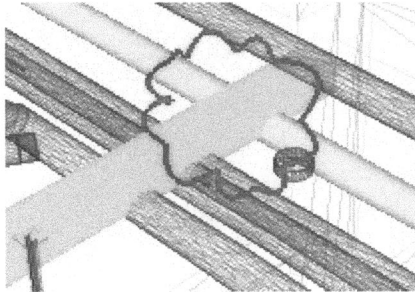

Figure 8-26 Clouds drawn around the clashing objects

3. Next, choose the **Text** tool from the **Redline** panel of the **Review** tab. Place the cursor inside the drawn cloud and click; the **Autodesk Navisworks Manage 2017** dialog box is displayed.

4. In this dialog box, enter **Clashed Items** in the **Enter Redline Text** text box, and choose the **OK** button; the text is added, as shown in Figure 8-27.

5. Repeat the procedure followed in steps 1 to 4 and save the other clashed views.

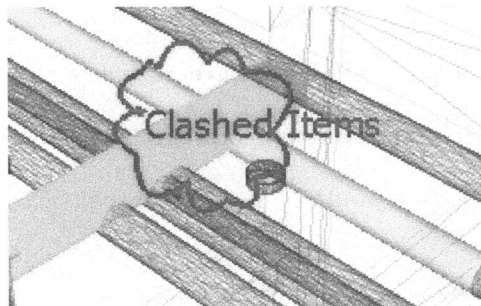

Figure 8-27 Text added to the clashing objects

Creating Clash Report

In this section, you will create the clash report.

1. Choose the **Report** tab in the **Clash Detective** window.

2. In the **Contents** and **Include these statuses** areas of the window, ensure that the check boxes are selected, as shown in Figure 8-28.

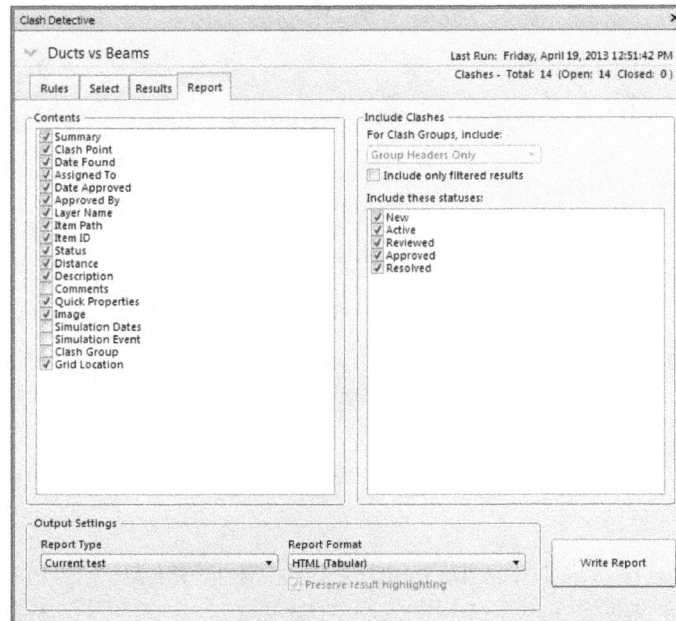

*Figure 8-28 Check boxes selected in the **Report** tab of the **Clash Detective** window*

3. Next, in the **Output Settings** area, ensure that the **Current test** option is selected in the **Report Type** drop-down list, refer to Figure 8-28.

4. Select the **HTML (Tabular)** option from the **Report Format** drop-down list, refer to Figure 8-28.

5. Next, choose the **Write Report** button; the **Save As** dialog box is displayed.

6. In this dialog box, browse to the following location: *C:\Navisworks_2017\c08_tutorial*.

7. Specify the name **Clash_Report_tut01** in the **File name** edit box and choose the **Save** button; the clash report is saved at the specified location.

8. Next, browse to the location on your system where the report is saved and click on the **Clash_Report_tut01** HTML document; the created report is displayed, as shown in Figure 8-29.

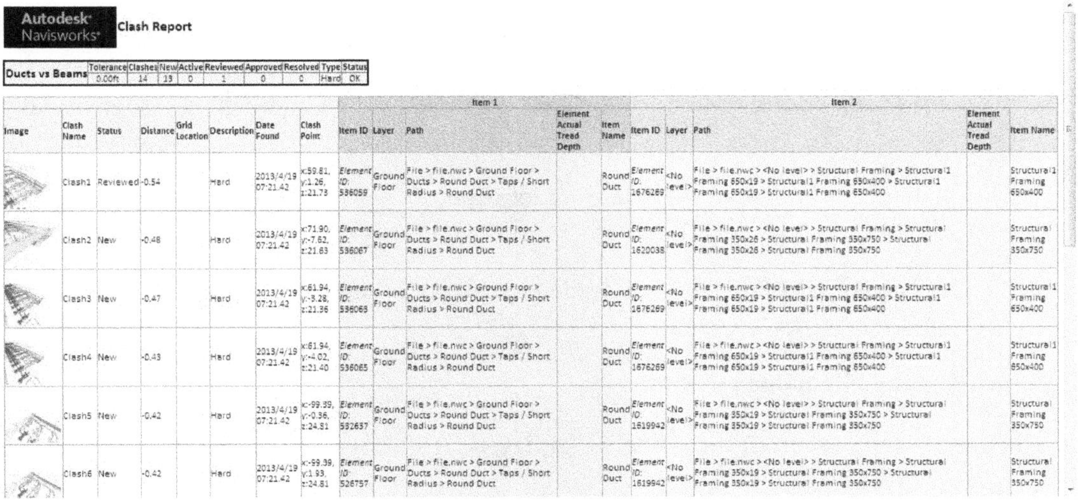

Ducts vs Beams	Tolerance	Clashes	New	Active	Reviewed	Approved	Resolved	Type	Status
	0.00ft	14	13	0	1	0	0	Hard	OK

								Item 1					Item 2				
Image	Clash Name	Status	Distance	Grid Location	Description	Date Found	Clash Point	Item ID	Layer	Path	Element Actual Tread Depth	Item Name	Item ID	Layer	Path	Element Actual Tread Depth	Item Name
	Clash1	Reviewed	-0.54		Hard	2013/4/19 07:21.42	x:59.81, y:1.26, z:21.73	*Element ID:* 536059	Ground Floor	File > file.nwc > Ground Floor > Ducts > Round Duct > Taps / Short Radius > Round Duct		Round Duct	*Element ID:* 1676269	<No level>	File > file.nwc > <No level> > Structural Framing > Structural1 Framing 650x19 > Structural1 Framing 650x400 > Structural1 Framing 650x19 > Structural1 Framing 650x400		Structural1 Framing 650x400
	Clash2	New	-0.48		Hard	2013/4/19 07:21.42	x:71.90, y:-7.62, z:21.63	*Element ID:* 536067	Ground Floor	File > file.nwc > Ground Floor > Ducts > Round Duct > Taps / Short Radius > Round Duct		Round Duct	*Element ID:* 1620038	<No level>	File > file.nwc > <No level> > Structural Framing > Structural Framing 350x26 > Structural Framing 350x750 > Structural Framing 350x26 > Structural Framing 350x750		Structural Framing 350x750
	Clash3	New	-0.47		Hard	2013/4/19 07:21.42	x:61.94, y:-3.28, z:21.36	*Element ID:* 536068	Ground Floor	File > file.nwc > Ground Floor > Ducts > Round Duct > Taps / Short Radius > Round Duct		Round Duct	*Element ID:* 1676269	<No level>	File > file.nwc > <No level> > Structural Framing > Structural1 Framing 650x19 > Structural1 Framing 650x400 > Structural1 Framing 650x19 > Structural1 Framing 650x400		Structural1 Framing 650x400
	Clash4	New	-0.43		Hard	2013/4/19 07:21.42	x:61.94, y:-4.02, z:21.40	*Element ID:* 536065	Ground Floor	File > file.nwc > Ground Floor > Ducts > Round Duct > Taps / Short Radius > Round Duct		Round Duct	*Element ID:* 1676269	<No level>	File > file.nwc > <No level> > Structural Framing > Structural1 Framing 650x19 > Structural1 Framing 650x400 > Structural1 Framing 650x19 > Structural1 Framing 650x400		Structural1 Framing 650x400
	Clash5	New	-0.42		Hard	2013/4/19 07:21.42	x:-99.39, y:-0.36, z:24.81	*Element ID:* 532637	Ground Floor	File > file.nwc > Ground Floor > Ducts > Round Duct > Taps / Short Radius > Round Duct		Round Duct	*Element ID:* 1619942	<No level>	File > file.nwc > <No level> > Structural Framing > Structural Framing 350x19 > Structural Framing 350x750 > Structural Framing 350x19 > Structural Framing 350x750		Structural Framing 350x750
	Clash6	New	-0.42		Hard	2013/4/19 07:21.42	x:-99.39, y:1.93, z:24.81	*Element ID:* 526757	Ground Floor	File > file.nwc > Ground Floor > Ducts > Round Duct > Taps / Short Radius > Round Duct		Round Duct	*Element ID:* 1619942	<No level>	File > file.nwc > <No level> > Structural Framing > Structural Framing 350x19 > Structural Framing 350x750 > Structural Framing 350x19 > Structural Framing 350x750		Structural Framing 350x750

***Figure 8-29** The clash report generated in the HTML Tabular format*

Saving the Project

In this section, you will save the project.

1. Close the **Clash Detective** window.

2. Select the **3D View** in the **Saved Viewpoints** window; the view is displayed in the Scene View.

3. Choose the **Save As** option from the Application Menu, the **Save As** dialog box is displayed.

4. Browse to *Navisworks_2017* folder and enter *c08_navisworks_2017_tut01* in the **File name** edit box. Next, choose **Navisworks File Set (*.nwf)** file format from the **Save as type** drop-down list, and choose the **Save** button; the project is saved.

Tutorial 2 Clash Detection Between Pipes and Walls

In this tutorial, you will perform clash detection to find the interferences between pipes and walls in a model. To perform this tutorial, you will use the *c08_navisworks_2017_tut2* file.

(Expected time: 20 min)

The following steps are required to complete this tutorial:

a. Open the *c08_navisworks_2017_tut2* file.
b. Add the test and select the test items.
c. Perform the clash test.
d. Review clash results.
e. Save the clashed views.
f. Create clash report.
g. Save the project.

Opening the File

In this section, you will open the file.

1. Choose the **Open** button from the Quick Access Toolbar; the **Open** dialog box is displayed on the screen.

2. In this dialog box, browse to the following location:
    C:\nws_2017\c08_nws_2017_tut

3. Select the **Navisworks File Set (*.nwf)** file format from the **Files of type** drop-down list.

4. Next, select the *c08_navisworks_2017_tut2* from the displayed list of files, the file name appears in the **File name** edit box.

5. Choose the **Open** button; the model is displayed, as shown in Figure 8-30.

Figure 8-30 Model displayed in the Scene View

Adding the Test and Selecting Test Items

In this section, you will add a test to the **Clash Detective** window and then select the test items used for clash testing.

1. Choose the **Clash Detective** tool from the **Tools** panel in the **Home** tab; the **Clash Detective** window is displayed.

2. Choose the **Add Test** button in the **Clash Detective** window; **Test 1** is added to the **Test** panel.

3. Rename **Test1** as **Pipes vs Walls** by double-clicking on it and press ENTER, refer to Figure 8-31.

4. Choose the **Select** tab in the **Clash Detective** window.

5. In the **Select** tab, select the **Sets** option from the drop-down lists available at the top in the **Selection A** and **Selection B** areas; the created selection sets will be displayed in both the areas of the window.

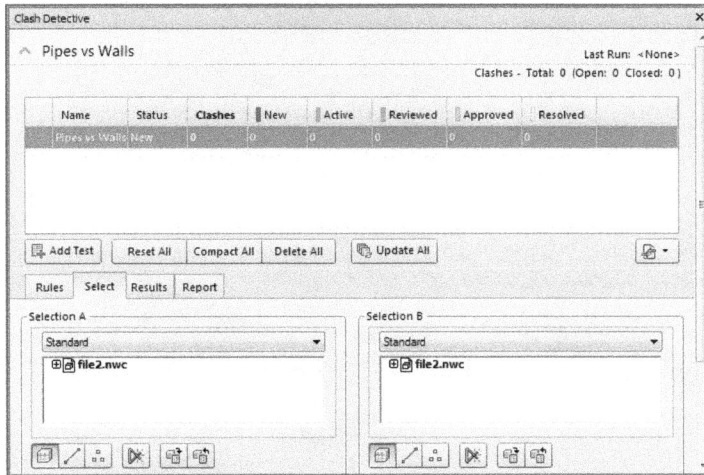

*Figure 8-31   Test added to the **Test** panel of the **Clash Detective** window*

6.   Select **Pipes** in the **Selection A** area and **Walls** in the **Selection B** areas, refer to Figure 8-32.

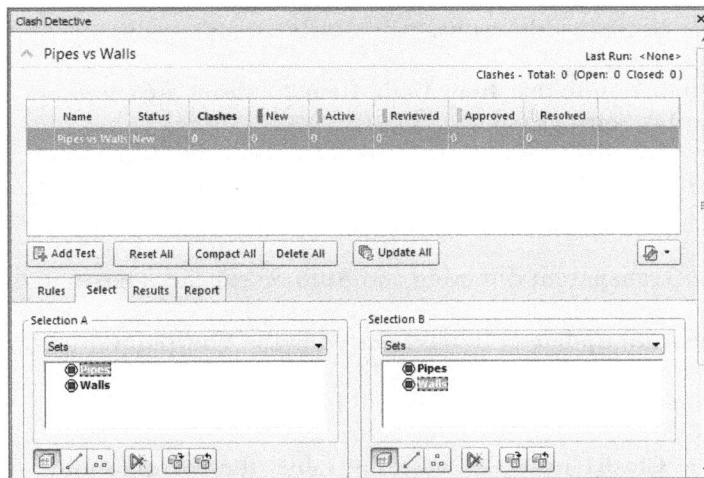

*Figure 8-32   Test items selected in the **Selection A** and **Selection B** areas*

Note
*Ensure that all check boxes in the **Ignore Clashes Between** area of the **Rules** tab are cleared.*

Performing the Clash Test
In this section, you will perform clash test between various building components.

1.   In the **Select** tab, ensure that the **Hard** option is selected in the **Type** drop-down in the **Settings** area.

2.   Choose the **Run Test** button in the **Clash Detective** window; the **Results** tab is displayed with all the clash results. Also, the summary is displayed in the **Test** panel, as shown in Figure 8-33.

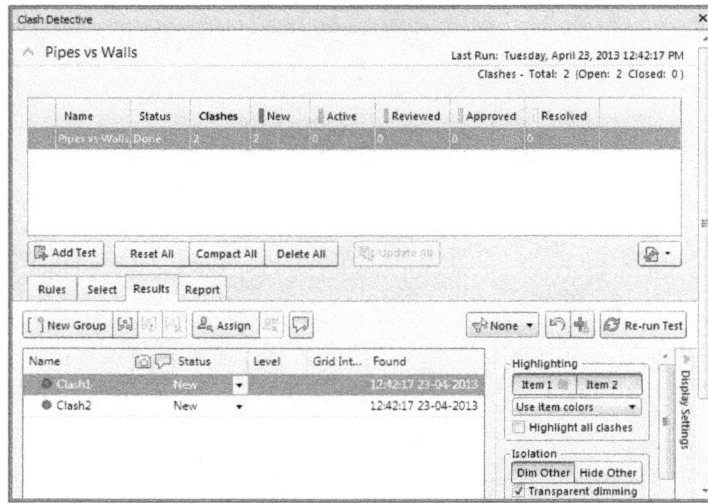

Figure 8-33 *Clash results and summary displayed in the window*

Reviewing Clash Results

In this section, you will view clash results.

1. In the **Results** tab, ensure that **Item 1** and **Item 2** buttons are chosen in the **Highlighting** area of the **Display Settings** panel.

2. Next, ensure that the **Dim Other** button is chosen in the **Isolation** area of the **Display Settings** panel.

3. Next, select the **Transparent dimming** and **Auto reveal** check boxes in the **Isolation** area.

4. Ensure that the **Auto-update** option is selected in the **Viewpoint** drop-down list.

5. Select the **Animate transitions** check box, if not selected by default.

6. Next, select the **Clash1** result in the result table; the corresponding clashed items are displayed in the Scene View, as shown in Figure 8-34.

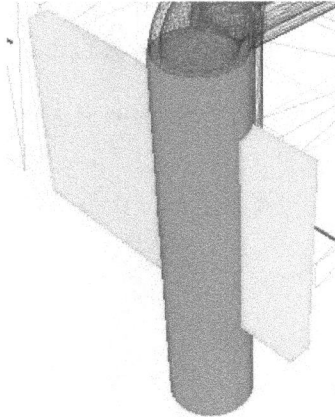

Figure 8-34 *Clash objects in the Scene View*

7. Click in the **Status** column corresponding to the **Clash1** result; a drop-down list is displayed.

8. Select the **Reviewed** option from this list. Next, select the **Use status color** option from the drop-down list in the **Highlighting** area of the **Display Settings** panel; the clashed items are highlighted with the color assigned for the clash status, as shown in Figure 8-35.

9. Repeat the procedure followed in step 7 and approve the **Clash2** result by selecting the **Approved** option from the **Status** drop-down list.

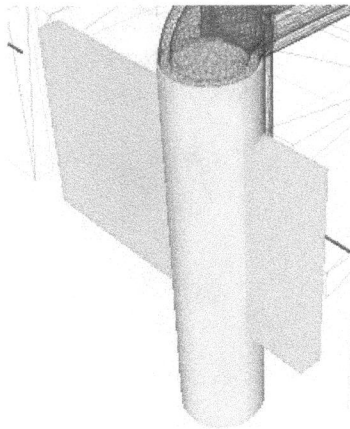

Figure 8-35 *Reviewed clash objects in the Scene View*

Saving the Clashed Views

In this section, you will highlight the clashed item and add texts to them.

1. Choose the **Ellipse** tool from **Review > Redline > Draw** drop-down; the cursor changes into a pencil cursor.

2. Draw the ellipse around the clashed objects of the **Clash1** result, as shown in Figure 8-36.

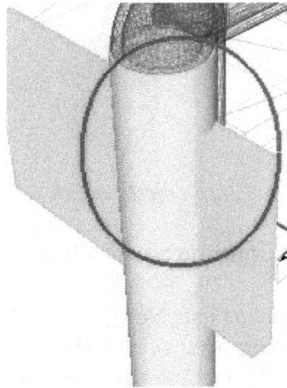

Figure 8-36 Ellipse drawn around the clashed objects

3. Next, choose the **Text** tool from the **Redline** panel of the **Review** tab. Place the cursor inside the drawn ellipse and click; the **Autodesk Navisworks Manage 2017** dialog box is displayed.

4. In this dialog box, enter the text **Clashed Objects** in the **Enter Redline Text** text box, and choose the **OK** button; the text is added, as shown in Figure 8-37.

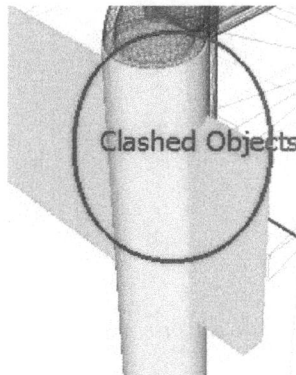

Figure 8-37 Text added to the clashing objects

5. Repeat the procedure followed in steps 1 to 4 and save the **Clash2** result view.

Creating Clash Report

In this section, you will create the clash report.

1. Choose the **Report** tab in the **Clash Detective** window.

2. In the **Contents** and **Include these statuses** areas, select the check boxes, as shown in Figure 8-38.

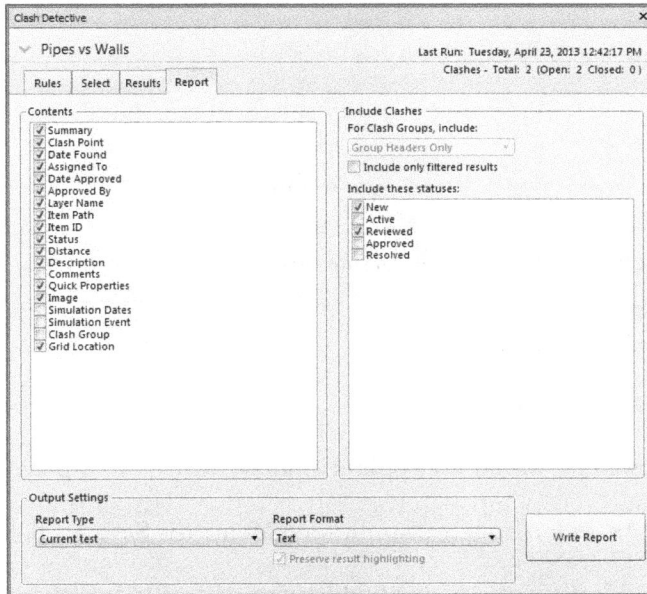

*Figure 8-38 The **Report** tab in the **Clash Detective** window*

3. Next, in the **Output Settings** area, ensure that the **Current test** option is selected in the **Report Type** drop-down list, refer to Figure 8-38.

4. Select the **Text** option from the **Report Format** drop-down list, refer to Figure 8-38.

5. Next, choose the **Write Report** button; the **Save As** dialog box is displayed.

6. In this dialog box, browse to the following location: *C:\nws_2017\c08_nws_2017_tut*.

7. Enter the name **Clash_Report_tut02** in the **File name** edit box and choose the **Save** button; the clash report is saved at the specified location.

8. Next, browse to the location on your system where the report is saved and click on the **Clash_Report_tut02** text document; the created report is displayed in the text format, as shown in Figure 8-39.

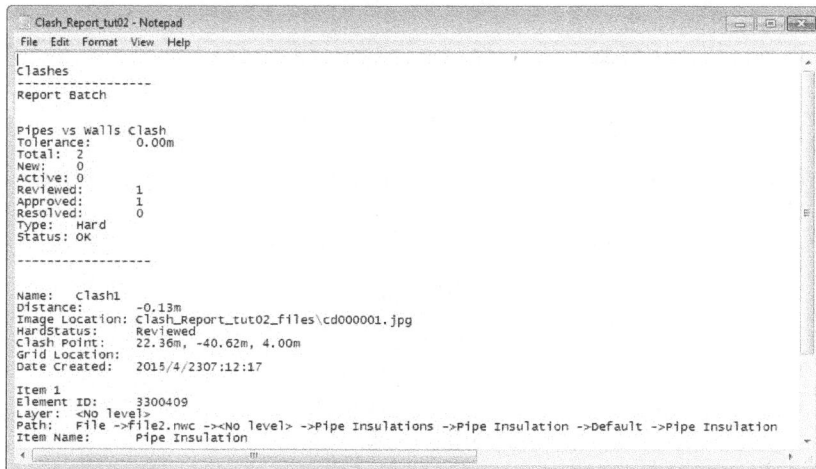

Figure 8-39 *Partial view of the clash report in the text format*

Saving the Project

In this section, you will save the project.

1. Close the **Clash Detective** window.

2. Select the **3D View** in the **Saved Viewpoints** window; the view is displayed in the Scene View.

3. Choose the **Save As** option from the Application Menu, the **Save As** dialog box is displayed.

4. Browse to *nws_2017\ c08_nws_2017_tut* folder and enter *c08_navisworks_2017_tut02* in the **File name** edit box. Next, choose **Navisworks File Set (*.nwf)** file format from the **Save as type** drop-down list, and choose the **Save** button.

Self-Evaluation Test

Answer the following questions and then compare them to those given at the end of this chapter:

1. You can reset the status of all the clash tests to **New** by choosing the _____ button.

2. You can use the _____ rule to ignore clashes between objects in the same layer.

3. You can include the surface geometries in the clash tests by choosing the _____ button.

4. The _____ button is used to ungroup the selected clash results.

5. You can fade the objects that are not involved in the clash test by choosing the _____ button.

6.   You cannot link a clash test to an object animation. (T/F)

7.   You cannot include point geometry in a clash test. (T/F)

8.   You cannot customize the rules available in the **Rules** tab. (T/F)

9.   The **Test** panel displays the summary of the performed clash test. (T/F)

10. You cannot ignore clashes between objects of the same selection set. (T/F)

Review Questions

Answer the following questions:

1.   Which of the following options is used to remove all the resolved clash tests?

   a) **Update All**   b) **Delete All**
   c) **Compact All**   d) **Reset All**

2.   Which of the following options is used to include line geometry in a clash test?

   a) **Point**   b) **Surfaces**
   c) **Self-Intersect**   d) **Lines**

3.   Which of the following options is used to perform a clash test?

   a) **Start**   b) **Play**
   c) **Run Test**   d) **Stop**

4.   Which of the following options is used to group the selected clash results?

   a) **Remove from Group**   b) **New Group**
   c) **Group Selected Clashes**   d) **Explode Group**

5.   Which of the following options is used to assign a clash or a clash group to a person or trade?

   a) **Unassign**   b) **Assign**
   c) **Delete**   d) **Reset All**

6.   You cannot save the viewpoint for the clash results. (T/F)

7.   You can highlight the objects involved in a clash by choosing the **Select in Scene** button.

8.   The **Unassign** button is used to clear the assignment for selected clashes. (T/F)

9.   You cannot add comments to the clashes. (T/F)

10. You cannot customize the display of clash results in the Scene View. (T/F)

EXERCISE

Exercise 1   Performing Clash Detection

Download the *c08_navisworks_2017_ex1.nwf* file from *http://www.cadcim.com* and then open it. Find clashes between model elements using the **Clash Detective** window. Figure 8-40 shows the model to be used for the clash detection. Use the parameters given below to complete this exercise.   **(Expected time: 45 min)**

Project parameters for Exercise 1:

Clash Test 1: **Ducts vs Walls**
Clash Test 2: **Pipes vs Walls**
Report format: **HTML (Tabular)**

The following steps are required to complete this exercise:

a. Open the model and display the **Clash Detective** window.
b. Add a test and select objects for the clash test.
c. Use the selection set defined in the **Sets** window for the clash detection.
d. Run the test and view the clash results.
e. Group clash results that are related to a single issue.
f. Assign the clash status as reviewed in both the clash tests.
g. Create a combined clash report with the name *Clash_Report_ex01*.
h. Save the project as *c08_navisworks_2017_ex01*.

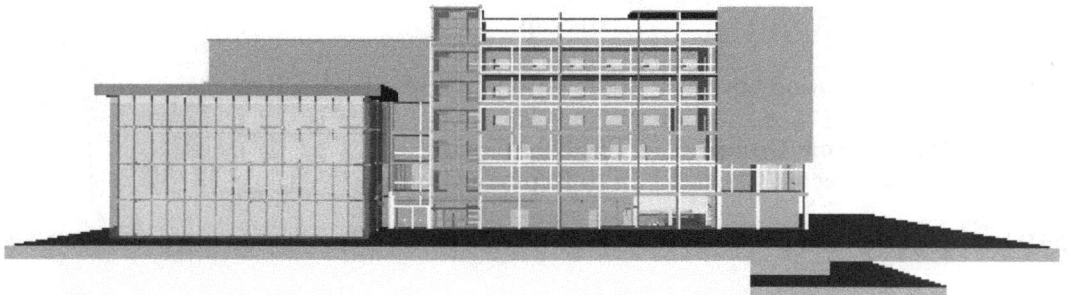

Figure 8-40 *The Conference Center*

Answers to Self-Evaluation Test
1. Reset All, 2. Items in Same Layer, 3. Surfaces, 4. Explode Group, 5. Dim Other, 6. F, **7.** F, **8.** F, **9.** T, **10.** F

Chapter 9

Autodesk Rendering in Navisworks

Learning Objectives

After completing this chapter, you will be able to:
- *Use Autodesk materials*
- *Use Autodesk lights*
- *Apply Sun, Sky, and Exposure effects*
- *Render using Autodesk Graphics*

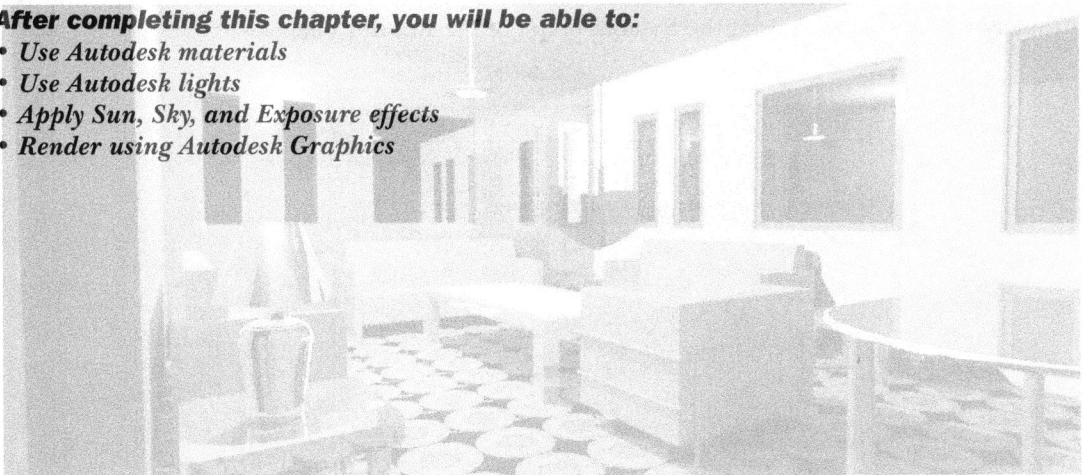

INTRODUCTION

The Autodesk Rendering feature in Navisworks allows you to achieve better photorealistic results with the help of the **Ray Trace** tool. In the language of computer graphics, ray tracing is a technique used for generating photorealistic image. This technique is performed by tracing the path of light through pixels in an image plane. This technique of rendering is capable of producing a very high degree of visual realism, usually higher than that of typical scanline rendering methods, but at a greater computational cost.

One of the advantages of using this Autodesk Rendering feature is that on opening a model created in any of the Autodesk products, all the materials assigned to the model will be displayed in Navisworks in a rendered view. You do not need to apply the materials again to the model in Navisworks.

Autodesk Rendering feature allows you to use existing material libraries, apply lightings, and adjust environment settings. You can also create custom libraries for a specific project using this feature.

THE Autodesk Rendering WINDOW

In Navisworks, the **Autodesk Rendering** window can be used to apply Autodesk material libraries and lightings. Also, you can change the environment settings of a project using this window. To invoke this window, choose the **Autodesk Rendering** tool from the **Tools** panel of the **Home** tab; the **Autodesk Rendering** window will be displayed, as shown in Figure 9-1. Alternatively, you can select the **Autodesk Rendering** check box from **View > Workspace > Windows** drop-down to display the **Autodesk Rendering** window. You can also invoke this window by choosing the **Autodesk Rendering** tool from the **System** panel of the **Render** tab.

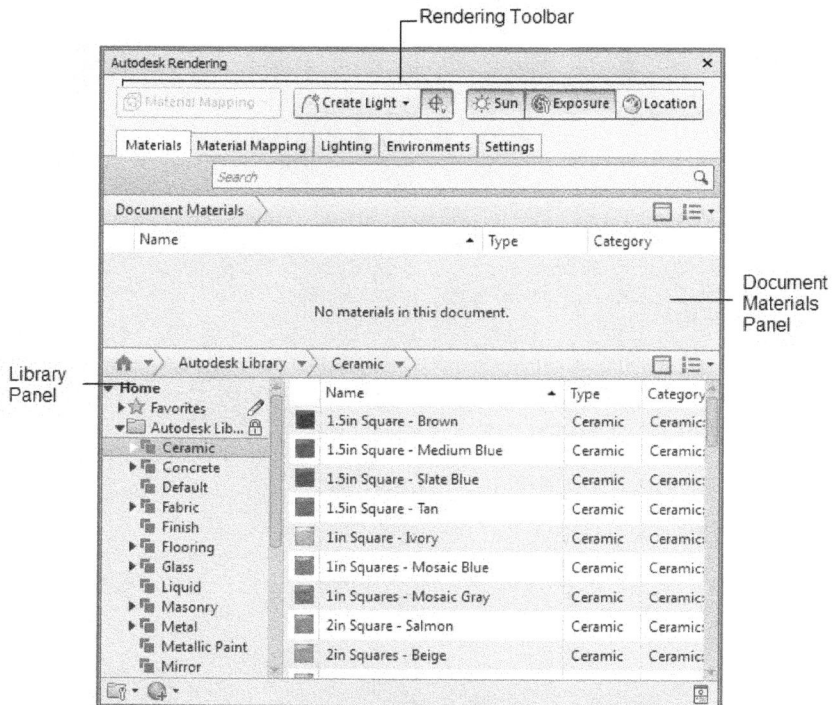

Figure 9-1 The Autodesk Rendering window

The **Autodesk Rendering** window comprises of the Rendering toolbar and five tabs: **Materials**, **Material Mapping**, **Lighting**, **Environments**, and **Settings**. The Rendering toolbar and these tabs are discussed next.

The Rendering Toolbar

The Rendering toolbar is located at the top of the **Autodesk Rendering** window. It contains various drop-down lists and buttons that are used for material mapping, creating and placing lights, adjusting exposure settings, and specifying geographical location of the model. These drop-down lists and buttons are discussed next.

Material Mapping

The **Material Mapping** drop-down list contains the options to be used for mapping the materials on the selected objects in the model. These options are: **Planar**, **Box**, **Cylindrical**, **Spherical**, and **Explicit**.

To apply material on flat faces such as floor, select the **Planar** option from the **Material Mapping** drop-down list. To apply material on a box like shape, select the **Box** option. When you apply this mapping type, the material is applied on all sides of the selected object. To apply material on cylindrical objects such as pipes, select the **Cylindrical** option. To map materials on spherical objects, select the **Spherical** option. Figure 9-2(a to d) show the same material applied but with different types of mappings. The **Explicit** option will be available only when the objects have explicit texture coordinates as part of the geometry.

(a) Planar mapping type   (b) Box mapping type   (c) Cylindrical mapping type   (d) Spherical mapping type

Figure 9-2 Materials applied using different mapping types

When you select an option from the **Material Mapping** drop-down list, various parameters related to the selected mapping type will be displayed in the **Material Mapping** tab which are discussed later in this chapter.

Note
*The **Material Mapping** drop-down list will be enabled only when you select an object in the model.*

Create Light

The **Create Light** drop-down list contains the options that are used for creating lights in the Scene View. These options are discussed next.

Point Light

The point light radiates light in all directions from a fixed location. It lights everything around it. To add a point light in the scene, select the **Point** light option from the **Create Light** drop-down list; a light glyph will be displayed along with the cursor. Specify a location for the light by clicking in the Scene View at the desired location; a gizmo will be displayed at the specified location. Use this gizmo for adjusting the light. Figure 9-3 shows a scene illuminated by the point light.

Note

*On selecting a light option from the **Create Light** drop-down list, the **Lighting** tab will be displayed in the **Autodesk Rendering** window, which will be discussed later in this chapter.*

Figure 9-3 *Scene illuminated by the point light*

Spot Light

The spot light is used for producing directional lights. It is used for lighting a particular area in the model. To add a spot light, select the **Spot light** option from the **Create Light** drop-down list; a torch will be displayed along with the cursor. Specify a location for the light and then specify the direction of the target area by clicking at the desired location in the Scene View that is to be illuminated by the spot light; a gizmo will be displayed at the specified location. Use this gizmo for adjusting the light and its direction. Figure 9-4 shows a scene illuminated by the spot light.

Figure 9-4 *Scene illuminated by the spot light*

Distant Light

A distant light emits uniform parallel light rays in one direction. It can be used for lighting a background uniformly. To add a distant light, select the **Distant** light option from the **Create Light** drop-down list; a light glyph will be displayed along with the cursor.

Specify a location for the light by clicking in the Scene View. Next, specify the direction of light by clicking in the required direction; a gizmo will be displayed. Use this gizmo for adjusting the light's position and direction. Figure 9-5 shows a scene illuminated by the distant light.

Figure 9-5 Scene illuminated by the distant light

Web Light

The web light is 3D representation of intensity distribution of a light source. It produces non-uniform distribution of light. To add a web light to a scene, select the **Web** light option from the **Create Light** drop-down list; a light glyph will be displayed along with the cursor. Specify location for the light in the Scene View. Next, specify the direction by clicking in the required direction; the light gizmo will be displayed. Use this gizmo for adjusting the light position and target.

Light Glyphs

The **Light Glyphs** button is used to toggle the light glyph displayed in the Scene View.

Sun

The **Sun** button is used to toggle the effect of sunlight in the current view. By default, this button is chosen, which shows the presence of sunlight effect in the scene. You can turn off the sunlight effect by choosing this button. Note that on choosing the **Sun** button, the **Environment** tab will be added to the window. This tab displays various parameters related to Sun, Sky, and Exposure effects.

Exposure

The **Exposure** button is used to toggle the exposure settings in the current view. By default, this button is chosen. Note that whenever you will turn on the exposure settings by choosing the **Exposure** button, the **Environment** tab will be invoked.

Location

The **Location** button is used to invoke the **Geographic Location** dialog box. In this dialog box, you can set the latitude, longitude, and position of the Sun in the Scene View, which will be discussed in detail later in this chapter.

Materials Tab

In the **Materials** tab, you can manage materials, apply materials to the selected objects in the model, and can create custom materials. The **Materials** tab comprises the Search area, Library panel, and the **Document Materials** panel, refer to Figure 9-1. These areas and panels are discussed next.

Search

In the Search area of the **Materials** tab, you can search for a specific material or a material category. To do so, enter the related keywords in the **Search** edit box and press ENTER; the searched result will be displayed in the Library panel.

Library Panel

The Library panel contains a list of categories of pre-defined materials available in the Autodesk Library. In the Library panel, expand the **Autodesk Library** node; various material categories will be displayed under this node. Select the required category in the left pane; a list of materials of the selected category will be displayed in the right pane of the Library panel. Figure 9-6 shows different ceramic tile materials in the Autodesk Material Library.

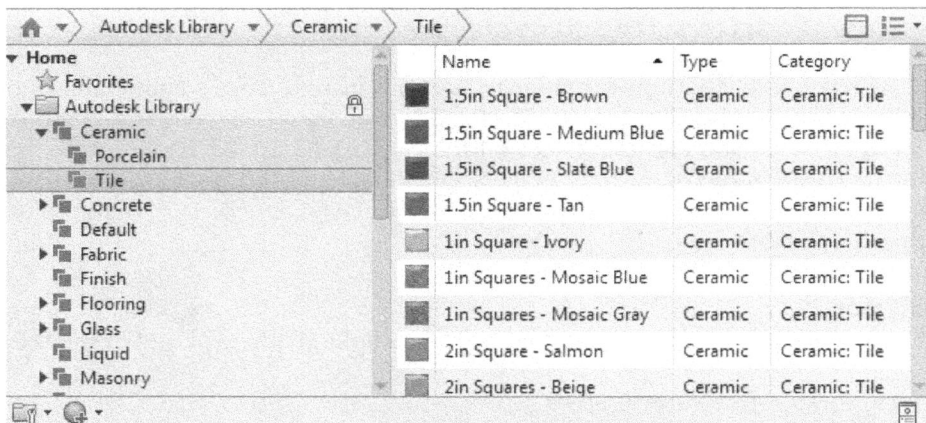

Figure 9-6 Different type of tiles displayed in the Library panel

To apply a material to the model, select an object in the model. Next, right-click on the desired material in the left pane of the Library panel and choose the **Assign to Selection** option from the shortcut menu displayed; the material will be applied to the selected object in the model. Note that whenever you assign a material to the model elements, the material is added to the **Document Materials** panel, which is discussed later in the chapter. Figure 9-7 shows the material applied on a roof.

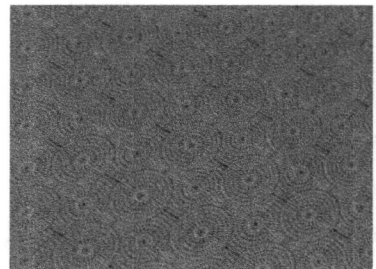

Figure 9-7 Material applied on the roof

In the Library panel, the **Favorites** node contains the user defined materials. By default, this node is empty and you can add materials to this node for a specific project. You can add materials directly to the **Favorites** node or under a category in it. To create a category in the **Favorites** node, right-click on the **Favorites** node; a shortcut menu will be displayed. Choose the **Create Category** option from the menu; a new category with a default name will be added. You can change the name of the category, if required. Next, to add a material under this category, right-click on the required material in the right pane of the Library panel and then choose **Add to > Favorites > Category**; the material will be added under the new category. Similarly, you can add materials directly under the **Favorites** node.

There are some buttons available in the Library panel, which are used for performing various functions. These buttons are discussed next.

Display

The **Display** button is used for displaying and filtering the list in the Library panel. When you choose the **Display** button, a drop-down list will be displayed, as shown in Figure 9-8. This list contains various options grouped in areas such as **Library**, **View Type**, **Sort**, and **Thumbnail Size**.

Show/Hides Library Tree

The **Show/Hide library tree** button is used to toggle the display of the Library panel.

Manage Libraries

The **Manage Libraries** button is used to create, open, and edit user defined libraries. When you choose this button, a drop-down list will be displayed. You can use the options from this list to open an existing library, create a new library, remove a library, and so on.

Figure 9-8 The drop-down list

Create Material

You can choose the **Create Material** button to create a new material from a pre-defined material or create a custom based material. When you choose this button, a drop-down list will be displayed. This list contains two areas: **New using type** and **New Generic Material**. The **New using type** area contains the pre-defined materials based on which you can create materials. The **New Generic Material** is used for creating new generic material. To create a new material from a pre-defined material, select the required material from the **New using type** area; the **Material Editor** dialog box will be displayed. In this dialog box, you can specify the material parameters. To create a new generic material, select the **New Generic Material** option; the **Material Editor** dialog box will be displayed. Specify the material parameters in this dialog box.

Material Editor

The **Material Editor** button is used to open/close the **Material Editor** dialog box.

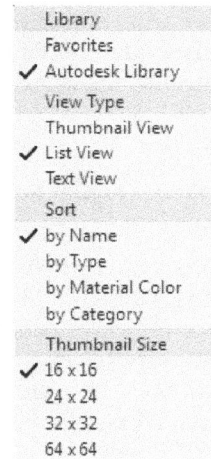

Document Materials Panel

The **Document Materials** panel contains a list of materials used in the model. If the Autodesk materials are assigned to a model in the native software in which it is created then all the assigned materials will be displayed in the **Document Materials** panel of the **Autodesk Rendering** window. If the materials are not assigned in the native software then this panel will be empty. When you assign a material from the Library panel to the model elements, the material will be added to the **Document Materials** panel.

The materials in the Library panel cannot be edited, so in this case you will add the required material to the **Document Materials** panel and then edit it. To add a material from the Library panel, place the cursor on the required material in the left pane of the Library panel; two buttons will be displayed in the material. Figure 9-9 shows the buttons displayed when the cursor is placed on the **Mosaic - Blue** material. You can choose the first button to add the material selected in the Library panel to the **Document Materials** panel. Alternatively, choose the second button to add the material to the **Document Materials** panel as well as to display the material parameters in the **Material Editor** dialog box. You can also add materials to the panel by right-clicking on the required material in the right pane of the Library panel. On doing so, a shortcut menu will be displayed. Choose **Add to > Document Materials** from the menu; the material will be added.

Figure 9-9 Buttons displayed while the cursor is placed on the material

After adding a material to the **Document Materials** panel, you can make changes in it. To do so, double-click on the required material in the **Document Materials** panel; the **Material Editor** dialog box will be displayed, as shown in Figure 9-10. Alternatively, to invoke this dialog box, place the cursor on the required material in the **Document Materials** panel; a button will be displayed. Next, choose this button; the **Material Editor** dialog box will be displayed. In this dialog box, there are two tabs: **Appearance** and **Information**. These two tabs contain various options which are discussed next.

Appearance Tab

The options in the **Appearance** tab are used to edit the material properties. The options displayed in the **Material Editor** dialog box vary for each material. The preview of the material will be displayed at the top in the dialog box. You can change the preview by selecting options from the **Preview** drop-down list. You can change the name of material by specifying a new name in the text box displayed below the **Preview** drop-down list. You can control the material properties such as color, glossiness, transparency, reflectivity, cutouts, self-illumination, bump, image fade, and so on from their respective areas in the **Material Editor** dialog box, refer to Figure 9-10.

Information Tab

The **Information** tab contains the information such as material name, its type, location, file path of the texture files, and so on of the material. You can change the name of material, its description, and keywords in their corresponding edit boxes.

You can assign materials to the model elements using the **Document Materials** panel. To do so, select the required model element and right-click on the desired material in the **Document Materials** panel; a shortcut menu will be displayed. Choose the **Assign to Selection** option; the material will be applied on the selected object in the Scene View. Choose the **Select Objects Applied to** option to highlight the objects on which the selected material is to be applied in the Scene View. Choose the **Edit** option to display the **Material Editor** dialog box. Choose the **Rename** option to rename the material. Choose the **Delete** option to remove the material from the **Document Materials** panel. Choose **Add to > Favorites** option to add the selected material to the **Favorites** node in the Library panel.

Material Mapping Tab

In the **Material Mapping** tab, you can customize the settings of the material mapping type that is selected from the **Material Mapping** drop-down list. Note that the **Material Mapping** tab will not display parameters unless you select individual objects, such as polygon or an instanced item of geometry, which are represented by ⬦ and ⬦ symbols in the **Selection Tree** window. When you select an object in the **Selection Tree** window, the material mapping parameters will be displayed according to the selected mapping type in the tab. These mapping types are discussed next.

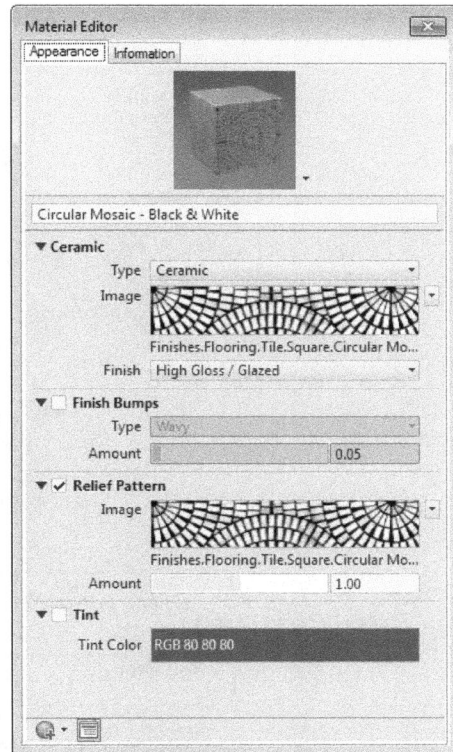

*Figure 9-10 The **Material Editor** dialog box*

Planar Mapping
The planar mapping is used for planar objects. Select an object such as floor in the **Selection Tree** window. Then select the **Planar** mapping type from the **Material Mapping** drop-down list; the options for the **Planar** mapping type will be displayed in the **Material Mapping** tab, as shown in Figure 9-11.

In this tab, you can specify the values for mapping the materials such as Translation, Rotation, and Scale, in the corresponding **X**, **Y**, and **Z** edit boxes.

Box Mapping
The box mapping is used for box type geometry such as roof. When you select the **Box** mapping type from the **Material Mapping** drop-down list, the options for the **Box** mapping type will be displayed in the **Material Mapping** tab, as shown in Figure 9-12.

In this mapping type, you can define the actual mapping for each face by selecting the options from the **Top**, **Bottom**, **Left**, **Right**, **Front**, and **Back** drop-down lists in the **UV Direction** area.

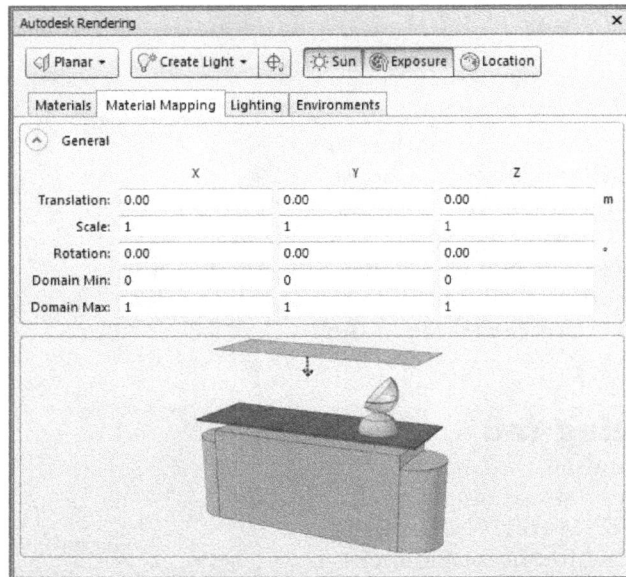

Figure 9-11 *The parameters displayed for the **Planar** mapping type*

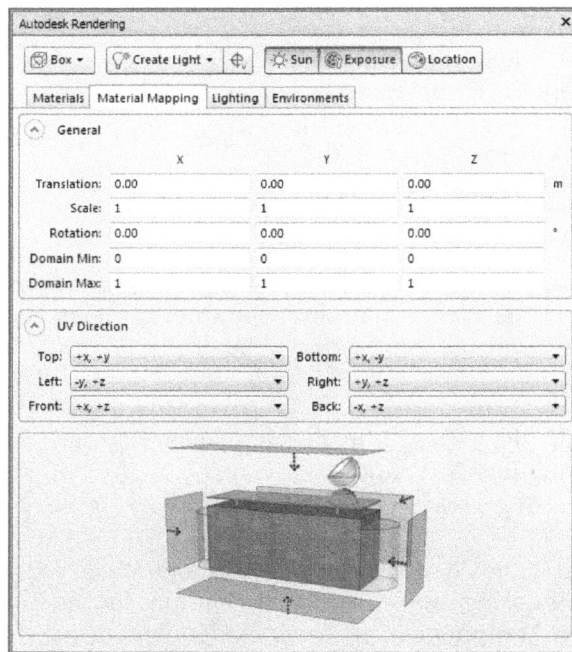

Figure 9-12 *The parameters displayed for the **Box** mapping type*

Cylindrical Mapping

The cylindrical mapping is used for cylindrical objects such as circular column. When you select the **Cylindrical** mapping type from the **Material Mapping** drop-down list, the options for the

Cylindrical mapping type will be displayed in the **Material Mapping** tab, as shown in Figure 9-13. In this mapping type, coordinates are mapped to the cylindrical surfaces. You can specify if the top and bottom of the cylinder are to be textured with the planar caps and not with the cylindrical side. To do so, select the **On** check box in the **Cap** area. You can define the angle between the point and cylindrical axis to decide whether the cap or side mapping should be used. To do so, specify an angle in the **Threshold** edit box in the **Cap** area.

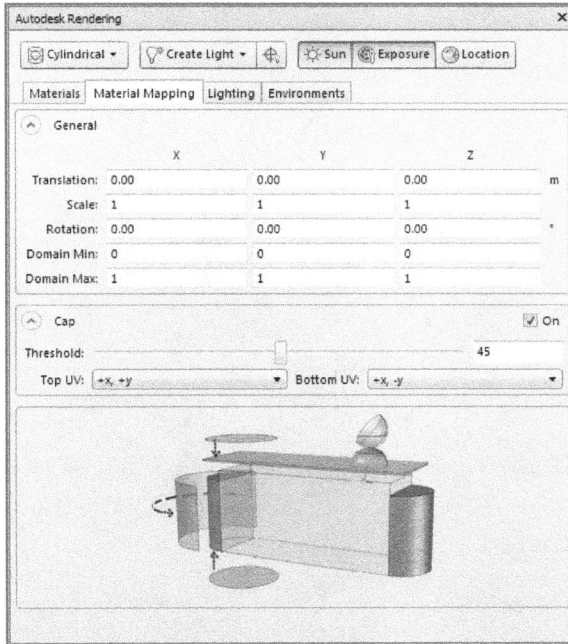

*Figure 9-13 The parameters displayed for the **Cylindrical** mapping type*

Spherical Mapping

The spherical mapping is used for spherical objects. When you select the **Spherical** mapping type from the **Material Mapping** drop-down list, the options for the **Spherical** mapping type will be displayed in the **Material Mapping** tab, as shown in Figure 9-14. In this mapping type, the texture coordinates are computed by spherical projection at origin.

Lighting Tab

In the **Lighting** tab, you can manage lights in a model. The **Lighting** tab is divided into two areas, the Lights view and the Properties view, refer to Figure 9-15.

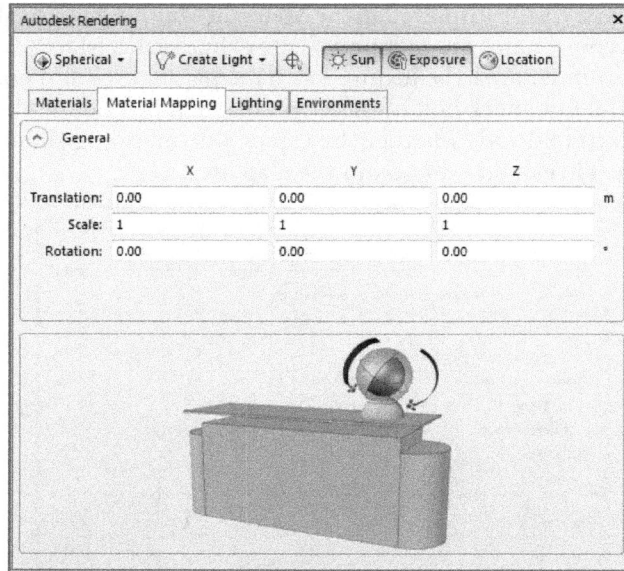

*Figure 9-14 The parameters displayed for the **Spherical** mapping type*

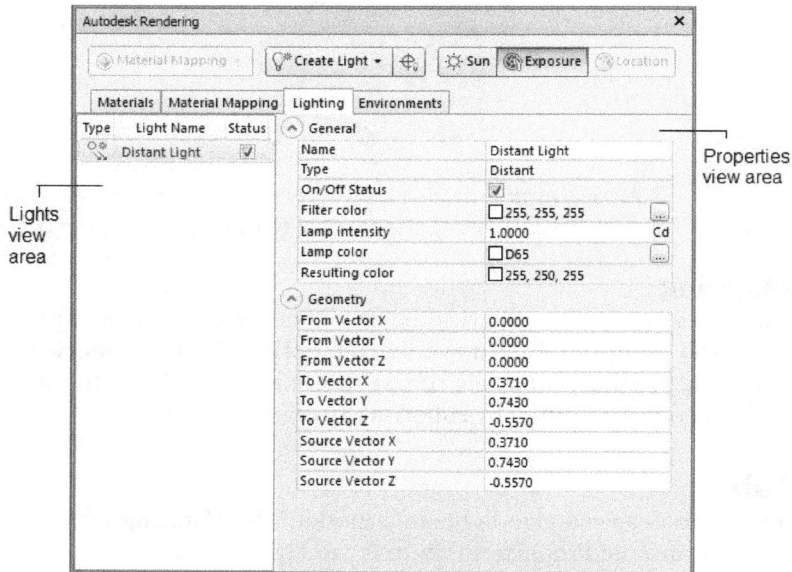

*Figure 9-15 The **Lighting** tab*

The Lights view area contains a list of lights used in the scene, with their name and type. You can turn the lights on and off by selecting the **Status** check box in the Lights view area.

The Properties view area shows the properties of the selected light. When you select a light in the Lights view area, its properties will be displayed in the Properties view area. Using these properties, you can change the color of light, its intensity, and so on.

The lights in the Lights view area will be displayed only after adding lights in the scene. For example, select the **Distant** light option from the **Create Light** drop-down list. Next, specify the location and direction of light by clicking on the required location in the model; the **Distant Light** will be added to the Lights view area and its properties will be displayed in the Properties view area. In the **General** area of the Properties view area, you can change the name, color, intensity, and so on. The **Geometry** area displays the position of light and its direction in the Scene View. You can change them by specifying the required values in the corresponding edit boxes.

Environments Tab

The options in the **Environments** tab are used to adjust the sun, sky, and exposure effects. Figure 9-16 shows the **Environments** tab in the **Autodesk Rendering** window. The various options in this tab are discussed next.

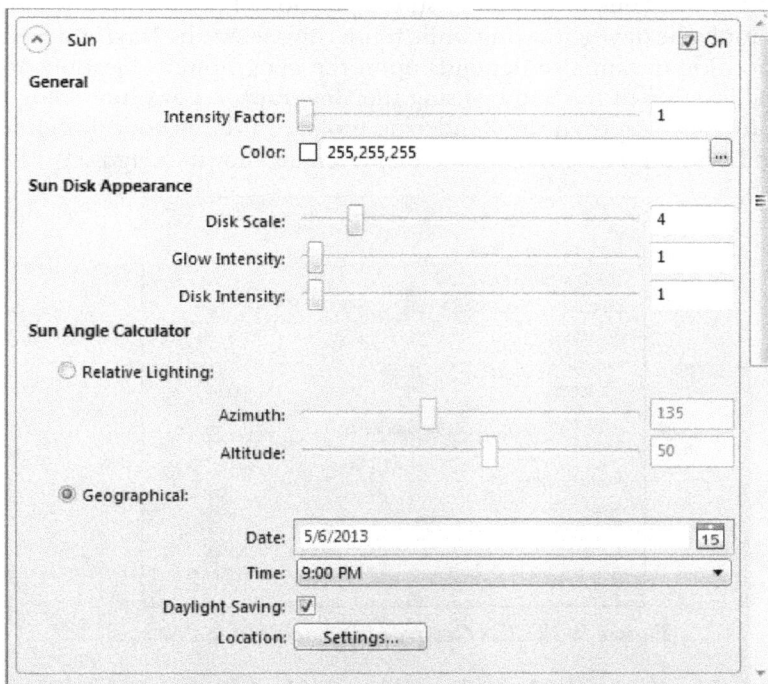

Figure 9-16 *Partial view of the **Environments** tab*

Sun

In the **Sun** area of the **Environments** tab, you can modify the properties of the Sun. To toggle the Sun effect, select the **On** check box. You can also enable the Sun effect by choosing the **Sun** button from the Rendering toolbar. On enabling this effect, various parameters in the **Sun** area of the **Environments** tab will be enabled, which are discussed next.

You can modify the general properties of the Sun such as its color and intensity factor. To adjust the intensity factor, drag the **Intensity Factor** slider. You can also specify the required value in the edit box displayed next to the **Intensity Factor** slider. Note that higher value will produce a brighter Sun effect. To change color of the Sun, click on the browse button displayed next to the **Color** edit box; the **Color** palette will be displayed. Select the required color from the palette and choose the **OK** button; the color is changed and is displayed in the **Color** edit box.

You can also control the appearance of the Sun disk. This will affect the background only. To change the size of the Sun disk, drag the **Disk Scale** slider in the left or right direction; the size of the Sun will change accordingly. To change the intensity of the Sun glow, drag the **Glow Intensity** slider. To change the intensity of the Sun disk, drag the **Disk Intensity** slider or specify a value in the edit box displayed next to the slider.

You can also change the position of the Sun according to the date and time. To specify the date for calculating Sun angle, first select the **Geographical** radio button. Next, click on the calendar icon displayed in the **Date** edit box; a calendar will be displayed. Select the required date from the displayed calendar; the selected date will be displayed in the **Date** edit box. Next, select the time from the **Time** drop-down list; the position of the sun and its effect will change accordingly. To apply the daylight saving while rendering, select the **Daylight Saving** check box. Note that the position of Sun also depends upon the geographical location. You can change the geographical location of the Sun by using the **Geographic Location** dialog box. To do so, choose the **Location** button from the Rendering toolbar in the **Autodesk Rendering** window; the **Geographic Location** dialog box will be displayed, as shown in Figure 9-17.

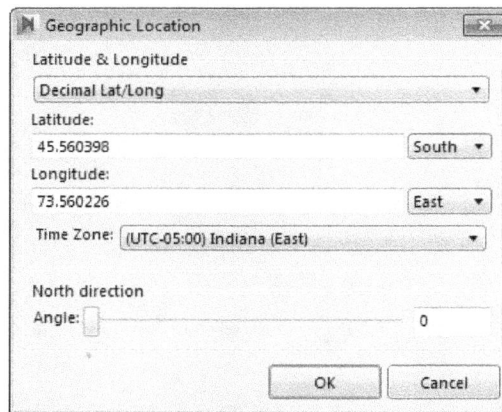

*Figure 9-17 The **Geographic Location** dialog box*

In this dialog box, you can set the latitude and longitude, and the angle of North direction of the geographic location in the model. In the **Latitude & Longitude** area, you can select the format for specifying latitude and longitude by selecting an option from the drop-down list, refer to Figure 9-17. Select the **Decimal Lat/Long** option to display the latitude and longitude in decimal values. Select the **Degree Minutes Seconds Lat/Long** option to display the latitude and longitude in degrees, minutes, and seconds. You can enter the latitude and longitude in the **Latitude** and **Longitude** edit boxes. You can specify the time zone by selecting the required option from the **Time Zone** drop-down list. You can control the position of Sun in the Scene View by using the **Angle** slider in the **North direction** area. After specifying all the parameters, choose the **OK** button; the dialog box will be closed and the changes will be applied in the Scene View.

Note
*When the Sun effect is on, you can view only the Perspective view of the model. The **Orthographic** option will be disabled in this case.*

Sky

In the **Sky** area of the **Environments** tab, you can adjust the sky properties such as intensity factor, haze, blur, and so on. The parameters in this area will be enabled when the sun effect is enabled. These parameters are discussed next.

Select the **Render Sky Illumination** check box to add the soft lighting effects caused by interaction between the Sun and atmosphere. To adjust the effect of sky light, specify a value in the **Intensity Factor** edit box. To adjust the scattering effect of sky light, drag the **Haze** slider or specify a value in the edit box displayed next to the slider. To change color of the night sky, choose the browse button displayed next to the **Night Color** edit box; a **Color** palette will be displayed. Choose the required color from the **Color** palette and choose the **OK** button; the color will be applied. To adjust location of the ground plane, drag the **Horizon Height** slider. Dragging the slider to the right direction will increase the height of the ground plane and dragging it to the left direction will decrease the height. To adjust the blur between ground plane and sky, drag the **Blur** slider. Note that dragging the slider in the right direction will decrease the blur. To change the color of the ground, choose the browse button next to the **Ground Color** edit box; a **Color** palette will be displayed. Choose the required color and then choose the **OK** button; the color will be applied.

Exposure

In the **Exposure** area, you can adjust the exposure related settings. You can turn the exposure on/off by selecting the **On** check box. When you clear this check box, the background will become white and the Sun and Sky effect will not be shown in the Scene View. To adjust the overall brightness, drag the **Exposure Value** slider. Note that dragging the slider in the right direction will decrease the brightness. The value of the exposure ranges from -6 to 16. You can adjust the light level for the brightest area of the scene by dragging the **Highlights** slider. Similarly, you can adjust the other exposure factors such as shadows, white point, saturation, and midtones.

Settings Tab

In the **Settings** tab, various rendering styles and quality options are available which can be used to render the model. Figure 9-18 shows the **Settings** tab in the **Autodesk Rendering** window. There are three areas in this tab: **Current Render Preset**, **Basic**, and **Advanced**. These areas are discussed next.

Current Render Preset

In the **Current Render Preset** area, a drop-down list is available which contains six rendering quality options. These options are: **Low Quality**, **Medium Quality**, **High Quality**, **Coffee Break Rendering**, **Lunch Break Rendering**, **Overnight Rendering**, and **Custom Settings**. The **Custom Settings** option is used to create custom rendering style. You can use these options to control the rendering quality and speed of rendered output. Here, the **Custom Settings** option is used to create a customized rendering style. Rest of the options are discussed next.

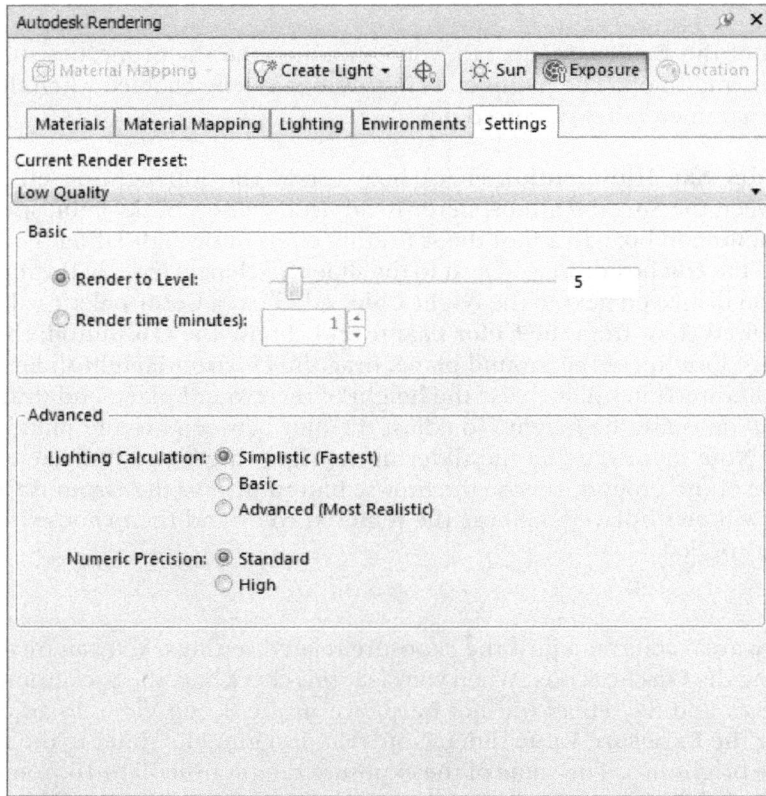

*Figure 9-18 The **Settings** tab in the **Autodesk Rendering** window*

Low Quality

The **Low Quality** option is used to quickly view the rendered scene. On selecting this option, the rendered output will be of low quality and may contain some inaccuracies and artifacts. This option avoids anti-aliasing during rendering.

Medium Quality

The **Medium Quality** option is used when you need to see the final preview of the scene. In this rendering style, anti-aliasing, sample filtering, and ray tracing are active with increased reflection depth setting. This option provides you satisfactory results with few errors.

High Quality

The **High Quality** option is used when you need to deliver good quality rendered output to the clients. This option will deliver high quality results with no errors. It includes all reflections, shadows, transparencies, and anti-aliasing on edges. However, such images will take a longer time to render.

Coffee Break Rendering

The **Coffee Break Rendering** option includes simplistic lighting calculation and standard numerical precision. When you select this option in the **Advanced** area of the **Autodesk**

Rendering window, the **Simplistic (Fastest)** and **Standard** radio buttons will be selected automatically. In this case, the rendering time is 10 minutes which will be filled automatically in the **Render time (minutes)** edit box in the **Basic** area.

Lunch Break Rendering

The **Lunch Break Rendering** option includes advanced lighting calculation which will give you more realistic output. Similar to **Coffee Break Rendering**, it uses standard numerical precision. The rendering time taken on selecting this option is 60 minutes. On selecting this option, the values of the related parameters will be filled automatically in the **Basic** and **Advanced** areas of the **Autodesk Rendering** window.

Overnight Rendering

Similar to **Lunch Break Rendering**, the **Overnight Rendering** option supports advanced lighting calculation. It uses high numerical precision method during rendering. The rendering time to be taken on selecting this option is 720 minutes. On selecting this option, the values of the related parameters will be filled automatically in the **Basic** and **Advanced** areas of the **Autodesk Rendering** window.

Custom Settings

The **Custom Settings** option is used to create custom render presets. On selecting this option, you can specify the rendering time, lighting calculation, and numeric precision parameters in the **Basic** and **Advanced** areas.

Note

*These options can also be selected from the **Ray Trace** drop-down list in the **Render** panel of the **Render** tab.*

Basic Area

In the **Basic** area, you can specify the basic parameters for rendering. Select the **Render to Level** radio button and specify the rendering level value by dragging the slider or by specifying a value in the corresponding edit box. The higher value will produce higher quality of rendering. Next, select the **Render time (minutes)** radio button and specify the rendering time in the corresponding edit box.

Advanced Area

In the **Advanced** area, you can specify the method for calculating the illumination in the scene and precision of its results. To specify the lighting calculation method, select the **Simplistic (Fastest)**, **Basic**, or **Advanced (Most Realistic)** radio button. Similarly, to specify the numeric precision, select the **Standard** or **High** radio button.

RENDERING USING AUTODESK GRAPHICS

Using Autodesk Graphics, you can produce highly detailed and photorealistic images. After selecting the desired render style type in the **Autodesk Rendering** window, choose the **Ray Trace** button from the **Interactive Ray Trace** panel of the **Render** tab; the rendering process will start and you can see the render progress indicator on the screen. While rendering, you can pan, orbit, and zoom the model. However, performing these actions will restart the rendering process.

You can control the rendering process by using the buttons available in the **Interactive Ray Trace** panel of the **Render** tab. Choose the **Pause** button to temporarily stop the rendering process. Choose the **Close** button to exit the rendering process. You can also save the rendered scene as an image. To do so, choose the **Save** button from the **Interactive Ray Trace** panel of the **Render** tab; the **Save As** dialog box will be displayed. In this dialog box, select the desired file type from the **Save As Type** drop-down list. Browse to the desired location to save the file. Enter the file name and choose the **Save** button; the image will be saved at the specified location.

RENDERING WITH AUTODESK 360

The Autodesk 360 is a cloud-based application that helps to render views online. You can use this application to access saved rendered images, change the render quality of existing rendered images, and use different background effects using the Autodesk render gallery. To render the scenes in Autodesk 360 cloud, choose the **Render in Cloud** tool from the **System** panel of the **Render** tab; the **Autodesk - Sign In** dialog box will be displayed, as shown in Figure 9-19.

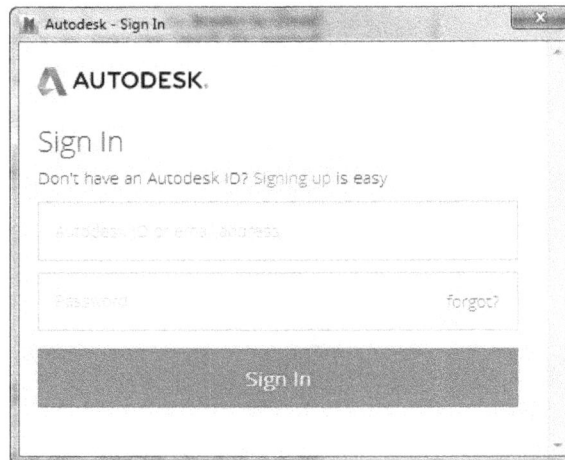

Figure 9-19 The Autodesk - Sign In dialog box

Enter your Autodesk Id or e-mail address in the **Autodesk ID or e-mail address** edit box and the required password in the **Password** edit box. Next, choose the **Sign In** button; the **Render in Cloud** dialog box will be displayed, as shown in Figure 9-20. In this dialog box, specify the required parameters and choose the **Start Rendering** button; the **Rendering in Cloud** progress bar will be displayed. Once the rendering process is complete, the rendered image will be uploaded in your Autodesk 360 account.

Now if you want to view the rendered images and in progress rendering, you can do so in the web browser. To do so, choose the **Render Gallery** tool from the **System** panel of the **Render** tab; the Autodesk 360 web page will be displayed with all the rendered images.

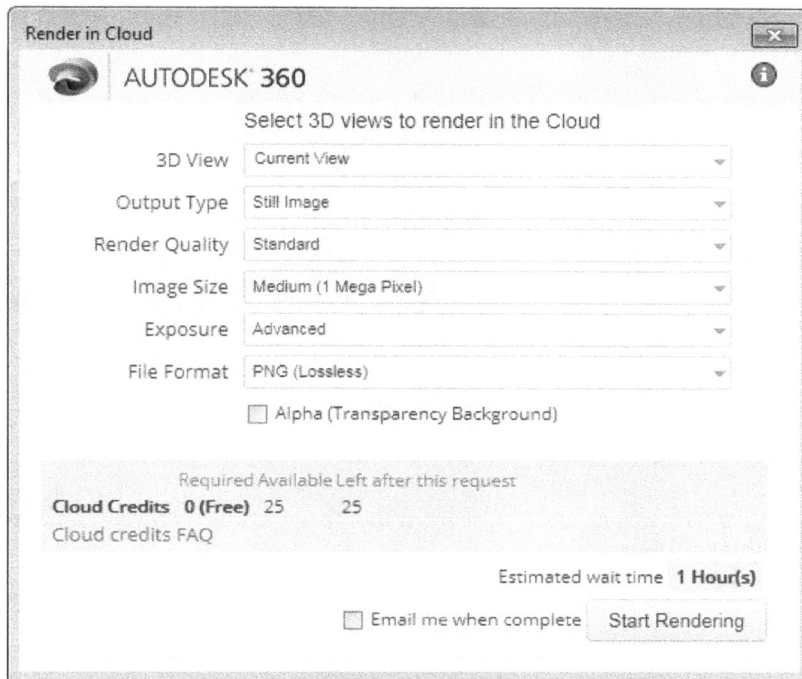

*Figure 9-20 The **Render in Cloud** dialog box*

TUTORIAL

General instructions for downloading tutorial files:

1. Download the *c09_nws_2017_tut* zip file for the tutorial from *www.cadcim.com*. The path of the file is as follows: *Textbooks > Civil/GIS > Navisworks > Exploring Autodesk Navisworks 2017.*

2. Now, save and extract the downloaded folder at the following location:
    C:\ nws_2017\ c09_nws_2017_tut

Tutorial 1	Rendering the Room

In this tutorial, you will open the *c09_navisworks_2017_tut1.nwf* file. Next, you will apply Autodesk materials and lights to the model elements such as furniture, walls, windows, and so on. After applying the materials and lights, you will render the scene using high quality rendering.

(Expected time: 45 min)

The following steps are required to complete this tutorial:

a. Open the file *c09_navisworks_2017_tut1.nwf*.
b. Invoke the **Autodesk Rendering** window.
c. Add materials to the **Document Materials** panel in the window.
d. Display the **Sets** window.

e.   Apply materials to the model elements.
f.   Add lights to the scene.
g.   Specify the location, date, and time.
h.   Render the model.
i.   Save the project.

Opening the File

In this section, you will open the file in Navisworks.

1.   Choose the **Open** button from the Quick Access Toolbar; the **Open** dialog box is displayed.

2.   In this dialog box, browse to the following location:
     C:\nws_2017\c09_nws_2017_tut

3.   Select **Navisworks File Set (*.nwf)** from the **Files of type** drop-down list.

4.   Select the file *c09_navisworks_2017_tut1* from the displayed list of files; the file name appears
     in the **File name** edit box.

5.   Choose the **Open** button in this dialog box; the model is displayed in the Scene View, as
     shown in Figure 9-21.

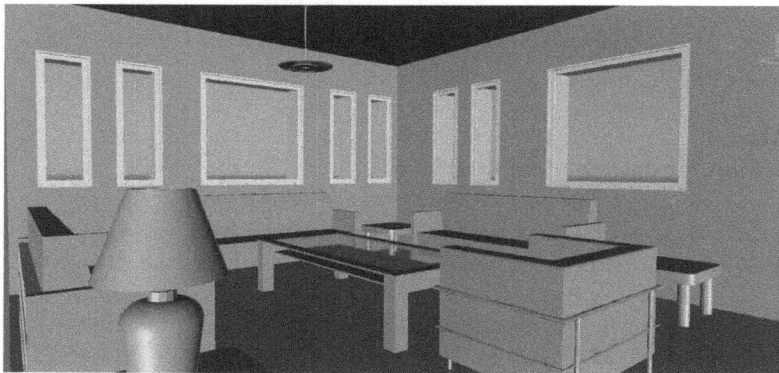

Figure 9-21 The interior view of the model

Invoking the Autodesk Rendering Window

In this section, you will invoke the **Autodesk Rendering** window.

1.   Choose the **Autodesk Rendering** tool from the drop-down in the **Tools** panel of the **Home**
     tab; the **Autodesk Rendering** window is displayed.

Note
*If materials are displayed in the **Document Materials** panel of this window, then delete them to
add new materials.*

Adding Materials to the Document Materials Panel

In this section, you will add materials to the **Document Materials** panel in the **Autodesk Rendering** window.

1. In the **Autodesk Rendering** window, ensure that the **Materials** tab is chosen. In this tab, expand the **Autodesk Library** node in the Library panel, if not expanded by default; various material folders are displayed.

2. Next, select the **Glossy** category under the **Autodesk Library > Wall Paint** node; various materials are displayed in the right pane of the Library panel.

3. In the right pane of the Library panel, scroll down and place the cursor on the **Maroon** wall paint; two buttons are displayed, as shown in Figure 9-22.

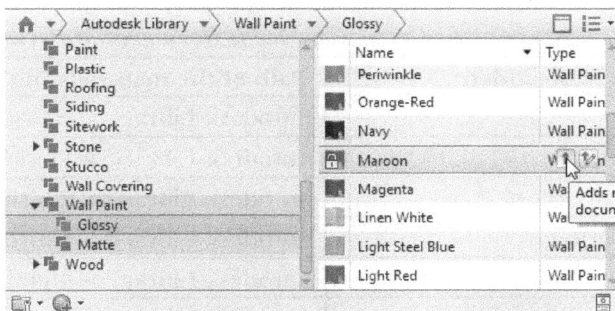

Figure 9-22 Buttons displayed on selecting the wall paint

4. Next, choose the **Adds material to document** button; the **Maroon** wall paint is added to the **Document Materials** panel of the **Autodesk Rendering** window, as shown in Figure 9-23.

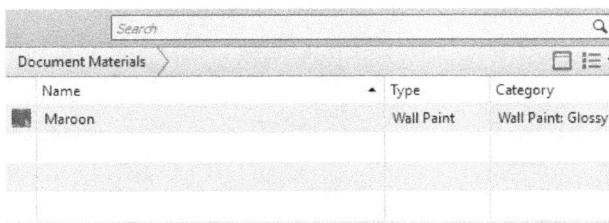

*Figure 9-23 The **Maroon** wall paint added to the **Document Materials** panel*

Next, you will add the materials to be used for window frames and glasses.

5. Select the **Wood** node from the left pane of the Library panel; various wood types are displayed in the right pane of the Library panel.

6. Select the **Teak - Stained Dark Medium Gloss** wood type and add it to the **Document Materials** panel, as explained in step 4. Similarly, add the **Teak - Natural Polished** wood type to the **Document Materials** panel.

7. Next, select the **Glass** node from the left pane of the Library panel; various glass types are displayed in the right pane.

8. Select the **Tinted - Blue** glass type and add it to the **Document Materials** panel.

9. Next, select **Glass > Glazing** from the left pane; different glass types are displayed in the right pane.

10. Select **Light Blue Reflective** glass type and add it to the **Document Materials** panel.

Repeat the procedure followed in the previous steps and add the fabric, carpet texture, and material for side table, lamps, and other fixtures respectively, as given in Table 9-1. Figure 9-24 show the partial list of the materials added to the **Document Materials** panel.

Table 9-1 List of materials to be added to the **Document Materials** panel

Materials to be added	Path of the materials in the Library panel
Pebbled - Mauve	Autodesk Library > Fabric > Leather
Rug - Braided	Autodesk Library > Flooring > Carpet
Wax - Black on Red Wood	Autodesk Library > Finish
Wax - Red on Light Wood	Autodesk Library > Finish
Titanium - Polished	Autodesk Library > Metal
Chrome - Satin	Autodesk Library > Metal

Figure 9-24 Materials added to the **Document Materials** panel

Displaying the Sets Window
In this section, you will display the **Sets** window.

1. Choose the **Manage Sets** tool from the **Home > Select & Search > Sets** drop-down; the **Sets** window is displayed, as shown in Figure 9-25.

Notice that in the **Sets** window, the model elements are saved as selection sets so that it is easier for you to apply the materials to the required model elements.

Applying Materials to the Model Elements

In this section, you will apply the added materials to the model elements such as windows, sofa, table, lamp, and so on.

1. In the **Sets** window, select **Wall**; all walls are highlighted in the Scene View.

2. Next, right-click on the **Maroon** wall paint in the **Document Materials** panel; a shortcut menu is displayed.

Figure 9-25 The Sets window

3. Choose the **Assign to Selection** option from the menu; the **Maroon** wall paint is applied on the walls.

4. Repeat the procedure followed in steps 1 to 3 and apply the materials to the model elements as given in Table 9-2. Figure 9-26 shows the model after applying all the materials.

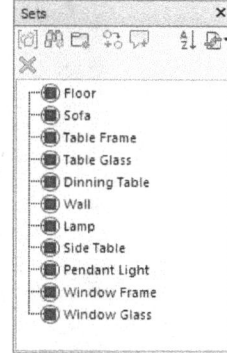

Table 9-2 List of objects and materials assigned to them

Objects in the Sets window	Materials to be Assigned
Window Frame	Teak - Stained Dark Medium Gloss
Window Glass	Light Blue Reflective
Table Frame	Wax Red on Light Wood
Table Glass	Tinted Blue
Dining Table	Teak - Natural Polished
Side Table	Wax Black on Red Wood
Lamp	Chrome - Satin
Sofa	Pebbled - Mauve
Floor	Rug - Braided
Pendant Light	Titanium - Polished

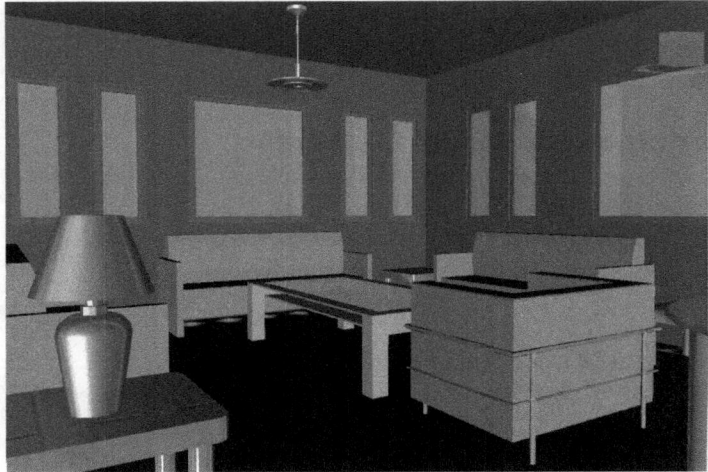

Figure 9-26 *View after applying all the materials*

Adding Lights to the Scene

In this section, you will add lights in the scene.

1. In the **Autodesk Rendering** window, select the **Spot** light option from the **Create Light** drop-down list in the Rendering toolbar; the **Lighting** tab is displayed in the window. Also, a torch is attached to the cursor.

    Now, you will specify the location and target of the light.

2. Click on the lamp, as shown in Figure 9-27; the location of the Spot light is specified.

Note
*If there are **Point** lights in the **Light View** area, then delete them.*

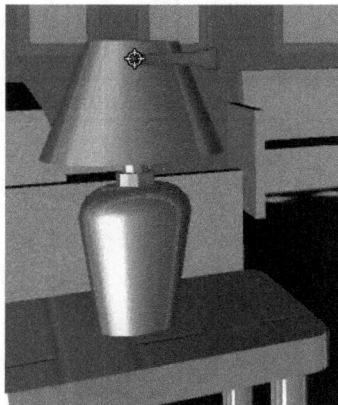

Figure 9-27 *Specifying the location for the Spot light*

3. Next, to specify the target for the spot light, move the cursor so that the torch points downward. Next, click in the drawing; the base of the Spot light is specified as the target. Note that the light gizmo is also displayed in the Scene View, as shown in Figure 9-28.

Figure 9-28 Light gizmo displayed in the Scene View

4. Next, specify the properties for the **Spot** light in the **Properties** view area, as shown in Figure 9-29.

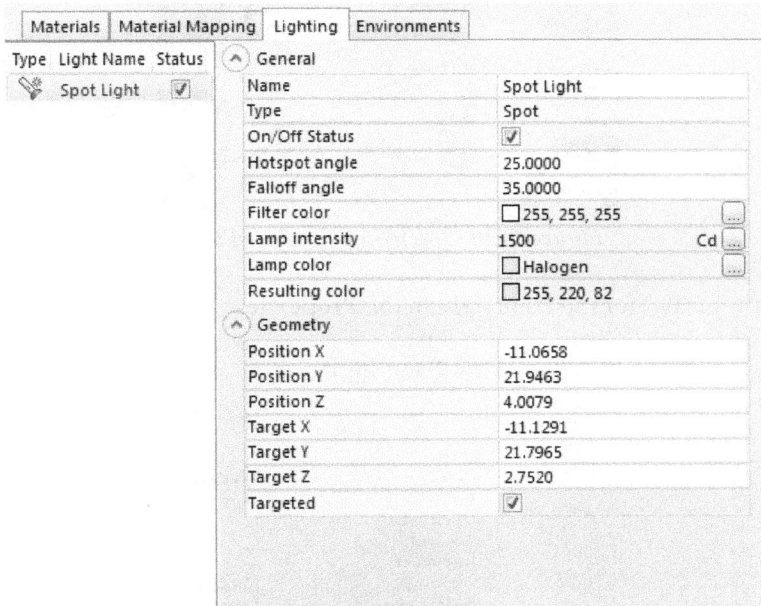

*Figure 9-29 Specifying parameters for the **Spot** light*

5. In the Properties view area of the **Lighting** tab, choose the browse button displayed in the field corresponding to the **Light Intensity** option in the **General** area; the **Lamp Intensity** dialog box is displayed.

6. In this dialog box, ensure that the **Intensity (Candela)** option is selected in the **Unit** drop-down list and enter **50** in the **Intensity** edit box.

7. Choose the **OK** button; the dialog box closes and the light intensity of the Spot light is changed accordingly.

Next, you will add point light to the scene.

8. Select the **Point** light option from the **Create Light** drop-down list in the Rendering toolbar; a light glyph is displayed along with the cursor.

9. Click to specify the position of the Point light in the Scene View, as shown in Figure 9-30; the gizmo is displayed in the Scene View. Also, the Point light is added in the **Properties** view area of the **Lighting** tab.

> **Tip**
> *Ensure that the **Full Lights** option is selected from the **Viewpoint > Render Style > Lighting** drop-down list. It will produce a high quality lighting.*

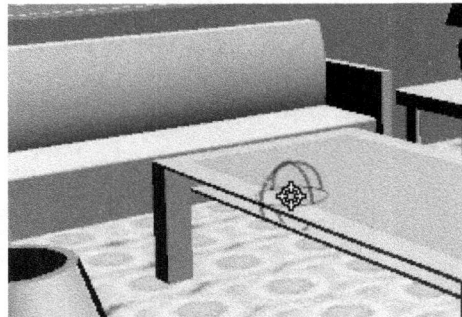

Figure 9-30 Specifying location of light

10. Specify the properties for the Point light in the **Properties** view area, as shown in Figure 9-31.

Next, you will change the color of the Point light.

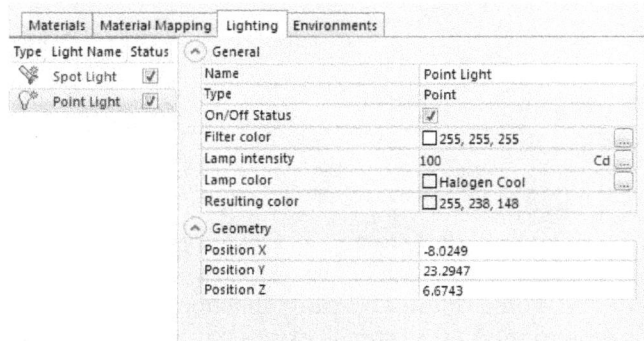

*Figure 9-31 Properties for the **Point** light in the **Properties** view area*

11. Click on the browse button displayed in the **Lamp color** edit box; the **Lamp Color** dialog box is displayed.

12. Select the **Halogen Cool** lamp color from the **Standard colors** drop-down list and choose the **OK** button; the selected color is applied to the lamp.

13. Choose the **Light Glyphs** button in the **Autodesk Rendering** window to hide the glyphs from the Scene View .

Specifying Location, Date, and Time of the Scene
In this section, you will set the time, date, and geographic location of the scene.

1. Choose the **Location** button from the Rendering toolbar; the **Geographic Location** dialog box is displayed.

2. In this dialog box, select the **Indiana (East)** option from the **Time Zone** drop-down list and then choose the **OK** button; the dialog box is closed.

3. Now, choose the **Environments** tab and specify the date and time in the **Sun Angle Calculator** area of the **Sun** area, as shown in Figure 9-32.

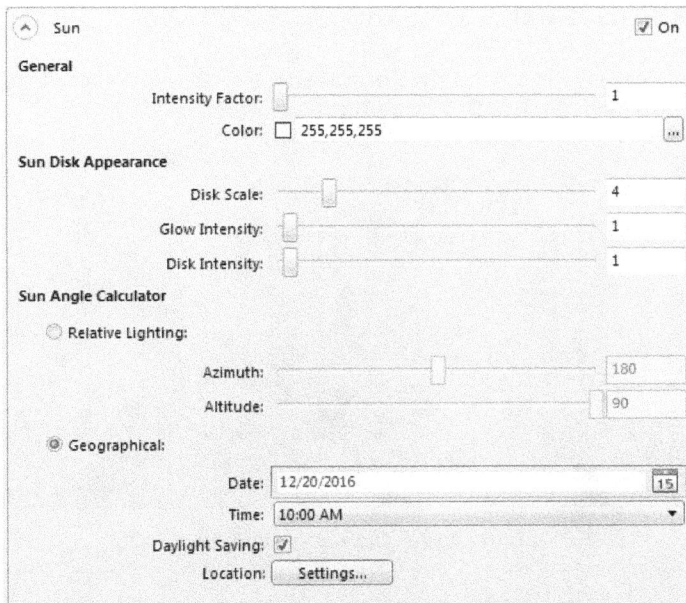

*Figure 9-32 Partial view of the **Environments** tab*

Figure 9-33 shows the view after changing the location and time.

Figure 9-33 *Model view after changing the location and time*

Rendering the Scene

In this section, you will render the scene with medium quality effects.

1. Close the **Autodesk Rendering** window and select the **Medium Quality** option from **Render > Interactive Ray Trace > Ray Trace** drop-down list.

2. Choose the **Ray Trace** button from the **Interactive Ray Trace** panel of the **Render** tab; the rendering process starts. Figure 9-34 shows the rendered model.

Figure 9-34 *View after rendering*

Similarly, select other rendering options to render the scene and compare the results. Figures 9-35 and 9-36 show a scene rendered using the **Coffee Break Rendering** and **Lunch Break Rendering** options respectively.

Saving the Rendered Scene as an Image

In this section, you will save the rendered scene as a JPEG image.

1. When the rendering is completed, choose the **Image** button from the **Export** panel of the **Render** tab; the **Save As** dialog box is displayed.

2. Save the image as a JPEG file with the name **c09_navisworks_2017_tut01**.

*Figure 9-35 Scene rendered by using the **Coffee Break Rendering** option*

*Figure 9-36 Scene rendered by using the **Lunch Break Rendering** option*

Saving the Project

In this section, you will save the project.

1. To save the project with the current view, choose **Save As** from the Application Menu; the **Save As** dialog box is displayed.

2. Browse to the *nws_2017/ c09_nws_2017_tut* folder and enter **c09_navisworks_2017_tut01** in the **File name** edit box. Now, select the **Navisworks File Set (*.nwf)** file format from the **Save as type** drop-down list, and choose the **Save** button.

Self-Evaluation Test

Answer the following questions and then compare them to those given at the end of this chapter:

1. You can invoke the **Autodesk Rendering** window using the _____ drop-down in the _____ panel of the _____ tab.

2. You can produce directional lights by using the _____ light option.

3. The _____ button is used to invoke the **Geographic Location** dialog box.

4. The _____ panel in the Autodesk Rendering window displays a list of Autodesk materials library.

5. The _____ option is used to assign material to the selected object.

6. In the **Autodesk Rendering** window, you can modify the materials in the Library panel. (T/F).

7. You can invoke the **Autodesk Rendering** window from the **Review** tab. (T/F)

8. The **Planar** mapping type is the best mapping type for circular objects. (T/F)

9. In the **Box** mapping type, the material is applied on all the sides of an object. (T/F)

10. There are five tabs in the **Autodesk Rendering** window. (T/F)

Review Questions

Answer the following questions:

1. Which of the following lights emits lights in all directions from a fixed location?

    a) **Point**   b) **Spot**
    c) **Distant**   d) **Web**

2. Which of the following mapping types is used for spherical objects?

    a) **Box**   b) **Cylinder**
    c) **Spherical**   d) **Planar**

3. Which of the following buttons is used to invoke the **Material Editor** dialog box?

    a) **Display**   b) **Material Editor**
    c) **Point**   d) **Material**

4. Which of the following options is used to produce a quick rendering output?

    a) **Medium Quality**   b) **High Quality**
    c) **Low Quality**   d) **Ray Trace**

5. Which of the following tabs contains the options for the sun, sky, and exposure effects?

    a) **Lighting**   b) **Environments**
    c) **Materials**   d) **Render**

6. In Autodesk Rendering, you cannot adjust the Sun's appearance. (T/F)

7. The **High Quality** option takes the shortest time for rendering. (T/F)

8. You can specify the geographical location by using the **Geographic Location** dialog box. (T/F)

9. You can save the rendered view as an image. (T/F)

10. You can change the light intensity and lamp color in the Properties view area of the **Lighting** tab. (T/F)

EXERCISE

Exercise 1   Rendering the House

Download and open the *c09_navisworks_2017_ex1* file from *www.cadcim.com*. Render the model using the Autodesk materials and effects. The model elements are saved as selection sets in the **Sets** window. Figure 9-37 shows the model to be used for this exercise. Use the parameters given below to complete this exercise. Figure 9-38 shows the final view of the model after applying materials and effects.   **(Expected time : 45 min)**

The following steps are required to complete this exercise:

a. Use Table 9-3 for adding materials.
b. Use Table 9-4 for applying materials.
c. Specify the Time Zone as **(UTC: 06:00) Central America** and time 09:00 AM.
d. Render the scene and export the rendered model as a JPEG image.
e. Save the file as *c09_navisworks_2017_ex01*.

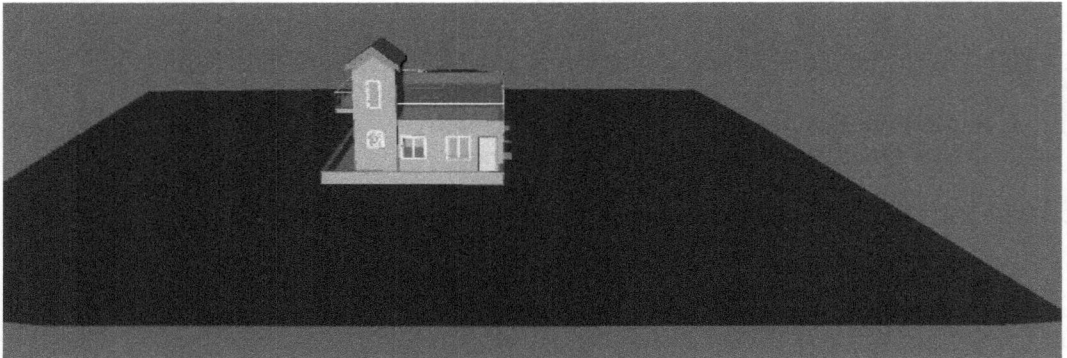

Figure 9-37 *Model displayed in the scene*

Table 9-3 *List of materials to be added to the* **Document Materials** *panel*

Materials to be added	Path of the materials in the Library panel
Adobe - Beige	Autodesk Library > Masonry > Brick
Weave - Brown	Autodesk Library > Masonry > Brick
Shake - Handsplit	Autodesk Library > Roofing
Mosaic	Autodesk Library > Flooring > Vinyl
Marble - White	Autodesk Library > Flooring > Stone
Paneling - Yellow Brown	Autodesk Library > Wood > Panels
Yellow Pine - Solid Natural Polished	Autodesk Library > Wood
Clear	Autodesk Library > Glass > Glazing

Table 9-4 *List of objects and the materials assigned to the objects*

Objects in the Sets window	Materials to be Assigned
Outer Walls	Adobe - Beige
Basic Roof 1 & Basic Roof 2	Shake - Handsplit
Roof Floor	Mosaic
Floor	Marble - White
Boundary Walls	Weave - Brown
Window & Door Frames	Yellow Pine - Solid Natural Polished
Door Panel	Panelling - Yellow Brown
Window Glass	Clear

Figure 9-38 *The final view of the model*

Answers to Self-Evaluation Test

1. F, **2.** F, **3.** F, **4.** T, **5.** F, **6. Windows, Workspace, View**, **7. Spot**, **8. Location**, **9.** Library, **10. Assign to Selection**

Index

T

U

V

W

X

Y

Z

This page is intentionally left blank

Other Publications by CADCIM Technologies

The following is the list of some of the publications by CADCIM Technologies. Please visit *www.cadcim.com* for the complete listing.

Exploring RISA 3D Textbook
• Exploring RISA-3D 14.0

Exploring Bentley STAAD.Pro Textbook
• Exploring Bentley STAAD.Pro V8i (SELECT series 6)

Raster Design Textbooks
• Exploring Raster Design 2017
• Exploring Raster Design 2016

Oracle Primavera Textbooks
• Exploring Oracle Primavera P6 R8.4
• Exploring Oracle Primavera P6 v7.0

Autodesk Revit Architecture Textbooks
• Exploring Autodesk Revit 2017 for Architecture, 13th Edition
• Autodesk Revit Architecture 2016 for Architects and Designers, 12th Edition
• Autodesk Revit Architecture 2015 for Architects and Designers, 11th Edition

AutoCAD Civil 3D Textbooks
• Exploring AutoCAD Civil 3D 2017, 7th Edition
• Exploring AutoCAD Civil 3D 2016, 6th Edition
• Exploring AutoCAD Civil 3D 2015, 5th Edition

AutoCAD Map 3D Textbooks
• Exploring AutoCAD Map 3D 2017, 7th Edition
• Exploring AutoCAD Map 3D 2016, 6th Edition
• Exploring AutoCAD Map 3D 2015, 5th Edition

Autodesk Revit Structure Textbooks
• Exploring Autodesk Revit 2017 for Structure, 7th Edition
• Exploring Autodesk Revit Structure 2016, 6th Edition
• Exploring Autodesk Revit Structure 2015, 5th Edition

Autodesk Revit MEP Textbooks
• Exploring Autodesk Revit 2017 for MEP, 4th Edition
• Exploring Autodesk Revit MEP 2016, 3rd Edition
• Exploring Autodesk Revit MEP 2015

Autodesk Navisworks Textbooks
- Exploring Autodesk Navisworks 2016, 3rd Edition
- Exploring Autodesk Navisworks 2015

AutoCAD MEP Textbooks
- AutoCAD MEP 2017 for Designers
- AutoCAD MEP 2016 for Designers
- AutoCAD MEP 2015 for Designers

SOLIDWORKS Textbooks
- SOLIDWORKS 2017 for Designers, 15th Edition
- SOLIDWORKS 2016 for Designers, 14th Edition
- SOLIDWORKS 2015 for Designers, 13th Edition

EdgeCAM Textbooks
- EdgeCAM 11.0 for Manufacturers
- EdgeCAM 10.0 for Manufacturers

CATIA Textbooks
- CATIA V5-6R2016 for Designers, 14th Edition
- CATIA V5-6R2015 for Designers, 13th Edition
- CATIA V5-6R2014 for Designers, 12th Edition

Creo Parametric and Pro/ENGINEER Textbooks
- PTC Creo Parametric 3.0 for Designers, 3rd Edition
- Creo Parametric 2.0 for Designers
- Pro/Engineer Wildfire 5.0 for Designers

Autodesk Alias Textbooks
- Learning Autodesk Alias Design 2016, 5th Edition
- Learning Autodesk Alias Design 2015

ANSYS Textbooks
- ANSYS Workbench 14.0: A Tutorial Approach
- ANSYS 11.0 for Designers

Creo Direct Textbook
- Creo Direct 2.0 and Beyond for Designers

Customizing AutoCAD Textbook
- Customizing AutoCAD 2013

AutoCAD LT Textbooks
- AutoCAD LT 2016 for Designers, 11th Edition
- AutoCAD LT 2015 for Designers, 10th Edition
- AutoCAD LT 2014 for Designers

AutoCAD Electrical Textbooks
- AutoCAD Electrical 2017 for Electrical Control Designers, 8th Edition
- AutoCAD Electrical 2016 for Electrical Control Designers, 7th Edition

3ds Max Design Textbooks
- Autodesk 3ds Max Design 2015: A Tutorial Approach, 15th Edition
- Autodesk 3ds Max 2017 for Beginners : A Tutorial Approach
- Autodesk 3ds Max 2016 for Beginners : A Tutorial Approach

3ds Max Textbooks
- Autodesk 3ds Max 2017: A Comprehensive Guide, 17th Edition
- Autodesk 3ds Max 2016: A Comprehensive Guide, 16th Edition
- Autodesk 3ds Max 2016 for Beginners: A Tutorial Approach, 16th Edition
- Autodesk 3ds Max 2015: A Comprehensive Guide, 15th Edition

Autodesk Maya Textbooks
- Autodesk Maya 2016: A Comprehensive Guide, 8th Edition
- Autodesk Maya 2015: A Comprehensive Guide, 7th Edition
- Character Animation: A Tutorial Approach
- Autodesk Maya 2014: A Comprehensive Guide

ZBrush Textbooks
- Pixologic ZBrush 4R7: A Comprehensive Guide
- Pixologic ZBrush 4R6: A Comprehensive Guide

Fusion Textbooks
- Blackmagic Design Fusion 7 Studio: A Tutorial Approach
- The eyeon Fusion 6.3: A Tutorial Approach

Flash Textbooks
- Adobe Flash Professional CC2015: A Tutorial Approach
- Adobe Flash Professional CC: A Tutorial Approach
- Adobe Flash Professional CS6: A Tutorial Approach

Computer Programming Textbooks
- Introduction to C^{++} programming
- Learning Oracle 11g
- Learning ASP.NET AJAX
- Introduction to Java Programming
- Learning Java Programming
- Learning Visual Basic.NET 2008
- Introduction to C^{++} Programming Concepts
- Learning C^{++} Programming Concepts
- Introduction to VB.NET Programming Concepts
- Learning VB.NET Programming Concepts

AutoCAD Textbooks Authored by Prof. Sham Tickoo and Published by Autodesk Press

- AutoCAD: A Problem-Solving Approach: 2013 and Beyond
- AutoCAD 2012: A Problem-Solving Approach
- AutoCAD 2011: A Problem-Solving Approach
- AutoCAD 2010: A Problem-Solving Approach
- Customizing AutoCAD 2010
- AutoCAD 2009: A Problem-Solving Approach

Textbooks Authored by CADCIM Technologies and Published by Other Publishers

3D Studio MAX and VIZ Textbooks

- Learning 3DS Max: A Tutorial Approach, Release 4
  Goodheart-Wilcox Publishers (USA)
- Learning 3D Studio VIZ: A Tutorial Approach
  Goodheart-Wilcox Publishers (USA)

CADCIM Technologies Textbooks Translated in Other Languages

SolidWorks Textbooks

- SolidWorks 2008 for Designers (Serbian Edition)
  Mikro Knjiga Publishing Company, Serbia
- SolidWorks 2006 for Designers (Russian Edition)
  Piter Publishing Press, Russia
- SolidWorks 2006 for Designers (Serbian Edition)
  Mikro Knjiga Publishing Company, Serbia

NX Textbooks

- NX 6 for Designers (Korean Edition)
  Onsolutions, South Korea
- NX 5 for Designers (Korean Edition)
  Onsolutions, South Korea

Pro/ENGINEER Textbooks

- Pro/ENGINEER Wildfire 4.0 for Designers (Korean Edition)
  HongReung Science Publishing Company, South Korea
- Pro/ENGINEER Wildfire 3.0 for Designers (Korean Edition)
  HongReung Science Publishing Company, South Korea

Autodesk 3ds Max Textbook

- 3ds Max 2008: A Comprehensive Guide (Serbian Edition)
  Mikro Knjiga Publishing Company, Serbia

AutoCAD Textbooks
- AutoCAD 2006 (Russian Edition)
  Piter Publishing Press, Russia
- AutoCAD 2005 (Russian Edition)
  Piter Publishing Press, Russia
- AutoCAD 2000 Fondamenti (Italian Edition)

AutoCAD Textbooks Authored by Prof. Sham Tickoo and Published by Autodesk Press
- AutoCAD: A Problem-Solving Approach: 2013 and Beyond
- AutoCAD 2012: A Problem-Solving Approach
- AutoCAD 2011: A Problem-Solving Approach

Textbooks Authored by CADCIM Technologies and Published by Other Publishers

3D Studio MAX and VIZ Textbooks
- Learning 3ds max5: A Tutorial Approach
  (Complete manuscript available for free download on www.cadcim.com)
- Learning 3ds Max: A Tutorial Approach, Release 4
  Goodheart-Wilcox Publishers (USA)

CADCIM Technologies Textbooks Translated in Other Languages

3ds Max Textbook
- 3ds Max 2008: A Comprehensive Guide (Serbian Edition)
  Mikro Knjiga Publishing Company, Serbia

SolidWorks Textbooks
- SolidWorks 2008 for Designers (Serbian Edition)
  Mikro Knjiga Publishing Company, Serbia
- SolidWorks 2006 for Designers (Russian Edition)
  Piter Publishing Press, Russia
- SolidWorks 2006 for Designers (Serbian Edition)
  Mikro Knjiga Publishing Company, Serbia
- SolidWorks 2006 for Designers (Japanese Edition)
  Mikio Obi, Japan

NX Textbooks
- NX 6 for Designers (Korean Edition)
  Onsolutions, South Korea
- NX 5 for Designers (Korean Edition)
  Onsolutions, South Korea

Pro/ENGINEER Textbooks
- Pro/ENGINEER Wildfire 4.0 for Designers (Korean Edition)
  HongReung Science Publishing Company, South Korea
- Pro/ENGINEER Wildfire 3.0 for Designers (Korean Edition)
  HongReung Science Publishing Company, South Korea

AutoCAD Textbooks
- AutoCAD 2006 (Russian Edition)
  Piter Publishing Press, Russia
- AutoCAD 2005 (Russian Edition)
  Piter Publishing Press, Russia

Coming Soon from CADCIM Technologies

- Exploring ArcGIS
- Mould Design using NX 11.0: A Tutorial Approach
- Autodesk Fusion 360: A Tutorial Approach

Online Training Program Offered by CADCIM Technologies
CADCIM Technologies provides effective and affordable virtual online training on animation, architecture, and GIS softwares, computer programming languages, and Computer Aided Design, Manufacturing, and Engineering (CAD/CAM/CAE) software packages. The training will be delivered `live' via Internet at any time, any place, and at any pace to individuals, students of colleges, universities, and CAD/CAM/CAE training centers. For more information, please visit the following link: http://*www.cadcim.com*.

www.ingramcontent.com/pod-product-compliance
Lightning Source LLC
Chambersburg PA
CBHW080705220326
41598CB00033B/5309